HUMAN GENETICS

The Jones and Bartlett Series in Biology

The Biology of AIDS
Hung Fan, Ross F. Conner, and Luis P. Villarreal,
all of the University of California-Irvine

Basic Genetics
Daniel L. Hartl, Washington University School of Medicine;
David Freifelder, University of California, San Diego; Leon
A. Snyder, University of Minnesota, St. Paul

General Genetics
Leon A. Snyder, University of Minnesota, St. Paul; David
Freifelder, University of California, San Diego; Daniel L.
Hartl, Washington University School of Medicine

Genetics
John R. S. Fincham, University of Edinburgh

Genetics of Populations
Philip W. Hedrick, University of Kansas

Human Genetics: A New Synthesis
Gordon Edlin, University of California, Davis

Microbial Genetics
David Freifelder, University of California, San Diego

Experimental Techniques in Bacterial Genetics
Stanley R. Maloy, University of Illinois-Urbana

Cells: Principles of Molecular Structure and Function
David M. Prescott, University of Colorado, Boulder

Essentials of Molecular Biology
David Freifelder, University of California, San Diego

Introduction to Biology: A Human Perspective
Donald J. Farish, California State University at Sonoma

Introduction to Human Immunology
Teresa L. Huffer, Shady Grove Adventist Hospital
Gaithersburg, Maryland, and Frederick Community College,
Frederick, Maryland; Dorothy J. Kanapa, National
Cancer Institute, Frederick, Maryland; George W. Stevenson,
Northwestern University Medical Center, Chicago, Illinois

Molecular Biology, Second Edition
David Freifelder, University of California, San Diego

The Molecular Biology of Bacterial Growth (a symposium volume)
M. Schaechter, Tufts University Medical School; F. Neidhardt,
University of Michigan; J. Ingraham, University of California,
Davis; N.O. Kjeldgaard, University of Aarhus, Denmark, editors

Evolution
Monroe W. Strickberger, University of Missouri-St. Louis

Molecular Evolution: An Annotated Reader
Eric Terzaghi, Adam S. Wilkins, and David Penny, all
of Massey University, New Zealand

Population Biology
Philip W. Hedrick, Unversity of Kansas

Virus Structure and Assembly
Sherwood Casjens, University of Utah College of Medicine

Cancer: A Biological and Clinical Introduction, Second Edition
Steven B. Oppenheimer
California State University, Northridge

Introduction to Human Disease, Second Edition
Leonard V. Crowley, M.D.
St. Mary's Hospital, Minneapolis

Handbook of Protoctista
Lynn Margulis, John O. Corliss, Michael Melkonian,
and David I. Chapman, editors

Living Images
Gene Shih and Richard Kessel

Early Life
Lynn Margulis, Boston University

Functional Diversity of Plants in the Sea and on Land
A.R.O. Chapman, Dalhousie University

Plant Nutrition: An Introduction to Current Concepts
A.D.M. Glass, University of British Columbia

*Methods for Molecular Cloning and Analysis of
Eukaryotic Genes*
Alfred Bothwell, Yale University School of Medicine; Fred
Alt and George Yancopoulous, both of Columbia
University

Medical Biochemistry
N.V. Bhagavan
John A. Burns School of Medicine,
University of Hawaii at Manoa

Vertebrates: A Laboratory Text, Second Edition
Norman K. Wessells, Stanford University and
Elizabeth M. Center, College of Notre Dame, editors

The Environment, Third Edition
Penelope ReVelle, Essex Community College
Charles ReVelle, The Johns Hopkins University

Medical Ethics
Robert M. Veatch, editor
The Kennedy Institute of Ethics--Georgetown University

Cross Cultural Perspectives in Medical Ethics: Readings
Robert M. Veatch, editor
The Kennedy Institute of Ethics--Georgetown University

*100 Years Exploring Life, 1888-1988, The Marine
Biological Laboratory at Woods Hole*
Jane Maienschein, Arizona State University

Writing a Successful Grant Application, Second Edition
Liane Reif-Lehrer, Tech-Write Consultants/ERIMON Associates

HUMAN GENETICS

A Modern Synthesis

GORDON EDLIN
University of California, Davis

JONES AND BARTLETT PUBLISHERS
BOSTON

Editorial, Sales, and Customer Service Offices
Jones and Bartlett Publishers
20 Park Plaza
Boston MA 02116

Printed in the United States of America
10 9 8 7 6 5 4 3 2 1

Library of Congress Cataloging-in-Publication Data

Edlin, Gordon, 1932–
   Human genetics: a modern synthesis / Gordon Edlin—2nd ed.
      p.   cm.
   Rev. ed. of: Genetic principles, © 1984.
   Includes bibliographies and index.
   ISBN 0-86720-112-6
   1. Genetics.   2. Human genetics.   I. Edlin, Gordon, 1932–
Genetic principles.   II. Title.
   [DNLM: 1. Genetics.   QH 431 E23g]
QH430.E.343 1989
575—dc19
DNLM/DLC
for Library of Congress   88–39688
                                                   CIP

ISBN: 0-86720-112-6

Production editor: Patricia Zimmerman
Text and cover designer: Rafael Millán
Illustrators and technical artists: Kim Fraley, Linda McVay, Pam Posey;
   James Rue Design/Kramer Design Assoc.

Cover illustration:
Polarized light photomicrograph of liquid crystalline DNA in a highly concentrated solution. At DNA concentrations close to those found in virus capsids, bacterial nucleoids, and sperm heads, liquid crystalline phases are observed that show microscopic textures comparable to those observed in smectic or columnar hexatic phases of small molecules. (Michael W. Davidson)

Science can be a tool for liberation, but it can also become a tool for oppression; in order to avoid this deflection, those responsible for its progress must share their knowledge by expressing themselves in such a way as to be understood by all.

ALBERT JACQUARD, *philosopher*
Endangered by Science

Preface

APPLICATIONS of new genetic discoveries now affect our daily lives in a multitude of ways. A multibillion dollar biotechnology industry based on the principles of genetic engineering has blossomed since restriction enzymes and recombinant DNAs were first isolated and synthesized in the 1970s. The manufacture of drugs and the breeding of crops and animals have been dramatically altered by the new techniques of gene cloning, *in vitro* fertilization, and embryo transplants. The diagnosis of disease-causing microorganisms in human beings and animals has been facilitated by the development of DNA probes that can identify specific viruses and bacteria in those infected. Today, one can hardly pick up a newspaper or magazine that does not contain an article on some aspect of genetics—on the potential profits of genetically engineered drugs, on advances in human *in vitro* fertilization, on prospects for human gene therapies, or on the potential benefits of mapping the complete human genome.

Indeed, it is in the area of human genetics that the most-dramatic and startling genetic advances have been made in the past decade. Hundreds of hereditary human diseases can now be detected *in utero*, giving prospective parents childbearing options that were not available to earlier generations. To date, more than four thousand human genes have been identified, most of which participate in some clinically recognized human disorder. In many instances, persons can be tested to determine whether they carry such recessive deleterious genes as those responsible for sickle-cell anemia, Huntington disease, Tay-Sachs, muscular dystrophy, and others.

A class of DNA probes has been used to help identify and convict suspects involved in criminal or civil litigation. For example, murder and rape suspects have been convicted through the analysis of DNA found in blood stains or in sperm samples removed from rape victims. It has been shown that every person (except identical twins) inherits a unique DNA "fingerprint." Tissue found at the scene of a crime can be analyzed for the cells' particular DNA pattern and compared with the DNA patterns of suspects.

Even the study of evolution and the relatedness of individuals has been revolutionized by new techniques for analyzing DNAs. Paternity suits can be resolved by analyzing DNA from cells of parents and their putative children. On a larger scale, the relatedness of human populations and their evolutionary histories can be determined by modern strategies of DNA-DNA hybridization, DNA cloning, and DNA base sequencing. The reconstruction of biological history on earth has been reduced in large measure to computer analyses of base sequences of DNA from different organisms.

Some grasp of the basic principles of genetics and their applications in today's world is essential if people are to function successfully and knowledgeably in a complex, technological society. It is the purpose of this text to inform readers with little or no scientific background about the principles of genetics. Having taught a course on heredity and evolution for nonscience majors for almost 20 years, I have continually been impressed with how much basic genetics students can learn and understand. Even more rewarding has been how students' attitudes regarding human hereditary diseases and human evolution change as their understanding of genetics increases.

The text begins with three chapters describing the fundamentals of classical genetics—Mendel's discoveries, mitosis and meiosis, and chromosome structure and functions. Understanding genetics, as is the case for all sciences, depends on mastering a new language of scientific terminology. Thus, key words are defined as they are introduced, and their definitions are also given at the end of each chapter, as well as in a glossary at the end of the book. Chapter 4 (which some instructors may choose to omit) contains a basic introduction to chemistry, particularly the molecules and reactions found in cells.

Molecular genetics deals with the chemistry of hereditary information (which is the rationale for Chapter 4) and specifically with the expression and regulation of genes in cells. Chapters 5 through 12 introduce the essential principles and discoveries of molecular genetics that, although elucidated in microorganisms, apply to human cells as well. Included in this section are discussions of how these molecular principles relate to understanding and treating cancer (Chapter 11) and to the development of the biotechnology industry (Chapter 12). Chapters 13 through 16 discuss specific aspects of human heredity — reproduction, hereditary diseases, genetic counseling and screening, and the detection of hereditary diseases by mapping and DNA probes. Chapter 17 introduces the subject of immunogenetics and antibodies. (This material also can be excluded at the instructor's option without any loss of continuity.) Chapters 18 and 19 present the principles of population genetics and evolution. Emphasis on the findings of molecular evolution at the DNA level puts the more-controversial aspects of human evolution into a molecular perspective. Finally, Chapter 20 discusses the rather sordid social history of genetic abuses in the past century. These topics include the worldwide eugenics movement, involuntary sterilization, restrictive immigration policies and miscegenation laws in the United States, genocide in Nazi Germany, the IQ controversy and racism, and the ongoing debate concerning the relative contributions of genes or environment or both to complex human behaviors such as alcohol and drug abuse, phobias, eating disorders, and so forth. The powerful new techniques that allow detection of particular human genes and particular genotypes make it imperative that we be alert to possible genetic abuses and infringement of basic human rights. The dismal history of the eugenics movement in this century demands that we proceed with care and compassion in any widespread application of the "new" genetics.

Each chapter in this book includes a summary of the concepts presented therein, a list of additional readings, study questions, and essay topics. Answers to the study questions can be found at the end of the book. Throughout the text, I have tried to engage the reader's interest by presenting controversial or thought-provoking material in boxes.

Many persons contributed valuable advice and criticism by reading part or all of this text. Any inaccuracies or misinterpretations that still remain are my responsibility. I am indebted to my previous editor, John Hendry, and to the present editor, Patricia Zimmerman. I also would like to thank my publishers, Don Jones and Art Bartlett, for their assistance and support. In particular, I would like to thank the following reviewers: Alan Atherly, Iowa State University; Kandy D. Baumgardner, Eastern Illinois University; Jon Beckwith, Harvard University School of Medicine; Ram Bhagavan, University of Hawaii School of Medicine; Willard Centerwall, University of California, Davis; A. John Clark, University of California, Berkeley; Cedric I. Davern, University of Utah; Wesley Ebert, California State University, Sonoma; David Freifelder, University of California, San Diego; Rachel Freifelder, University of California, Berkeley; Jon Gallant, University of Washington, Seattle; Eric Golanty, University of California, Davis; George A. Hudock, Indiana University; Gerard O'Donovan, Texas A & M University; David Prescott, University of Colorado, Boulder; Barton Slatko, New England Biolabs, Inc., Beverly, Massachusetts; and W. Dorsey Stuart, University of Hawaii.

Gordon Edlin

Davis, California

Contents

List of Topics

List of Boxes

1

Mendel's Laws

The Rules of Inheritance

As failures go, attempting to recall the past is like trying to grasp the meaning of existence. Both make one feel like a baby clutching at a basketball: one's palms keep sliding off.

JOSEPH BRODSKY, *poet*

BIOLOGY IS the branch of science that studies living organisms—from simple single-celled bacteria to complex human beings composed of at least a hundred thousand billion cells. Biology is concerned with all aspects of living organisms, including their development (embryology), their interactions with other organisms and with the environment (ecology), and the hereditary information that gives each individual organism its unique characteristics (genetics). In many respects, genetics is the most fundamental of all the branches of biology because information carried by the genes determines how an organism will develop and function. That information dictates the similarities and differences between parents and progeny, and it orchestrates the chemistry of every cell in every organism. For example, every cell in a person's body, from a skin cell to a nerve cell in the brain, contains complete copies of the genetic information. It is how that genetic information is expressed in different cells that accounts for the functions of various tissues and organs.

Genetics refers to all aspects of heredity in organisms. Because genetics encompasses so many different aspects of heredity, it is usually subdivided into more-specialized categories. Classical genetics consists of studies similar to those carried out by Gregor Mendel on peas, population genetics investigates the frequencies of various genes in individual members of a population, evolutionary genetics studies how gene frequencies change in populations through time (the history of living and extinct organisms), and molecular genetics deals with the biochemical and physical properties of genetic information. **Genes** are the discrete hereditary units located at specific positions (**loci**) in a **chromosome**, which is the microscopically visible structure in cells that contains the genetic information. In classical genetics, a gene affects one or more observable traits of an organism; the sum of all traits is the organism's **phenotype**. Correspondingly, the **genotype** of an organism refers to the sum of all of its genes, or genetic information.

For thousands of years, people have been aware that "cats beget cats." They have also been aware that not all cats are alike, nor are all the kittens in a litter identical. Not until recently, however, have we been able to explain the biological similarities and differences between organisms. It is only within the past hundred years or so that we have come to understand the principles of inheritance—how genetic information is passed from parent to progeny and how it may become altered from one generation to the next. And only in the past fifty years or so have we come to understand the chemical nature of that genetic information—how the molecules of heredity encode, express, and regulate the genetic information contained within cells. All genetic information in all organisms is contained in chemically identical molecules called DNA (deoxyribonucleic acid). DNA molecules are chemically identical, but the genetic information that they carry is infinitely variable (discussed in Chapter 6).

Finally, to the astonishment of some people (and to the concern of others), in the past fifteen years or so scientists have learned how to manipulate genes so as to create new arrangements of genetic information and even novel plants and animals that do not exist in nature. Genetic engineers are now capable of creating unique organisms and of moving genetic information from one species to another (Figure 1-1).

FIGURE 1-1 A sheep-goat chimera that is a mixture of cells from an Alpine sheep and a Nubian goat. Embryonic cells are mixed together *in vitro* and the reconstituted embryo is implanted into a receptive female animal. Chimeric animals are but one example of the dramatic changes in the breeding of plants and animals since Mendel's experiments. (Courtesy of Gary Anderson, University of California, Davis).

Scientific Ideas

Jacob Bronowski, an eminent historian of science, described three creative ideas that seem to be crucial to all branches of science: the idea of order, the idea of cause and effect, and the idea of chance.

Isaac Newton was impelled by the idea of order when he realized that all objects in the universe, small objects on earth as well as stars and planets, exert a force on one another called gravity. His mathematical definition of that force became the organizing principle for explaining the motions of all objects. At about the same time, the biological world was thought to be organized generally into plant and animal kingdoms. Then, in the eighteenth century, a scientist named Linnaeus reclassified organisms into five kingdoms (Table 1-1). He proposed that each kingdom be further classified into phylum, class, order, family, genus, and species. That system of classification revealed the fundamental similarities and differences between organisms and gave other scientists a framework for experimental study. According to Bronowski, this "ability to order things into likes and unlikes is the foundation of human thought."

The biological revolution prompted by Linnaeus's classification of organisms led to the discovery of cause-and-effect relations between organisms and their environment. In the nineteenth century, for example, scientists discovered that light causes plants to assimilate carbon dioxide from the environment and to give off oxygen, that microorganisms cause the souring of wine and the onset of human diseases such as cholera and typhoid fever, and that biochemical defects cause certain inherited diseases such as albinism in human beings.

It was the idea of chance (statistical probabilities), however, that led to even more revolu-

TABLE 1-1 All living organisms are classified into five kingdoms.

Kingdom	Kinds of organisms
Monera	Bacteria and simple algae
Protista	Amoebas, paramecia, and diatoms
Fungi	Molds and mushrooms
Plantae	Grasses, ferns, trees, and flowering plants
Animalia	Worms, insects, mollusks, fishes, birds, and mammals

FIGURE 1-2   Charles Darwin, whose works established the scientific principles of evolution. (Courtesy of the National Library of Medicine.)

tionary theories. In physics, the theory of quantum mechanics (based on probabilities) yielded a new understanding of the physical world. In biology, the theory of evolution formulated by Charles Darwin supplied the unifying concept for understanding the diversity of nature (Figure 1-2). Darwin proposed that evolution has two key components—variation and selection. Genetic variation ultimately is the result of random mutations; natural selection affects the survival and reproduction of better-adapted organisms. In the same period (Darwin's book *On the Origin of Species by Means of Natural Selection* was published in 1859), an Augustinian monk, Gregor Mendel (Figure 1-3), discovered the laws of inheritance by studying how particular traits in pea plants are passed from one generation to the next. We begin our exploration of the principles of genetics by examining the remarkable discoveries of Gregor Mendel, who is regarded as the "father" of genetics.

Classical Genetics

The discoveries of Gregor Mendel and his successors concerning the patterns of inheritance of specific traits among sexually reproducing plants and animals constitute the period of *classical genetics*—a period extending from 1866, when Mendel first published his findings, to about 1940. During this period, scientists conducted their experiments with no precise knowledge of the chemical nature of the genetic information that is transmitted from one generation to the next. Modern molecular genetics began with the discovery of the structure of DNA by James Watson and Francis Crick in 1953, although many experiments bearing on the chemical properties of genes and chromosomes had been performed much earlier.

By analyzing the results of crosses between pea plants having different traits, Mendel discovered the universal rules of inheritance. He was the first to explain the phenomenon of

FIGURE 1-3   Gregor Mendel, founder of the science of genetics. (Courtesy of the National Library of Medicine.)

hybrid plants—that is, progeny plants having combinations of traits that are different from those of the parent plants. Moreover, he proposed three new concepts that revolutionized the study of genetics:

1 Traits are determined by discrete hereditary factors that are transmitted intact from one generation to the next. (The term gene was not introduced until about forty years after Mendel's experiments.)
2 Those discrete hereditary factors are present as pairs in every individual organism and the members of each pair can differ; one factor is usually **dominant** (determines the trait) and the other is **recessive** (does not contribute to the trait).
3 Progeny plants with particular traits are produced in predictable numbers according to statistical rules from one generation to the next.

For thousands of years before Mendel undertook his revolutionary investigations, a very different explanation of the inheritance of traits had been widely accepted. That explanation was *pangenesis*. Its basic idea was that each part of an adult organism produces a tiny replica of itself, called a *pangene*, and that all the individual replicas are collected in the "seed" of that organism and are transmitted to the next generation. It was Mendel's bold and brilliant insights that helped disprove that notion (see Box 1-1).

Gregor Mendel entered the Augustinian monastery in Brünn, Austria (now Brno, Czechoslovakia), in 1843. This was a teaching order and Mendel was sent to study at the University of Vienna, where he took courses in plant science and physics. After several years of study, he twice failed to pass his examinations. Finally, he had to leave the university without receiving a degree. Some time after he returned to the monastery in 1853, he began his breeding experiments with the common garden pea, *Pisum sativum*.

Although we will never know just what motivated Mendel to undertake his experiments, the ideas of evolution and natural selection were generating lively interest and controversy at the time. (Darwin did not know about Mendel's work, but Mendel had read Darwin's *Origin of Species* by the time he published his own findings.) Moreover, plant breeders had puzzled over the question of hybrids for centuries. How could new combinations of traits that had never been observed in either parent plant arise in later generations of plants? Like plant breeders before him, Mendel must have been perplexed by this question.

After ten years of careful breeding experiments, Mendel announced his conclusions in 1865 and had them published a year later. His research placed the study of inheritance on a sound scientific basis and marked the beginnings of the science of genetics. For thirty-five years, however, the scientific community totally ignored Mendel's findings. Although Mendel did not live long enough to receive recognition for his discoveries, he understood the significance of his own accomplishments. In the introduction to the original paper titled "Experiments on Plant Hybrids," he wrote:

among the numerous experiments not one has been carried out to an extent or in a manner that would make it possible to determine the number of different forms in which hybrid progeny appear, permit classification of these forms in each generation with certainty, and ascertain their numerical interrelationships. It requires a good deal of courage, indeed, to undertake such a far-reaching task; however, this seems to be the one correct way of finally reaching the solution to a question whose significance for the evolutionary history of organic forms must not be underestimated.

From Pangenes to Genes
A Journey of 2,000 Years

BOX 1-1

Since the beginning of recorded history, people have wondered how organisms reproduce and how traits are inherited. Why do the phenotypes of each species remain recognizable generation after generation—why do cats always beget cats and not mice or dogs? When plants and animals reproduce, the progeny are never exactly like their parents in all respects, but they are invariably similar. Some traits appear to be identical with parental traits: others are distinctly different.

One theory that attempted to explain heredity was *pangenesis* (*pan* = all, *genesis* = origination). Among scientists and philosophers who believed in this theory were such famous figures as Aristotle, Hippocrates, and Darwin. The basic idea of pangenesis, particularly as it related to human reproduction, was that each part of the body produced tiny particles that, taken all together, carried all of a person's traits. The invisible particles, variously referred to as pangenes, gemmules, or plastidules, were carried by the blood to the organs of reproduction. As the different particles from each of the body's parts accumulated in the female womb, they became capable of reproducing all of the parts and traits of a new human being.

It was thought that the pangenes contained miniature replicas of each organ, tissue, or other body part. When the male and female pangenes came together in the womb, the particles grew and were modified, and a new person with a mixture of both parents' traits was produced. Such were the ideas that persisted for thousands of years and that were used to explain why children resemble their parents.

Pangenesis also explained how parents' defects or diseases could be passed on to their sons and daughters. If the father was mentally defective, the particular pangenes copied from his diseased brain would also be defective, and the child would inherit the defects of the father. Or if the mother had contracted cancer or tuberculosis, the pangenes made from her cancerous organ or diseased lungs would also be diseased (or at least prone to disease).

Of course, people had noticed that children did not always turn out like their parents. This was supposedly because defective male particles could be modified in the mother's womb during pregnancy and because the mother's contribution was not considered to be nearly as great as the father's. (Male chauvinism has a long history, particularly in matters of inheritance; even today most couples express a preference for having sons.) The ideas of pangenesis persisted through centuries and supported prejudices toward male lineage and maintenance of "royal blood lines" (because the pangenes were thought to be carried in the blood).

Leonardo da Vinci was one of the few creative geniuses of the past who believed that men and women contribute equally to the traits of their offspring. In the late fifteenth century he wrote: "The blacks of Ethiopia are not the products of the sun; for when black gets black with child in Scythia, the offspring is black; but if a black gets a white woman with child, the offspring is grey in color. Now this proves that, with regard to the embryo, the seed of the mother has the same vigor as that of the father."

Gregor Mendel seems to have been the first scientist to realize that the theory of pangenesis was all wrong and to correctly describe how traits are transmitted from generation to generation. It was Mendel who conceived of and documented the existence of discrete hereditary factors, which later came to be called genes. The word *gene* was derived by geneticist Wilhelm Johannesen from the term *pangenesis* many years after Mendel's discoveries. In 1909, Johannesen wrote, "We will simply speak of the 'gene' and the 'genes' instead of 'pangene' and the 'pangenes.' The word 'gene' is completely free from any hypothesis; it expresses only the evident fact that, in any case, many characteristics of the organism are specified in the germ cells." The discrete hereditary factors of Mendel eventually became the genes in chromosomes of classical genetics and, almost a century later, the sequences of bases in DNA.

Reading: F. H. Portugal and J. S. Cohen, *A Century of DNA*. MIT Press, 1979.

Although Mendel lived and worked as a monk in a strict religious order, he was well aware of the revolutionary import of his experiments.

True-breeding Plants

Mendel began his experiments by establishing true-breeding strains of peas. By repeatedly crossing plants that have a common trait, such as wrinkled seeds, with themselves for several generations, he eventually developed plants that produced only wrinkled seeds—that is, plants that were true-breeding for that trait generation after generation. Over time, he developed true-breeding strains for seven pairs of distinguishable traits (Figure 1-4). We do not know whether he conceived of the idea that some traits were dominant and others were recessive before he began the experiments or whether the idea occurred to him after he had analyzed the results. Once he had established true-breeding strains for the seven pairs of traits, however, he began to cross-pollinate plants with different pairs of traits. A discussion of his experiments requires the use of several key genetic terms, which are defined in Table 1-2. Knowing what these terms mean is essential to grasping the significance of the rules of inheritance that Mendel derived from his experiments.

The Law of Segregation

Pea plants normally inbreed—that is, the eggs in a pea flower are fertilized by pollen from other flowers on the same plant, or **self-fertilized**. In experimenting with **crosses**, Mendel carefully mated plants with different but related traits by dusting flowers of one true-breeding type—such as a plant with round seeds—with pollen from the flowers of a related true-breeding type—a plant with wrinkled seeds. He prevented self-pollination of these plants by "emasculating" the recipient flowers.

The techniques of artificial pollination had been known since antiquity (Figure 1-5), but Mendel was the first to apply them with his particular purpose and the first to analyze the number of plants in successive generations carrying one or the other of the original pairs of traits. After collecting the seeds produced from the parent plants that had been crossed, he planted them and counted the number of plants in the first filial (called the F_1) generation that exhibited each phenotype—that is, either round or wrinkled seeds. He found that all the seeds produced by the F_1 generation plants were round; he observed no wrinkled seeds. Next he permitted adult F_1 plants to self-fertilize and produce the second filial, or F_2, generation. Then

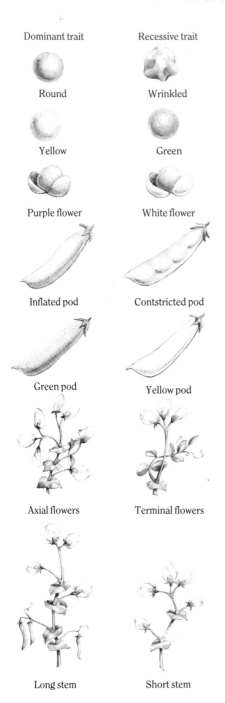

FIGURE 1-5 Artificial pollination has been performed for thousands of years. This Assyrian relief, dating from the ninth century B.C., shows masked priests pollinating date palms. (Courtesy of Hans Stubbe, Akademie der Wissenschaften der DDR.)

FIGURE 1-4 The seven different pairs of traits that Mendel used in his experiments with pea plants. From observing the traits in progeny plants, Mendel deduced the laws of genetics, including the concept of dominant and recessive traits.

he counted the number of F_2 plants that exhibited each trait.

After counting hundreds of plants in the F_2 generation, Mendel realized that, for every plant bearing wrinkled seeds, there were approximately three plants bearing round seeds: *on the average*, seed-shape traits appeared in the F_2 generation in the proportion of three round to one wrinkled (Figure 1-6). Through the years, Mendel performed many crosses with other pairs of the seven traits and always observed

TABLE 1-2 Definitions of genetic terms.

Chromosome	Threadlike structure observable in the nucleus of a plant or an animal cell that carries genetic information.

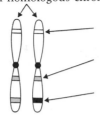

A pair of homologous chromosomes

A *genetic locus* is the position of a gene on a chromosome.

Identical alleles; an individual organism with identical alleles at a locus is said to be *homozygous* for that trait.

Different alleles; an individual organism with different alleles at a locus is said to be *heterozygous* for that trait. One allele may be dominant; the other recessive.

DNA (deoxyribonucleic acid)	The long, double-helical molecule in a chromosome. The chemical structure of this molecule contains genetic information (genes).
Gene	In Mendelian terms, a discrete hereditary factor situated at a particular locus (site) on a chromosome. A Mendelian gene determines an observable trait.
Allele	An alternative state of a gene. An allele can be either dominant or recessive. An individual organism can have identical alleles on both chromosomes of a homologous pair (the organism is then *homozygous* for that allele or trait) or the alleles can be different (in which case the organism is *heterozygous*). Although a diploid plant or animal can have only two different alleles at a genetic locus, a population (many individuals) may contain many different alleles (for example, the gene for the ABO blood type in people has three different alleles—A, B, and O—that are present in any pairwise combination in an individual person).
Phenotype	Observable traits in an individual organism that result from its genotype.
Genotype	The total genetic constitution of an individual organism or of those traits that have been identified.
Diploid (2N)	The chromosomal state in which each different chromosome (except for the sex chromosome) is present in two copies (2N), each parent having contributed one haploid set of chromosomes.
Haploid (N)	Half of the diploid number of chromosomes. Reproductive cells, or *gametes* (sperm, pollen, eggs), contain the haploid number of chromosomes. When sperm and egg combine to form a *zygote*, the first cell of a new offspring, the diploid number of chromosomes is restored.

this same 3 : 1 pattern of inheritance. Thus, in a cross of true-breeding plants that differed in a single trait, only one of the parental traits would appear in plants of the F_1 generation. However, in the F_2 generation, both traits would reappear, always in the ratio of 3 to 1.

Figure 1-7 summarizes the results of Mendel's experiments with a cross between F_1

FIGURE 1-6 Mendel's cross between true-breeding parent plants (P) having either round or wrinkled seeds. In the first (F_1) generation, all plants produced round seeds. When F_1 generation plants were crossed, the plants in the F_2 generation produced round and wrinkled seeds in the ratio of three round-seed plants to one wrinkled-seed plant.

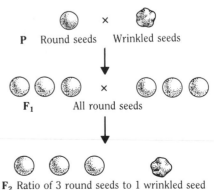

P Round seeds   ×   Wrinkled seeds

F_1 All round seeds

F_2 Ratio of 3 round seeds to 1 wrinkled seed

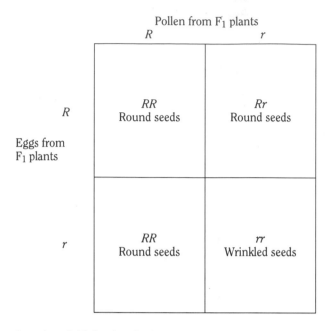

Pollen from F$_1$ plants

Eggs from F$_1$ plants

	R	r
R	RR Round seeds	Rr Round seeds
r	RR Round seeds	rr Wrinkled seeds

Round : wrinkled ratio — 3 : 1

R = Round phenotype
r = Wrinkled phenotype

FIGURE 1-7 A Punnett-square diagram analyzing the cross shown in Figure 1-6.

Genotypes	Phenotypes
RR	round seeds
Rr	round seeds
rr	wrinkled seeds

Ratio of phenotypes: 3 plants with round seeds to 1 with wrinkled seeds.

plants to produce F$_2$ plants having round or wrinkled seeds. This method of analyzing a genetic cross, known as the **Punnet square**, was devised by geneticist Reginald Crundall Punnett many years after Mendel's death. In this diagram, the capital letter stands for the dominant allele (R = round seed) and the small letter stands for the recessive allele (r = wrinkled seed). The squares in any Punnett diagram show the genotypes (and hence the pheno-

types) of the progeny that arise by combining gametes (and hence alleles) from the male and female parents in all possible combinations.

Mendel realized that the true-breeding parental plants must contain a pair of identical alleles for seed shape. Thus, the plant with round seeds would be **homozygous** (having identical alleles at a genetic locus on a pair of chromosomes) for alleles RR and the plant with wrinkled seeds would be homozygous for alleles rr. All progeny plants in the F$_1$ generation must be **heterozygous** (having different alleles at a genetic locus) because they receive the R allele from one parent and the r allele from the other. (Note that the results of the cross are the same irrespective of which allele is carried by the male or the female gamete.) Figure 1-7 shows clearly why a 3 : 1 ratio of plants with round or wrinkled seeds was observed by Mendel — only one plant in four on average will carry both recessive rr alleles.

To explain the 3 : 1 ratio of round/wrinkled seeds (he found the same result for all seven pairs of traits shown in Figure 1-4), Mendel hypothesized the existence of two discrete hereditary factors. To describe them, he introduced the concepts of *dominant* (able to mask) factors and *recessive* (able to be masked) factors. For the first time in history, insofar as we know, the transmission of traits from generation to generation had been correctly explained. The transmission of traits in the characteristic 3 : 1 ratio in the F$_2$ generation represents Mendel's first law: the law of segregation.

This law may be formally stated as follows. Genetic factors governing traits segregate from one another in the transmission of particular alleles (the genetic factors) from generation to generation.

The Law of Independent Assortment

Mendel's second discovery, the law of independent assortment, is merely a logical extension of the law of segregation, but it is an important one. Suppose that a cross is performed for *two* pairs of traits simultaneously, such as a cross between true-breeding plants that produce round, yellow seeds and true-breeding plants that produce wrinkled, green seeds. Figure 1-8 shows the genotypes and phenotypes of plants produced in the F_1 and F_2 generations. If we assume, as Mendel did, that the discrete factors (the dominant and recessive alleles governing each trait) are transmitted to progeny independently of one another, then we should observe the same 3 : 1 ratio for each pair of traits analyzed separately. However, because the two pairs of traits are inherited independently of one another, *four* different phenotypes are observed in the F_2 progeny plants in a ratio that, again on average, should be 9 : 3 : 3 : 1 because 16 different genotypes are possible. We will see in the next chapter that this result is due to the fact that the pairs of alleles are on different chromosomes that are randomly packaged into gametes. Mendel, however, had no knowledge of chromosomes.

The logic of the experiment outlined in Figure 1-8 is quite simple. If both traits from each parent were inherited together — that is, if traits that appeared together in the parent plants always appeared together in the offspring plants — then the pattern of inheritance would be identical with the pattern observed for the inheritance of a single trait. Thus, in the F_2 generation, we should find the expected 3 : 1 ratio for both traits inherited as a single unit. In other words, three-fourths of the plants would produce round, yellow seeds and one-fourth would produced wrinkled, green seeds. But that pattern does not appear. Instead, four different classes of plants appear in the ratio of 9 : 3 : 3 : 1, and each of the two related pairs of traits (round seed/wrinkled seed and green seed/yellow seed) appears in the 3 : 1 ratio: twelve plants produce

P	Round, yellow seeds × Wrinkled, green seeds

P: Round, yellow seeds RR GG × Wrinkled, green seeds rr gg

F_1: All round, yellow seeds Rr Gg

F_2:

	RG	Rg	rG	rg
RG	RRGG Round, yellow	RRGg Round, yellow	RrGG Round, yellow	RrGg Round, yellow
Rg	RRGg Round, yellow	RRGG Round, green	RrGg Round, yellow	Rrgg Round, green
rG	RrGG Round, yellow	RrGg Round, yellow	rrGG Wrinkled, yellow	rrGg Wrinkled, yellow
rg	RrGg Round, yellow	Rrgg Round, green	rrGg Wrinkled, yellow	rrgg Wrinkled, green

Ratio: 9 round, yellow : 3 round, green : 3 wrinkled, yellow : 1 wrinkled, green

12 round : 4 wrinkled = 3 : 1
12 yellow : 4 green = 3 : 1

FIGURE 1-8 A Punnett-square diagram showing how Mendel derived the law of independent assortment. Four combinations of phenotypes are observed in the ratio 9 : 3 : 3 : 1, which is the product of the individual 3 : 1 ratios. (The product of two quantities indicates their mathematical independence.)

round seeds and four produce wrinkled seeds; twelve plants produce yellow seeds and four produce green seeds. This characteristic ratio of 9 : 3 : 3 : 1 for two independently inherited pairs of factors defines Mendel's second law: the law of independent assortment.

This can be formally stated as follows. One pair of alleles (genetic factors) will segregate independently of a second pair if they are on different chromosomes (or otherwise unlinked).

To appreciate Mendel's insight, we must be aware that the 9 : 3 : 3 : 1 ratio is obtained *only* if the two different factors (genes) are inherited independently of each other and *only* if both traits are expressed completely in a dominant and recessive manner. If the pairs of genes for the different traits Mendel studied had been located close to each other on the same chromosome or if they had influenced each other's affect on the trait, Mendel would *not* have observed the $9:3:3:1$ pattern in the F_2 generation.

Mendel knew nothing about chromosomes, genes, or the positions of genes on chromosomes. And yet his reasoning and conclusions regarding the inheritance of traits were correct. Gregor Mendel, an academic failure whose work was long ignored had discovered the basic rules of inheritance. Several unresolved mysteries still surround Mendel's data and discoveries, however; they are discussed in Box 1-2.

Other Forms of Gene–Gene Interaction

Although it is true that genes (the discrete hereditary factors) ultimately determine an organism's traits, most alleles (the various alternative forms of a gene) are not expressed in the clear-cut dominant-recessive manner that marked Mendel's pea experiments. More often than not, the characteristics produced by genes interact both with one another and with the environment. Both interactions alter the expression of a gene—that is, the actual effect of the gene on the observed trait.

If one allele at a locus does not completely mask the expression of the other allele at that locus, the alleles are said to show **incomplete dominance**. For example, when certain true-breeding plants that produce white flowers are crossed with true-breeding plants that produce red flowers, only pink flowers will appear in the F_1 generation. If the alleles at a given locus are

expressed equally, they are said to show **codominance**. In human beings, the A and B alleles that determine the AB blood type are codominant alleles. A person with both an A allele and a B allele has AB type blood.

Yet another form of gene–gene interaction, called **overdominance**, is also commonly observed, especially in plants. Gene–gene interaction is overdominant when a heterozygous organism (*Aa*) shows a more-extreme phenotype than either homozygous type (*AA* or *aa*). For example, suppose allele *T* determines that a plant will be tall (about 2 feet in height) and allele *t* determines that it will be short (about 1 foot in height). The *T* allele is dominant and the *t* allele is recessive in conformity with the classical Mendelian definition. One would predict that plants with the genotype *TT* or *Tt* would be tall and plants with the genotype *tt* would be short. However, it is sometimes observed that plants with a *Tt* genotype are generally considerably taller (say, 3 feet) than plants with either the *TT* or the *tt* genotype. This phenomenon of overdominance, also referred to as **hybrid vigor**, plays an important role in plant breeding (Figure 1-9). Seed companies, for example, produce hybrid seeds from carefully maintained stocks of parent plants. Farmers buy these seeds every year because the hybrid plants that they produce are more productive than plants produced from seeds collected in their own fields.

In another sort of gene–gene interaction, an allele at one locus on a chromosome may completely mask the expression of alleles at a distant locus. This masking effect, or action-at-a-distance, is called **epistasis**. (Do not confuse epistasis with dominance, which is the masking of the expression of one allele by another at the same locus.) A classic example of epistasis in human beings is the so-called Bombay phenotype. In a person having this phenotype, the usual (ABO) blood type is completely masked by

FIGURE 1-9 An example of hybrid vigor, or overdominance, in corn. Hybrid corn (middle ear) is significantly larger and more productive than ears from either of the parental plants (at the left and at the right). (Courtesy of A.R. Hallauer, Corn Research Project, Iowa State University.)

the recessive allele *h*, located at a different locus from the *ABO* locus. An A-, B-, or AB-type person who is also homozygous for the *h* allele expresses an O-type blood phenotype. Normally, the O phenotype is expressed only if a person is homozygous for the recessive alleles *oo*. In the Bombay phenotype, the *A* and *B* alleles are present but are not expessed in the presence of the homozygous *hh* alleles.

Gene–Environment Interactions

Genes are affected not only by the expression of other genes, but also by such environmental factors as temperature, chemicals, infectious microorganisms, hormones, and other agents.

When an identical pair of alleles carried in separate cells of an individual organism are expressed in different ways, they are said to exhibit **variable expressivity**. A familiar example is the dark color at the tip of the tail, ears, and paws of Siamese cats, which otherwise have a uniformly light coat color (Figure 1-10). Although all the skin cells of these cats carry the same alleles for coat color, the temperature of each part of the body during development influences the degree of pigmentation. Because the tips of an animal's tail, nose, paws, and ears are generally cooler than the main part of its body, the fur that develops there is darker. Variable expressivity affects many traits in human beings, from baldness to disease susceptibility. Failure to understand the significance of variable expressivity (and other aspects of gene–environment interactions that will be discussed in later chapters) contributes to the mistaken, but widely held, belief that all persons

FIGURE 1-10 A Siamese cat showing the dark color at the extremities. Lower body temperature changes the function of an enzyme responsible for pigmentation. (Courtesy of Arthur and Nancy Bartlett.)

The Mysteries of Mendel's Experiments

BOX 1-2

No one questions the remarkable insights Mendel had into the mechanisms of inheritance. His discoveries are all the more astonishing because he worked alone in a remote monastery in what is now Czechoslovakia, with no collaborators or support for his ideas or his experiments. Even when he communicated his ideas to other scientists in letters, his research was either ignored or regarded as having no merit.

The importance of Mendel's discoveries has been discussed in this chapter—his view of a discrete particle, or factor, of inheritance; the bold concept of dominant and recessive factors; and the realization that precise statistical analysis of phenotypes can correctly indicate genotypes. There is no mystery regarding these accomplishments them-

selves. However, there is considerable mystery surrounding the source of his ideas and the remarkable precision of the data he obtained.

The first mystery concerns his selection of traits to study, not to mention how he chose the pea plant over all other plants. Keep in mind that, at the time Mendel began his experiments, nothing was known about the existence of chromosomes or genes. Yet Mendel chose to study seven traits—and it just so happens that peas have seven haploid chromosomes that segregate independently in mieosis. Even more fortunate for Mendel, four of these traits are determined by genes that lie on different chromosomes, and the other three

genes are so far apart on the chromosomes that they behave as if they assort independently because the genes become randomly separated by recombination during mieosis (see Chapter 2).

How did Mendel manage to choose seven traits that would all behave as if they were unlinked and segregate randomly to produce 3 : 1 ratios? Was it luck, or did he do preliminary experiments to determine which traits would behave "correctly"? If the latter, then Mendel must have divined the answer to his experiments even before he began them.

Then there is the mystery of how Mendel chose seven traits that unambiguously expressed dominant or recessive characteristics. Was this also luck, or did he deliberately exclude from his experiments any traits that did

carrying a particular disease-causing gene will inevitably develop symptoms of the disease and will suffer more or less to the same degree.

A single gene (or a single allele) may produce multiple effects on the phenotype, a phenomenon called **pleiotropy** (literally, "many turnings"). Whereas the genes that Mendel studied affected only one characteristic (a single phenotype such as seed color, shape, and so forth), many genes in animals affect several aspects of a phenotype. Pleiotropy is particularly evident in persons who carry defective alleles

that cause disease. *Phenylketonuria* (PKU), for example, is a human hereditary disease caused by a pair of recessive alleles, each of which produces a defective enzyme that is unable to chemically process the amino acid phenylalanine. (*Enzymes* are proteins that catalyze specific chemical reactions. They will be discussed in Chapter 4.) As a result of that enzyme deficiency, people who have PKU will be mentally retarded if they are not restricted to a diet low in phenylalanine immediately after birth. If untreated, they have abnormally small heads, de-not behave in the desired way? And how did he conceive of the idea of heredity being determined by discrete factors whose expression is always dominant or recessive?

Although this question cannot ever be answered with any assurance, one speculation is that Mendel got his brilliant ideas by thinking about sex. He may have noticed that individuals are always either male or female, never a mixture of the two. Because male and female characteristics are not blended in progeny but remain unchanged generation after generation, Mendel might have reasoned that this should be true for other traits as well.

Finally, how did Mendel obtain such good data? The numbers of plants of each phenotype that Mendel reported in 1865 are extremely close to the expected 3 : 1 ratio (see the accompanying table)—too close, according to recent statistical analysis of the data. If the same experiments are repeated today and comparable numbers of plants of each phenotype scored, the ratios rarely approximate the 3 : 1 ratio as closely as Mendel reported for his experiments.

We will never know how Mendel came upon his genetic insights. The monk who became head abbot of the monastery after Mendel's death destroyed all his notebooks and papers.

READING: H. Stubbe, *A History of Genetics.* MIT Press, 1968.

Mendel's data.

Parental characteristics	F₁	F₂	F₂ ratio
Round × wrinkled seeds	All round	5,474 round : 1,850 wrinkled	2.96 : 1
Yellow × green seeds	All yellow	6,022 yellow : 2,001 green	3.01 : 1
Gray × white seedcoats	All gray	705 gray : 224 white	3.15 : 1
Inflated × pinched pods	All inflated	882 inflated : 299 pinched	2.95 : 1
Green × yellow pods	All green	428 green : 152 yellow	2.82 : 1
Axial × terminal flowers	All axial	651 axial : 207 terminal	3.14 : 1
Long × short stems	All long	787 long : 277 short	2.84 : 1

fective tooth enamel, various skin problems, and a tendency toward convulsions and seizures. All of these pleiotropic effects result from the presence of a particular pair of recessive alleles and the absence of a normal functional enzyme, the lack of which can have serious consequences. However, management of the diet (that is, the environment) can alleviate almost all of the effects of the defective genotype.

Sometimes disease-causing genes may be present in an organism and yet produce no symptoms whatsoever. The nonexpression of a gene that ordinarily produces a particular phenotype is called **incomplete penetrance**. The failure of a gene to produce the same effects in different individual organisms results from genetic or environmental influences, usually a combination of both. In human beings, for example, certain members of the same family may exhibit a condition known as *polydactyly* (extra digits), whereas other members do not. Why the defective gene produces an abnormal phenotype in some persons but not in others is not understood.

Three Human Traits Governed by Single Dominant-Recessive Genes

BOX 1-3

Numerous traits in plants—including height at maturity; the color of flowers, leaves, and seeds; and the shape of leaves and fruits—are determined by single genes that are often expressed in a dominant or recessive manner. Some traits in animals also are governed by simple dominant-recessive genes. In human beings, many traits are governed by complex interactions of several genes with each other and with the environment. Thus, they do not obey the simple rules of Mendelian inheritance. Although the rules of inheritance for people *are* the same as for plants, human traits often appear to be inherited in a more-complex manner because of the ways in which genes are expressed and the ways in which they interact with other genes and with the environment.

However, three easily recognized human traits *are* determined by single genes that are expressed in a dominant or recessive fashion. The ability to roll one's tongue is determined by a dominant allele. If you possess

one or both copies of the dominant allele, and about 85 percent of people do, you can roll your tongue without any effort (see the accompanying photograph). The product of this gene controls transverse muscles in the tongue. If you possess only recessive alleles for this trait, no amount of effort will enable you to roll your tongue.

Another normal human trait is pattern baldness, which also appears to be determined by a single gene. This trait, however, is said to be *sex-limited* because the expression of the gene and the

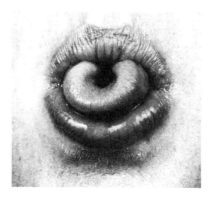

degree of baldness is affected by the presence or absence of particular male and female sex hormones. Men who carry either one or both of the dominant pattern-baldness alleles lose their hair (regardless of what magazine and newspaper advertisements say about preventing baldness and restoring hair). In women, on the other hand, even if both dominant alleles are present, there may be some thinning of the hair but rarely complete baldness. This example demonstrates how cellular chemistry may affect gene expression.

A third human trait that is the result of a single dominant-recessive gene is the ability to taste the bitterness of the chemical phenylthiocarbamide (PTC). About 70 percent of white Americans and 90 percent of black Americans are tasters—that is, they experience a bitter taste if a speck of PTC is placed on their tongue. Because the gene for PTC taste behaves in a Mendelian fashion, a taster has at least one dominant allele of this gene; nontasters have two recessive alleles.

Although several thousand so-called Mendelian traits—traits determined by single gene loci—have been identified in human beings, most human traits, especially behavioral traits,

are determined by many genes, each of which has a small effect on the overall phenotype (see Box 1-3). Intelligence, for example, is a trait that is determined by many genes interacting with many environmental factors. No single gene, and not even a small number of genes, determines intelligence. Complex human traits such as intelligence, musical ability, or personality are determined by complex gene–environment interactions and are generally referred to as polygenic-multifactorial traits. (Much more will be said about the inheritance of these kinds of traits in later chapters.) For example, we know that malnutrition in the course of fetal development and in early childhood adversely affects brain development and intelligence, though we do not understand the precise mechanism. The measurement of small phenotypic differences caused by the cumulative effects of many genes is called *quantitative genetics*, a science that has contributed greatly to the breeding of improved varieties of plants and domestic animals.

As we have seen, Mendel's insights and experiments laid the foundation for the science of genetics and the study of inheritance. However, many traits in plants and most traits in animals are not determined altogether by the rules of inheritance discovered by Mendel. Before proceeding to more-complex examples, however, we must examine the structure and transmission of chromosomes and the mechanisms by which genes are expressed and regulated.

Summary

Classical genetics refers to the study of patterns of inheritance undertaken before the chemical nature of the gene was known. Through breeding experiments with the common garden pea, *Pisum sativum*, Gregor Mendel, the father of genetics, discovered that each individual plant contains a pair of discrete hereditary factors (now called genes) that determine a given trait, such as seed color. Alternative states of genes (called alleles) can be dominant or recessive.

Mendel formulated two laws governing inheritance: in accord with the law of segregation, different alleles governing a given trait segregate from one another in the transmission of genetic information from generation to generation; in accord with the law of independent assortment, one pair of alleles segregates *independently* of another pair of alleles if the loci of the two pairs are on different chromosomes.

Since Mendel's time, interactions between genes (incomplete dominance, codominance, overdominance, and epistasis) and between genes and the environment (variable expressivity, pleiotropy, and incomplete penetrance) have been found to affect the phenotypic expression of genes.

Key Words

alleles   Alternative functional states of the same gene in homologous chromosomes.

chromosomes   Structures in the nuclei of eukaryotic cells that carry the genetic information in the form of DNA molecules.

codominance   Two alleles that are expressed equally.

diploid   The chromosomal state of a cell or organism in which each different chromosome is present in two copies (2N).

DNA   Deoxyribonucleic acid, the macromolecule carrying the hereditary information in the chromosomes of cells.

dominant allele   The allele in a heterozygous organism that determines the phenotype.

epistasis   The masking of the expression of a gene by a gene located elsewhere in the genome.

gene   A discrete hereditary unit located at a specific position (locus) on a chromosome; also a sequence of bases in DNA that codes for a protein.

genotype   The particular set of genes present in the chromosomes of an organism.

haploid   The chromosomal state of a cell in which each different chromosome is present in one copy; sperm and eggs are haploid cells in animals.

heterozygous   Having different alleles at the same locus on homologous chromosomes.

homozygous   Having identical alleles at the same locus on homologous chromosomes.

hybrid plant   A progeny plant with a different combination of traits from those observable in either parent.

incomplete dominance   Neither of two alleles is completely dominant or recessive.

incomplete penetrance   The failure of a gene to produce the same effects in different individuals.

overdominance   Also called hybrid vigor; heterozygous plants have a more vigorous phenotype than do homozygous plants.

pangenesis   The concept that each part of an adult organism produces a tiny replica of itself that is collected in the "seed" of the organism and then transmitted to offspring.

phenotype   The observable characteristics (traits) of an organism that result from the interaction of its genotype with the environment.

phenylketonuria (PKU)   An inherited human disease caused by a defect in the gene that converts phenylalanine into tyrosine.

pleiotropy   The ability of a gene to affect several unrelated traits in an individual organism.

Punnett square   A checkerboard method devised by R.C. Punnett showing the types of gametes in a cross and the phenotypes produced.

recessive allele   The allele in a heterozygous individual organism that is not observable in the phenotype.

variable expressivity   The range of phenotypes expressed by a given genotype in a defined environment or range of environments.

Additional Reading

Crow, James F. "Muller, Dobzhansky, and Overdominance." *Journal of the History of Biology*, Fall, 1987.

Falk, R. "The Gene in Search of an Identity." *Human Genetics*, 68, 195–204 (1984).

Keller, E. "McClintock's Maize." *Science 81*, August, 1981.

Lewontin, R.C. "Polymorphisms and Heterosis: Old Wine in New Bottles and Vice Versa." *Journal of the History of Biology*, Fall, 1987.

Mendel, G. "Experiments in Plant Hybridization." In C.I. Davern, ed., *Genetics*. W.H. Freeman, 1981.

Miller, J.A. "Mendel's Peas: A Matter of Genius or of Guile?" *Science News*, February 18, 1984.

Olby, R. *Origins of Mendelism*. University of Chicago Press, 1985.

Study Questions

1   What important fact can you deduce from knowing that a trait segregates in a 3 : 1 ratio in the F_2 generation?

2   What is the minimum number of alleles required to demonstrate Mendel's law of independent assortment?

3   How many different traits in peas did Mendel study?

4   Is a given gene expressed to the same degree in all individual organisms that carry that gene?

5   List four different ways in which the expression of a gene can be affected by other factors.

6   Describe three major areas of genetics that scientists study.

7   What is meant when we refer to a person's genotype? To his or her phenotype?

Essay Topics

1   Discuss why the ideas of pangenesis persisted for so many centuries.

2   Discuss the most-important scientific contributions of Mendel and Darwin.

3   Explain the difference between classical genetics and molecular genetics.

4   Discuss one or more ways in which Mendel might have been lucky in his experiments.

2

Mitosis and Meiosis

Chromosomes Carry Genetic Information

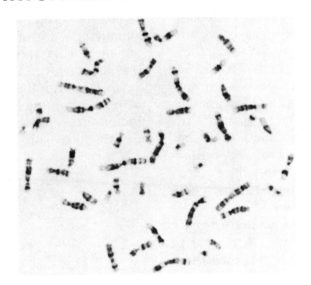

Without the person of outspoken opinion, without the critic, without the visionary, without the nonconformist, any society of whatever degree of perfection must fall into decay.

BEN SHAHN, *artist*

ENDEL had to hypothesize the existence of "discrete hereditary factors" in the paper that he published in 1866. Earlier, in the *Origin of Species*, published in 1859, Charles Darwin also had to hypothesize the existence of "hereditary factors" to explain how organisms evolve through time by *natural selection*. Darwin described that process as the differential reproduction of genetically different organisms in response to their ability to survive and reproduce in a particular environment. The reason that Mendel and Darwin were obliged to invoke hypothetical "hereditary factors" is that chromosomes had not yet been identified as the carriers of genetic information.

In 1590, in Holland, a primitive form of the light microscope was invented. Three hundred years later, German scientists managed to improve the microscope to the point at which they could observe chromosomes in the nuclei of dividing cells. Accompanying that advance was the discovery of chemicals that could be used to stain chromosomes and other components of cells. In 1882, Walther Flemming coined the term *chromatin* (from the Greek *chroma*, meaning color) to describe the chemically stained material that he observed while studying cells under the microscope (Figure 2-1). As he watched cells grow and divide, he noticed that the chromatin split longitudinally into two parts just before the cells divided. He proposed that each part might be transmitted to a new cell resulting from the division, in a process he called *mitosen* (from the Greek *mitos*, meaning thread), now called mitosis. In the growth cycle of most plant and animal cells, **mitosis** is the process by which chromosomes that have doubled divide and separate into two identical daughter cells.

All multicellular organisms are made up of two classes of cell. In plants and animals, the major class consists of **somatic cells**, which include all cells except reproductive cells. The **reproductive cells** in male animals are in the reproductive organ called the **testis** (pl. testes), which produces male sex cells called sperm. (In plants, the male sex cells are called pollen). In female plants and animals, the reproductive organ is the **ovary**, which produces cells called eggs (ova). Sperm, pollen, and ova are single-celled entities called **gametes**. Each gamete contains a haploid number (one complete set) of chromosomes. Because a gamete contains only a single copy of each chromosome, a process must exist by which plants and animals can reduce the diploid number of chromosomes in somatic cells to the haploid number found in gametes. That process is called **meiosis**. Through meiosis, the proper set of chromosomes is transmitted to the next generation.

To maintain the fidelity of the genetic information that is passed on to the next generation, meiosis must be highly accurate. If a chromo-

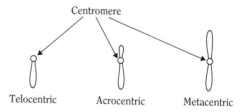

FIGURE 2-1 Three basic chromosomal structures. If the centromere is located at the extreme end, the structure is said to be telocentric. If the centromere is near the middle, the structure is metacentric, and if the centromere is noticeably off center, the structure is acrocentric. The centromere location is important in identifying particular human chromosomes and for recognizing chromosome defects.

some is altered or damaged in the course of meiosis, or if the wrong number of chromosomes is segregated into the gametes, the offspring may suffer disastrous consequences (see Box 2-1). However, a normal aspect of meiosis is the exchange of genetic information between homologous chromosomes. This exchange, called **crossing-over**, results in new combinations of genes (**recombination**).

When single-celled eukaryotes reproduce by dividing and when multicelled plants and animals grow from fertilized eggs into organisms, they do so by cell division. The series of steps in the production of two genetically identical cells (apart from the occasional mutation) from one cell constitute the *cell cycle*. The cell cycle includes cell growth, DNA replication (and thereby chromosome duplication), separation of the daughter chromosomes by the process of mitosis into daughter nuclei, and finally cell division.

This chapter deals with meiosis and mitosis, the fundamental processes in plants and animals by which chromosomes are distributed into newly synthesized cells during development, growth, and reproduction.

Mitosis

All somatic cells reproduce their chromosomes by mitosis; only reproductive cells reproduce their chromosomes by meiosis. Because, in some respects, mitosis is a simpler process than meiosis, the phases of mitosis that can be observed under the microscope will be considered first. Again, the function of mitosis is to transmit two identical sets of genetic information (contained in the chromosomes) to a pair of dividing cells. Because the parent cell is diploid, each of the two daughter cells must also be diploid.

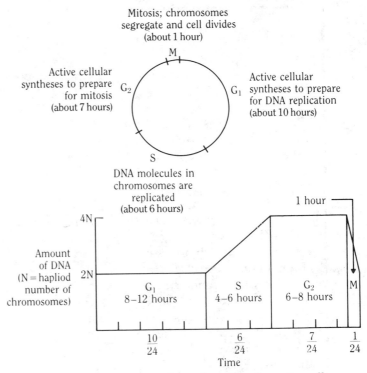

FIGURE 2-2 The growth cycle of eukaryotic cells. Growth occurs in two stages, called G_1 and G_2. DNA synthesis occurs during the S stage. Mitosis (M) is the shortest period of the cell cycle; this is when the cell divides and chromosomes are separated.

Mitosis is only one part of the full cycle of cell growth and division. For many human and other animal cells, that cycle takes about 24 hours to complete (Figure 2-2A). The complete cycle consists of four periods: G_1, S, G_2, and M. During G_1 (for gap-1), the cell undergoes normal metabolism and prepares for the replication of DNA (the synthesis of new strands). Replication takes place in the S (synthesis) period. This process results in two identical DNA molecules (two identical copies of the genetic information) that remain attached at a small unreplicated region until the midpoint of mitosis. The period of DNA synthesis is followed by another period of cellular growth and changing metab-

The Fragile X Chromosome

BOX 2-1

A goal of genetics is to be able to correlate observable chromosomal abnormalities with heritable human diseases: the classic example is the correlation between an extra chromosome 21 and Down syndrome. A *cytogeneticist* studies chromosomes by examining chromosomal structure at different stages in a cell's life cycle. Metaphase is the most useful stage at which to examine chromosomes because this is when they have condensed in the nucleus and are most visible.

Some chromosomal aberrations—for example, the presence of an extra chromosome, the loss of an entire chromosome, or the translocation of large chromosomal segments between nonhomologous chromosomes—are easily detected. But the detection of chromosomal abnormalities becomes increas-ingly difficult as the genetic defects become physically smaller. However, modification of the conditions in which the cells are grown in the laboratory may assist in the detection of small chromosomal aberrations that otherwise go unnoticed.

By varying cell-culture conditions, cytogeneticists in several laboratories have been able to identify a human chromosomal abnormality known as the *fragile X chromosome*, which is found in the cells of certain mentally retarded males (see the accompanying illustrations). Observe that this X chromosome has two faintly stained knoblike structures that are situated at the end of the long arm of the chromosome but that are only faintly connected to it—hence, the name *fragile X*. The significance of these tiny structures at the end of the X chromosome is not known, but they are not present in normal X chromosomes. Because some (but not all) mentally retarded males have this abnormal X chromosome, it is assumed that they inherited it from their mothers. Some of the mothers are normal, but others are slightly mentally retarded, indicating that the normal X chromosome in some women may, in some way, mask the effects of having an abnormal X chromosome.

As better cytological techniques are developed, it may become possible to detect this fragile X chromosome in men and women who are carriers. Genetic counseling and amniocentesis could then be offered to those women who are at high risk for having mentally retarded offspring. It should be emphasized that as yet having a fragile X chromosome does not guarantee that mental retardation is inevitable because some form of activation also seems to be required.

READING: G.R. Sutherland, "The Enigma of the Fragile X Chromosome," *Trends in Genetics.* April, 1985.

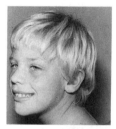

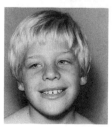

The mentally retarded boy in these photographs carries a fragile X chromosome. (Courtesy of G.R. Sutherland. From *Fragile Sites on Human Chromosomes* by G.R. Sutherland and F. Hecht, Oxford Monographs Medical Series, No. 13. New York, Oxford University Press, 1985.)

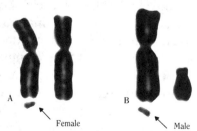

One of the maternal X chromosomes in part A is a fragile X chromosome (indicated by the arrow), as is the X chromosome in part B from a male karyotype. (Courtesy of G.R. Sutherland. From *Chromosome Abnormalities and Genetic Counselling* by R.J.M. Gardner and G.R. Sutherland. New York, Oxford University Press, 1988.)

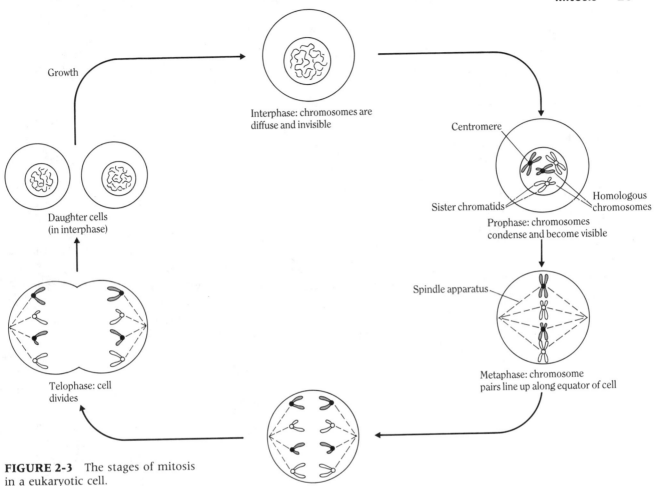

Growth

Interphase: chromosomes are diffuse and invisible

Centromere

Homologous chromosomes

Sister chromatids

Prophase: chromosomes condense and become visible

Spindle apparatus

Metaphase: chromosome pairs line up along equator of cell

Daughter cells (in interphase)

Telophase: cell divides

Anaphase: chromosomes separate

FIGURE 2-3 The stages of mitosis in a eukaryotic cell.

olism, G_2 (for gap-2), during which the cell is prepared for mitosis. These first three stages of the cell cycle — G_1, S, and G_2 — constitute what is called **interphase**. The last stage in the cell cycle, M (for mitosis), is the period during which the identical chromosome complement that existed in the parent-cell nucleus is distributed into each of the two newly formed daughter-cell nuclei. The cell cycle is completed with cytokinesis, division of the parent cell's cytoplasm between the two new cells. Cell growth therefore includes DNA replication and mitosis (nuclear division) followed by cytokinesis (cell division).

As mitosis begins, the parent cell is still diploid (2N); however, because each chromosome consists of two strands of newly replicated DNA, the nucleus contains twice the amount of DNA as was present in the early interphase nucleus. During the first phase of mitosis, called **prophase**, the chromosomes become visible as they gradually condense. The double nature of the chromosomes becomes apparent. Each chromosome consists of sister **chromatids** (the replicated DNA strands complexed with proteins), which are held together at a region of the chromosome called the **centromere** (Figure 2-3). In the next phase, called

metaphase, the sister chromatids line up at the center (the equator) of the cell. Structures called spindle fibers, which reach to the opposite poles of the cell, now become attached to opposite sides of the centromere. Once the sister chromatids have aligned with one another, **anaphase** begins. The centromeres and the chromatids separate and a copy of each chromosome is pulled to one pole or the other of the cell. (By this time, the chromatids have become independent chromosomes.) During the final phase of mitosis, called **telophase**, the chromosomes begin to decondense, the cell cytoplasm divides (cytokinesis), and a complete complement of chromosomes (2N) becomes encased in a new nuclear membrane in each of the two daughter cells. As the newly formed cells enter interphase, the cell cycle is completed.

The four phases of mitosis are not as well defined as Figure 2-3 implies. Mitosis is a continuous process, and the end of one phase shades gradually into the beginning of the next. Moreover, mitosis is more readily observed in some kinds of cells than in others. It is certain, however, that mitosis occurs whenever plant and animal cells divide.

Meiosis

Meiosis differs from mitosis in four respects: (1) meiosis takes place in reproductive cells but not in somatic cells; (2) meiosis reduces the number of chromosomes in the gametes from the diploid number (2N) to the haploid number (N); (3) **homologous chromosomes** (those containing the same linear sequence of genes) pair with precise alignment during meiosis but not during mitosis; (4) during meiosis, homologous pairs of chromosomes recombine to create new combinations of genetic information (Figure 2-4).

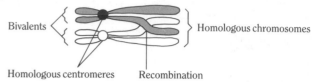

FIGURE 2-4   Formation of a bivalent during meiosis. The pairing of homologous chromosomes carrying different genetic information allows new combinations of information by recombination.

The function of meiosis is to reduce the chromosome number in gametes to the haploid number and to maintain the diploid number of chromosomes in plants and animals from one generation to the next. If it were not for meiosis, each generation of organisms would have *twice* as many copies of each chromosome as the preceding generation—clearly a situation that would soon get out of hand. Most species of plants and animals are diploid, although the number of different chromosomes varies greatly from one species to another (Table 2-1). Errors in meiosis sometimes occur, leading to eggs or sperm that have deficiencies or duplications of particular chromosomes. Embryos derived from such defective gametes often have severe defects. Many of these severely defective embryos abort before term, but some are born with birth defects (see Box 2-1).

Meiosis does much more than preserve single copies of the two chromosome sets that combine at fertilization to produce a new individual. In meiosis, chromosomes recombine to produce a chromosome set that is different from the gene combinations carried on either the paternal chromosome set or the maternal set.

The following calculation will give you some idea of the variety of gametes that can be generated by meiosis. Human cells contain 23 different pairs of chromosomes; the diploid (2N) number is 46 and the haploid (1N) number is 23. Any sperm or egg may receive the chromo-

TABLE 2-1 The number of chromosomes differs greatly among species. Note that the number of chromosomes in cells does not bear any relation to the complexity of the organism. For example, turkeys have 82 chromosomes in their cells, whereas human beings have 46; both potatoes and chimpanzees have 48 chromosomes.

Animal	Diploid chromosome number	Plant	Diploid chromosome number
Turkey, *Meleagris galiopavo*	82	Upland cotton, *Gossypium hirsutum*	52
Chicken, *Gallus domesticus*	78	Potato, *Solanum tuberosum*	48
Dog, *Canis familiaris*	78	Tobacco, *Nicotiana tabacum*	48
Horse, *Equus caballus*	64	Bread wheat, *Triticum aestivum*	42
Donkey, *Equus asinus*	62	Cherry, *Prunus cerasus*	32
Chimpanzee, *Pan troglodytes*	48	Tomato, *Solanum lycopersicum*	24
Rhesus monkey, *Macaca mulatta*	48	White oak, *Quercus alba*	24
Human being, *Homo sapiens*	46	Yellow pine, *Pinus ponderosa*	24
Rat, *Rattus norvegicus*	42	Rice, *Oryza sativa*	24
House mouse, *Mus musculus*	40	Bean, *Phaseolus vulgaris*	22
Cat, *Felis domesticus*	38	Radish, *Raphanus sativus*	18
Frog, *Rana pipiens*	26	Garden pea, *Pisum sativum*	14
House fly, *Musca domestica*	12	Barley, *Hordeum vulgare*	14
Fruit fly, *Drosophila melanogaster*	8	Rye, *Secale cereale*	14
Mosquito, *Culex pipiens*	6	Cucumber, *Cucumis sativus*	14

some for any one of the 23 homologous pairs from either the father or the mother. So the number of different gametes that can be produced by independent assortment of chromosomes into sperm or eggs alone is 2^{23}, or more than 8 million. And, if we consider the number of different gene combinations in chromosomes produced by recombination during meiosis,

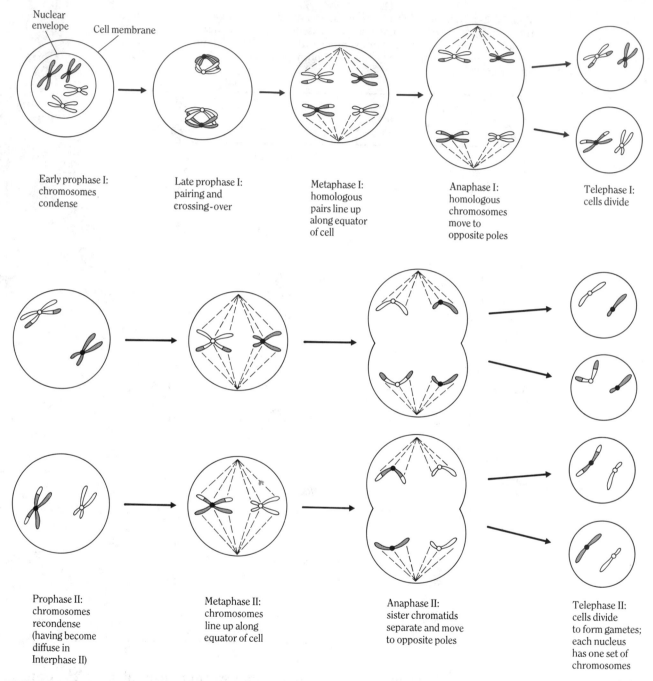

FIGURE 2-5 The stages of meiosis I (the first cell division) and the stages of meiosis II (the second cell division). Each stage is defined by the characteristic appearance of the chromosomes.

probably no two gametes synthesized in an organism are ever genetically identical.

Meiosis proceeds through eight phases that can be observed microscopically (Figure 2-5). In an early phase (prophase I), the replicated pairs of homologous chromosomes line up side by side and join together to form a tetrad of chromatids in a process known as **synapsis** (Figure 2-6). Shortly after synapsis, information is exchanged between the chromosomes in the process of recombination, which is also called crossing-over. (This term should not be confused with a genetic cross, which is a mating between genetically different individuals.) It is at this time that segments of DNA in the paired chromosomes are physically exchanged, producing new gene combinations in the chromosomes that end up in the gametes. Subsequent phases of meiosis resemble the phases of mitosis and have the corresponding names. The following discussion summarizes the essential features of the eight phases of meiosis (interphase I and II are not counted as part of the eight phases).

Interphase I As in mitosis, interphase I comprises periods G_1, S, and G_2 of the cell cycle. The most significant genetic event that takes place in this phase is the duplication of DNA and the doubling of the chromosome number during the S period. The cell is now 4N.

Prophase I Early in prophase I, the chromosomes condense and become visible. As the phase continues, the homologous chromosome pairs line up side by side (synapsis) to produce a structure called a **bivalent**, which contains four chromatids called a **tetrad**, as shown in Figure 2-6. Identical chromosomes in each bivalent contain sister chromatids; nonidentical chromosomes contain nonsister chromatids. Toward the end of prophase I, both sister and

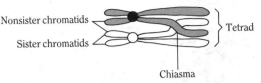

FIGURE 2-6 Synapsis of homologous chromosomes. The two pairs of sister chromatids line up alongside one another to form a tetrad. They are held together by at least one chiasma, where crossing over occurs. This figure shows additional features of the bivalent structure in Figure 2-4.

nonsister chromatids form structures called chiasmata (sing. **chiasma**), in which gene combinations change through crossing-over—that is, by the physical breaking and rejoining of DNA molecules. More than one chiasma may be formed between sister and nonsister chromatids in a tetrad, resulting in multiple recombination events. However, chromosomes can also be segregated having the same genetic information because a chiasma between sister (identical) chromatids does not cause any genetic change. In the event that crossing-over does take place between nonsister chromatids, the genetic information in each recombinant chromosome differs from that in the other (Figure 2-7). Prophase I ends when spindle fibers first appear and when the nuclear membrane of the cell disappears.

Metaphase I By metaphase I, the chromosomes are highly condensed and bivalents have lined up along the equator of the cell. The spindle fibers, which are now visible, become attached to the centromeres of the chromosomes and to one or the other pole of the cell.

Anaphase I In anaphase I, pairs of identical chromosomes containing sister chromatids separate and move to opposite poles of the cell. The centromeres do not split apart, as they do in the anaphase of mitosis. Moreover, because of crossing-over, the paired chromatids may no

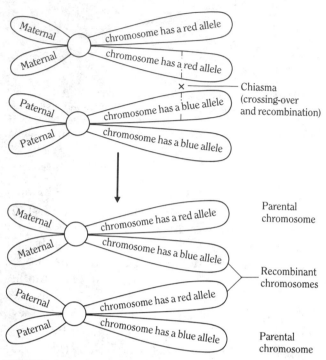

FIGURE 2-7 Recombination of DNA molecules in meiosis changes the genetic information that is passed from parents to progeny. Recombination accounts for the fact that progeny have different combinations of traits from those of their parents.

longer carry genetic loci that are identical with those in the parental chromosomes. In mitosis, as we have seen, crossing-over does not occur under normal circumstances.

Telophase I In telophase I, the cell divides into two daughter cells. Each daughter cell contains a haploid set of doubled chromosomes (two chromatids). Some of those chromosomes are identical with parental types; others are recombinant owing to the exchange of fragments of maternal and paternal chromosomes differing in genetic information. By the end of telophase I, meiosis I has been completed.

Interphase II Before meiosis II begins, the chromosomes in each daughter cell usually un-

coil briefly and become invisible. However, no DNA duplication (synthesis of a new DNA molecule) occurs during interphase II, as it does during interphase I.

Prophase II In prophase II, the chromosomes again condense and become visible. There is no pairing of homologous chromosomes, however, and no chiasma formation or recombination. New spindles begin to form and become attached to the centromeres. The stages of meiosis II are basically the same as those of mitosis.

Metaphase II In metaphase II, the chromosomes are again aligned along the equator of the cell, as in metaphase I.

Anaphase II In anaphase II, the centromere divides and the individual chromosomes are pulled to opposite poles of the cell.

Telophase II In telophase II, a nuclear envelope forms around each haploid set (1N) of chromosomes, which again become invisible. The cell divides and the second meiotic division is completed.

Meiosis performs three essential functions in the reproduction of organisms. First, it maintains the diploid number from generation to generation by reducing the number of chromosomes from diploid to haploid in the gametes. Second, it ensures genetic diversity in succeeding generations by means of the random segregation and independent assortment of chromosomes into gametes. Third, through the formation of chiasmata and recombination, it provides the gametes with new combinations of genes that are different from the combinations in either of the parents. Such genetic differences may enable the offspring to survive and reproduce more successfully in a particular environment. Moreover, they form the basis for evolution and, as Darwin proposed more than a century ago, may lead to the emergence of new species.

Human Chromosomes

As geneticists examined more and more cells under the microscope early in this century, it became apparent that somatic cells always contained a complete diploid set of chromosomes. The number, size, and shapes of chromosomes are unique for each species of plant and animal. Eventually, techniques were developed that allowed chromosomes to be stained with dyes and to be photographed. The chromosomes in animal cells in particular could be arranged and numbered according to their size.

Human chromosomes can be stained and photographed in cells that have been fixed during metaphase of mitosis, revealing 23 distinct pairs in cells from normal persons. Display of a cell's chromosomes allows geneticists to detect chromosomal abnormalities. Dividing white blood cells are used for this **karyotype** analysis. Red blood cells *cannot* be used for a karyotype analysis because, as the red blood cell matures and is released from bone marrow into the circulation, it loses its nucleus. Also, not all human cells divide normally to the same extent. For example, aging may be due to a failure in cell division of some cells (see Box 2-2). Figure 2-8 shows the karyotype of a normal male human being, with the pairs of chromosomes arranged according to size. The practiced eye can detect all of the different human chromosomes, and good chromosome preparations and experience make karyotype analysis a fairly routine procedure for detecting chromosomal abnormalities.

A universally recognized numbering system allows human genes to be assigned to specific

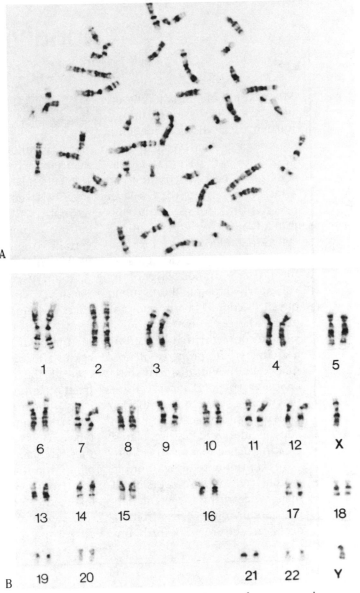

FIGURE 2-8   The ordered display of human chromosomes is called a karyotype analysis. The nuclei of cells are examined at the stage of mitosis when they are most visible—that is, at metaphase. The chromosomes are stained and photographed under the light microscope, as shown in part A. Pairs of homologous chromosomes are then arranged according to a standard numbering system from 1 to 22 and X, Y, as shown in part B. Karyotype analysis reveals any gross chromosomal abnormalities. (Courtesy of Patricia Jacobs, Director, Wessex Regional Cytogenetics Unit.)

Nondividing Cells
The Cause of Aging?

BOX 2-2

Many kinds of animal cells are now grown in tissue cultures—nutrient-rich liquid mediums contained in Petri plates or in flasks. Unlike bacteria, animal cells grow best when they can attach to a surface; thus, a layer one cell thick forms on the bottom of a Petri plate as animal cells grow and divide. Most animal cells stop dividing after forming a continuous layer of tissue, at which point all cells are in physical contact with one another. The cessation of cell division on contact with other cells is a characteristic property of normal cells. (One distinguishing characteristic of cancer cells is that they lose the ability to stop growing on contact and instead pile up in tissue-culture plates, yielding dense masses of cells.)

To continue to grow normal animal cells *in vitro* (outside of living organisms), researchers put nondividing cells from a dormant culture into fresh nutrient medium. These few cells again attach to the bottom of the Petri plate and begin growing. This sequential transfer of animal cells from a nondividing culture to new mediums allows certain strains of animal cells to be grown almost indefinitely.

About twenty years ago, two scientists discovered that cells from a human embryo can be grown and transferred in this manner but that cell division ceases after about fifty doublings, even if the cells are placed in fresh mediums and are not physically touching. More surprising was the finding that the same type of cells taken from adult tissues are capable of *fewer* doublings and that the older the person from whom the cells are taken, the fewer the number of cell divisions *in vitro*. Similarly, cells from other animals that have average life spans longer or shorter than the average human life span have been shown to have a range of cell doublings that correlates with these animals' maximum life spans. Some examples are listed in the accompanying table.

It appears, then, that body cells might be genetically programmed to undergo a finite number of cell divisions and then die. If this is the case, then the process of aging must be largely under genetic control. The idea that the rate of aging and the maximum life span of any animal species is controlled by a kind of "genetic clock" has some evidence to support it. By comparing different strains of mice. Roy Walford, a scientist at the University of California School of Medicine at Los Angeles, has been able to identify a group of mouse genes that affect the aging process. In human beings, genes on chromosome 6 correspond to the same group of genes in mice. Further studies may determine how significant the genetic contribution to aging is—and perhaps whether the human aging process can be changed.

READINGS: Leonard Hayflick, "The Cell Biology of Human Aging," *Scientific American*. January, 1980. Richard Connitt, "Living Longer," *Next Magazine*. May-June, 1981.

Species	Number of cell doublings	Estimated maximum life span (years)
Galapagos tortoise	90–125	175
Human being	40–60	110
Chicken	15–35	30
Mouse	14–28	3.5

locations on the chromosomes. For example, the allele that causes the severe hereditary disease *Lesch-Nyhan syndrome* is located at Xq26. This means that it is on the long arm of the X chromosome in segment 2, band 6. (A detailed description of how human genes are numbered and mapped is in Chapter 16.) More than 150 genes have been assigned to the X chromosome because sex-linked traits (those determined by genes on the X chromosome) are easiest to identify and map (discussed in Chapter 3). Until quite recently, only very rough estimates could be made of the total number of functional human genes.

From a knowledge of *Drosophila* genetics, it was thought that fruit flies have between 5,000 and 6,000 genes. Because human beings have fifteen times as much DNA as do flies, a human chromosome could contain about 90,000 genes. Estimates based on mouse genetics and other calculations suggest the number of human genes to be closer to 50,000. In the past few years, there has been an explosion in our understanding of human genetics. Almost 2,000 genes have been mapped to specific chromosomes, many to specific sites. This knowledge, in turn, has made it easier to detect disease-causing genes *in utero* and to screen for some recessive alleles that cause genetic disorders in susceptible families. Discussion of these topics, however, must be deferred until we learn more about the organization of chromosomes and genes.

Summary

Chromosomes, structures in the nuclei of most cells, are the carriers of genetic information. All plants and animals contain reproductive cells — those that produce gametes (eggs and sperm or pollen), which have the haploid number (1N) of chromosomes — and somatic cells — all nonreproductive cells, which have the diploid number (2N) of chromosomes.

Mitosis is the process by which a somatic cell distributes identical sets of genetic information to two daughter cells, each of which is diploid. It is one of four stages of somatic-cell division: gap-1 (G_1), DNA synthesis (S), gap-2 (G_2), and mitosis (M), followed by cytokinesis. The first three stages (G_1, S, and G_2) constitute interphase; the fourth (M) is subdivided into prophase, metaphase, anaphase, and telophase.

Meiosis is the process by which a reproductive cell distributes genetic information to four daughter cells, or gametes, each of which is haploid. To produce gametes that have the haploid number of chromosomes, reproductive cells must undergo two meiotic cell divisions. The first consists of interphase I (G_1, S, and G_2, as in somatic-cell growth), prophase I, metaphase I, anaphase I, and telophase I. The second consists of interphase II (in which the already duplicated chromosomes are not replicated), prophase II, metaphase II, anaphase II, and telophase II. In meiosis, unlike mitosis, homologous chromosomes recombine to create new combinations of genetic information.

Key Words

anaphase   The phase of mitosis following metaphase; chromosomes move toward opposite poles of the cell.

bivalent   A pair of homologous, synapsed chromosomes.

centromere   The attachment site on chromosomes for spindle fibers, which participate in the segregation of chromosomes in mitosis and meiosis.

chiasma (pl. chiasmata)   The visible physical crossing-over between sister or nonsister chromatids of homologous chromosomes at meiosis.

chromatid   One of the two daughter strands of a duplicated chromosome.

chromatin   The substances of which chromosomes are

composed; for example, nucleic acids, proteins, histones, and so forth.

crossing-over   The exchange of segments between pairs of homologous chromosomes in meiosis.

gamete   Male or female reproductive cell; in animals, it is called sperm or egg.

genetic map   The assignment of genes to specific locations on chromosomes.

homologous chromosomes   The two members of a pair of chromosomes, one member from each parent. Homologous chromosomes pair up (synapse) at meiosis.

interphase   The interval in the cell cycle between mitoses.

karyotype   Visual arrangement of all of the chromosomes from a single cell so that they can be identified and counted.

meiosis   Process by which the haploid set of chromosomes ends up in gametes (reproductive cells).

metaphase   The phase of mitosis following prophase; chromosomes line up along the equatorial plane of the cell.

mitosis   Process of chromosome segregation and cell division.

natural selection   The differential reproduction of genetically different individual organisms as a result of their adaptation to a particular environment.

prophase   The first phase of mitosis following interphase; chromosomes contract and become visible.

recombination   The breaking and rejoining of genetically different DNA molecules; the appearance of traits in progeny that were not observed in parents.

somatic cells   All cells other than reproductive cells in plants and animals.

synapsis   The pairing of homologous chromosomes in meiosis.

telophase   The phase of mitosis following anaphase; the chromosomes unwind and begin to return to their interphase condition.

tetrad   Four homologous chromatids (two in each bivalent).

X chromosome   The sex chromosome normally present in two copies in women and one copy in men.

Y chromosome   The sex chromosome normally present in one copy in men.

Additional Reading

Lloyd, A.T. "Pussy Cat, Pussy Cat, Where Have You Been?" *Natural History*, July, 1986.

Myers, E.H. "The Chromosome Connection." *The Sciences*, May-June, 1980.

Snyder, L.A., D. Freifelder, and D.L. Hartl. *General Genetics*. Jones and Bartlett Publishers, Inc., 1985.

Suzuki, D.T., A.J.F. Griffiths, J.H. Miller, and R.C. Lewontin. *An Introduciton to Genetics*. W.H. Freeman, 1986.

White, R., and D. Drayna. "The Genetic Linkage Map of the Human X Chromosome." *Science*, 230, 753–758 (1985).

Yunis, J.J. "High Resolution of Human Chromosomes." *Science*, 191, 1268–1270 (1976).

Study Questions

1   What molecules or structural features are used to distinguish between chromatin, chromosomes, and chromatids?

2   What is the purpose of performing a karyotype analysis of human chromosomes?

3   List the primary differences that distinguish mitosis from meiosis.

4   Describe the cellular processes that characterize the four stages of the cell cycle, G_1, S, G_2, and M.

5   What is the function of the centromere in chromosomes?

6   What are the three essential functions of reproduction that are accomplished in meiosis?

7   What is meant by a sex-linked trait? By a fragile X chromosome?

Essay Topics

1   Explain how meiosis and Mendel's laws are related.

2   Review the changes that take place in chromosomes during the different phases of mitosis.

3   Discuss how genetic diversity is generated by the process of meiosis.

3

Chromosomes
Normal and Abnormal

> Unpredictability is in the nature of the scientific enterprise. If what is to be found is really new, then it is by definition unknown in advance.
>
> FRANÇOIS JACOB, *biologist*

IN STUDIES of genetics, it was assumed right along that genes are located on chromosomes and that chromosomes segregate into gametes at random. Today, we know that those assumptions are correct, but for many years they were never rigorously tested. Proof of their correctness emerged from genetic studies of the fruit fly *Drosophila melanogaster* performed at the beginning of this century by Thomas Hunt Morgan, Calvin B. Bridges, and other "fruit fly" geneticists. From examination of the chromosomes in cells derived from male and female animals, it was eventually realized that the sex of an animal is determined by a special pair of chromosomes called the **sex chromosomes**.

In human beings, these are the X and Y chromosomes—XY people are male, XX are female. Chromosomes in animals other than the sex chromosomes are called **autosomes**. Thus, human beings have 22 pairs of autosomes and 2 sex chromosomes. Sex determination in human beings and other animals is more complex than simply the presence of X or Y chromosomes (discussed further in Chapter 13).

Morgan discovered that genes located on the X chromosome affect the inheritance patterns of flies in predictably different ways from those affected by the genes (factors) studied by Mendel in peas. The early *Drosophila* geneticists studied how eye color could be inherited in crosses between male and female flies that had either red eyes or white eyes. By means of these genetic crosses, they showed conclusively that (1) the hereditary information *is* carried by chromosomes and (2) the genes that determine eye color are located on the X chromosome. Their achievements preceded the discovery of the chemical nature of genes by many years.

Genes Are Located in Chromosomes

Before the genetics of *Drosophila* became known, geneticists had been unable to prove that the "discrete hereditary factors" that came to be called genes were located in chromosomes. As they examined the chromosomes in fruit-fly cells under the microscope, however, they noticed an occasional odd-shaped X chromosome. And they established unambiguously that the allele determining white eyes is always associated with that odd-shaped X chromosome. This discovery demonstrated for the first time that genes are located in chromosomes and, in particular, that the gene for eye color in fruit flies is located in the X chromosome.

Genetic Crosses Between Mutant and Wild-type Flies

In these classic experiments, the geneticists crossed true-breeding white-eyed female flies with true-breeding red-eyed male flies. The alleles that determine red eye color in flies are dominant; the alleles that determine white eyes are recessive. A mutation in the eye-color gene was discovered when rare white-eyed male mutant flies appeared in a population of wild-type red-eyed male and female flies. A **mutation** is any structural change in a gene, specifically in the chemical composition of DNA. A **mutant** refers to the organism that carries the mutation (altered gene). Organisms that predominate in nature are referred to as **wild-type**; those that are found only very rarely are called mutants.

We now know that male flies (and male human beings) have an X chromosome and a Y chromosome and that females have two X chromosomes. If the mutant white-eyed male flies are crossed with the wild-type red-eyed female flies, the eye color of progeny flies in both the F_1 and F_2 generations does not conform to what is predicted by Mendel's law of segregation. Recall that all individuals in the F_1 generation in Mendel's pea crosses had the same phenotype and that a 3 : 1 ratio of phenotypes was observed in the F_2 generation.

Figure 3-1 shows the results of crossing white-eyed (mutant) male flies with red-eyed (wild-type) female flies, assuming that the gene for eye color is on the X chromosome. All the male and female flies of the F_1 generation are red-eyed. In the F_2 generation, all the female flies have red eyes but half the males are red-eyed and half white-eyed. The Punnett squares show the genetic analysis of the crosses on the assumption that the alleles for eye color reside on the X chromosome. We now know that, if a female fly has either RR or Rr alleles, her eyes will be red; because males have only one X chromosome, a single allele determines their eye color. Repeated crosses of flies exhibiting **sex-linked traits**—that is, traits determined by genes located on an X chromosome—showed that the pattern of inheritance of such traits is different from that of the traits studied by Mendel.

If a gene is located on an autosome, it makes no difference which parent carries either one of the two alleles for that gene. Reciprocal crosses give identical results—that is, the recessive alleles can be present in either the male or the female parent. However, if a gene is located on the X chromosome, reciprocal crosses produce a different (non-Mendelian) pattern of inheritance. For example, the reciprocal cross of that illustrated in Figure 3-1 is between white-eyed female flies and red-eyed males, as shown in Figure 3-2. All the F_1 female flies are red-eyed and all the F_1 males are white-eyed. In the original cross (Figure 3-1), both males and fe-

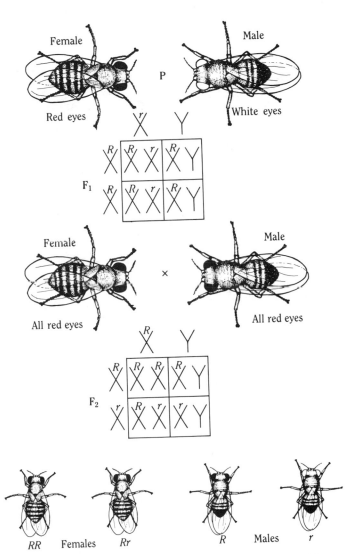

FIGURE 3-1 A genetic cross between red-eyed (wild-type) female flies and white-eyed (mutant) male flies. In the F_1 generation, all flies have red eyes. In the F_2 generation, however, all female flies have red eyes, whereas half of the males have red eyes and half have white eyes.

males in the F_1 generation are red-eyed. The pattern of inherited eye color is also different in the F_2 generations (compare Figures 3-1 and 3-2). When reciprocal crosses produce different results, we can confidently assign the gene to

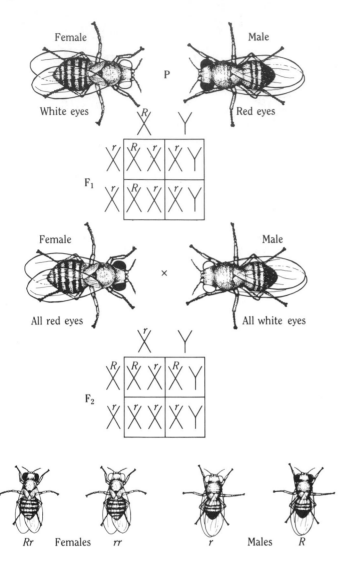

the X chromosome. (In fruit flies and human beings, the Y chromosome contains no genes that contribute to traits other than one gene for sex determination.)

Genes in Somatic Cells Are Not Inherited

We have just seen that patterns of inheritance of genes carried on X chromosomes is different from those of genes on autosomes. It is also important to remember that genes are inherited only if they are carried in sex cells—*male gametes* (sperm or pollen) or in *female gametes* (eggs). It is the union of male and female gametes and the growth of a new individual organism that accounts for the transfer of genetic information from generation to generation. However, as mentioned in Chapter 2, most of the billions of cells in plants and animals are *somatic cells*, which refers to all cells in an organism other than its sex cells.

Mutations are just as likely to occur in genes carried in the chromosomes of somatic cells as they are in genes carried in sperm or eggs. However, mutations that arise in somatic cells are *never* passed on to progeny—that is, somatic mutations are not inherited and do not cause genetic (hereditary) diseases. A mutation that arises in a gene carried in a lung cell, for example, may affect the growth of that cell, eventually causing it to grow and develop into a tumor. A person with such a tumor may be diagnosed as having cancer, but the disease cannot be passed on to future generations. Lung cancer and most other forms of cancer are not hereditary diseases, although hereditary predisposition to certain cancers is possible (discussed in Chapter 11).

FIGURE 3-2   A genetic cross between white-eyed female flies and red-eyed males. In the F₁ generation, all females have red eyes and all males have white eyes. If the F₁ generation males and females are crossed, the expected 3 : 1 ratio is not observed in the F₂ generation. Rather, red-eyed flies and white-eyed flies occur in a ratio of 1 : 1 among males and females. This deviation from the expected Mendelian pattern of inheritance led to the discovery that eye-color genes are located on sex chromosomes and are inherited in different patterns from those of genes on autosomes.

Nondisjunction of Chromosomes

In performing crosses between red-eyed and white-eyed flies, the early *Drosophila* geneticists examined the eyes of thousands of fruit flies. On rare occasions, they observed than an unexpected phenotype would crop up in a cross that otherwise gave the expected results. For example, on the basis of what was then known about the patterns of inheritance, in a cross between white-eyed females and red-eyed males, the F_1 female progeny should *always* have red eyes and the F_1 males should *always* have white eyes, as shown in Figure 3-2. And yet, among the many thousands of flies produced in such a cross, they observed an occasional white-eyed female or red-eyed male. In genetics, the rare exceptions to predictable patterns are often a clue to some new discovery.

After studying these exceptional flies, Bridges eventually realized that the white-eyed females must have inherited, in some abnormal fashion, both of their mother's X chromosomes, and that the red-eyed males must have inherited their single X chromosome from their father instead of from their mother. But how could the X chromosome be inherited in this unconventional manner?

If any two homologous chromosomes fail to separate from each other during meiosis—in an event called **nondisjunction**—gametes will be produced that have more or less than the normal number of chromosomes. For example, if the two X chromosomes of a female fail to separate in prophase I of meiosis, the gametes will be abnormal. Consequently, when a female fly carrying the abnormal eggs mates with a normal male, there will be four abnormal combinations of sex chromosomes in the progeny (Figure 3-3). The white-eyed female flies resulting from this mating will be genetically XXY (also suggesting to an observant geneticist that

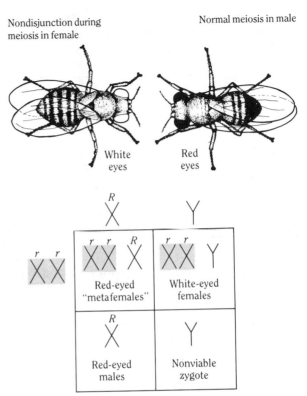

FIGURE 3-3 Nondisjunction of homologous X chromosomes during meiosis in the formation of fly eggs. When such eggs are fertilized by normal sperm, abnormal flies result that have too few or too many X chromosomes. The gray box indicates that the two X chromosomes are physically joined together; the pair segregates together in meiosis.

the Y chromosome does not always determine maleness in fruit flies), whereas the red-eyed males will have only a single X chromosome. When geneticist Bridges examined the chromosomes from cells of the male and female flies, he found an abnormal number of chromosomes. He found female flies with three X chromosomes—"metafemales"—and eventually discovered some nonviable embryos that contained the single Y chromosome, proving that the genetic analysis in Figure 3-3 is correct.

Nondisjunction of X chromosomes, and of autosomes as well, occurs during meiosis in

TABLE 3-1   The phenotypic effects of abnormal X-chromosome constitution in animals of different species.

Sex-chromosome constitution	Human	Mouse	Drosophila
XX	Normal female	Normal female	Normal female
XY	Normal male	Normal male	Normal male
XXY	Sterile male	Sterile male	Fertile female
XYY	Fertile male	Semisterile male	Fertile male
XO	Sterile female	Fertile female	Sterile male

human beings of both sexes, in other animal species, and in plants. In human beings, the X chromosome is most frequently affected by nondisjunction. The presence of an abnormal number of X chromosomes affects both the sex and the fertility of various species in different ways (Table 3-1). For example, a single X chromosome (XO) in human beings results in a sterile female, whereas fruit flies with only one X chromosome become males. And in mice a single X chromosome results in a fertile female. The presence of an extra X chromosome (XXY) in the cells of male human beings and in male mice results in sterility, but in fruit flies the presence of an extra X chromosome (XXX or XXY) results in a fertile female. Infertility may also arise from the abnormal development of sperm or eggs even though their chromosomal make-up is normal (Box 3-1).

Abnormalities of Sex Chromosomes in Human Beings

The first response made to a newborn infant is to determine whether it is male or female. Sex assignment is based on examination of the external genitals: scrotum and penis for males, vagina and clitoris for females. However, abnormal sex genotypes that are not detectable at birth by examination of sex organs are rather frequent both in human beings and in other animals. If nondisjunction of the X chromosome occurs during the formation of the mother's egg or the father's sperm, offspring with either XO or an XXY sex-chromosome constitution may be conceived and will survive. At birth, infants with such genotypes are usually indistinguishable from normal infants. Other sex-chromosome abnormalities include males with an XYY complement and females with three or more X chromosomes. Such abnormalities result from abnormal partitioning of the sex chromosomes into the male or female gametes during meiosis.

In some people, an abnormal X chromosome constitution has little noticeable effect; in others, however, development does not proceed normally, and the phenotypic effects of the abnormality may appear later in life. For example, most women with *Turner syndrome* (XO) do not become aware of their condition until they try to become pregnant and cannot. Sterility is a major characteristic of Turner syndrome, although a variety of other characteristics result from it. (A **syndrome** is a group of symptoms or traits that tend to appear together and that serve to characterize and diagnose a particular disease.) Many women who have Turner syndrome are shorter than average, show immature develop-

"Why Can't We Have a Baby?"

BOX 3-1

Millions of couples who want children have not been able to conceive any. Such human infertility has many causes. The first step in helping an infertile couple is to determine whether the reproductive problem exists in the man or in the woman.

In testing for male fertility, sperm are examined for the number produced, for normal appearance, and for the capacity to swim vigorously. If any of these properties are markedly abnormal, it may mean that the sperm are not sufficiently active or are not present in sufficient numbers to fertilize an egg.

Until quite recently, it was not possible to measure the capacity of human sperm for fertilization, but now tests for doing so are available. Using eggs obtained from hamsters or guinea pigs, researchers can observe whether or not a sample of human sperm is able to fertilize eggs from another mammalian species and, by inference, to fertilize human eggs, too.

Normally, sperm from one animal species are unable to fertilize eggs of another species. However, this interspecies barrier can be removed in the laboratory by stripping the zona pellucida—the outer layer—from the hamster or guinea pig eggs. Without this gelatinous barrier, normal human sperm can attach to the egg's surface, penetrate the egg, and actually stimulate a few cell divisions before the egg aborts, indicating that the sperm have the capacity for fertilization. Defective sperm—even those that appear normal by other tests—are unable to fertilize the hamster or guinea pig egg. This test, then, either assigns a couple's infertility problem to a defect in the sperm or indicates that the woman may have a reproductive problem. If the sperm prove normal, the woman may elect to undergo tests to determine whether her reproductive process is blocked at a certain stage and whether the problem can be corrected.

READING: Laurence, E. Karp, "Genetic Crossroads," *Natural History*. October, 1978.

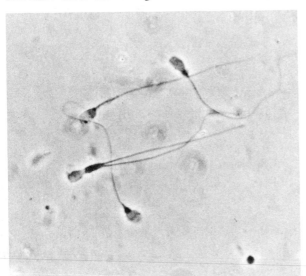 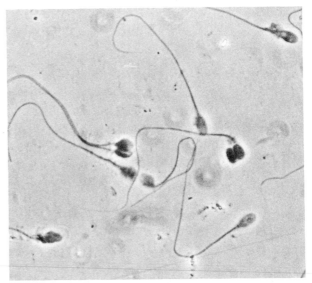

Abnormal human sperm: two tails (left) or two heads (right). (Courtesy of Jane Rogers, Vanderbilt University.)

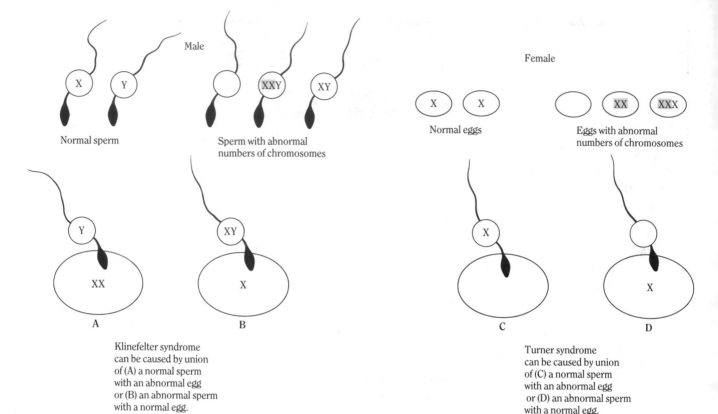

Klinefelter syndrome
can be caused by union
of (A) a normal sperm
with an abnormal egg
or (B) an abnormal sperm
with a normal egg.

Turner syndrome
can be caused by union
of (C) a normal sperm
with an abnormal egg
or (D) an abnormal sperm
with a normal egg.

FIGURE 3-4   Nondisjunction or abnormal segregation of the sex chromosomes in the formation of human gametes can produce progeny with abnormal genotypes.

ment of breasts and external genitals, and age prematurely.

Turner syndrome is thought to result from the fertilization of an egg that lacks an X chromosome as a consequence of meiotic nondisjunction (Figure 3-4). If such an egg is fertilized by a sperm carrying the father's X chromosome, a female will be produced whose cells contain only a single X chromosome. Such a woman is sterile, because the female reproductive organs have failed to develop in the embryo. Approximately one of every five thousand infants classified as female at birth has an XO karyotype (Figure 3-5).

In 1942, Harry F. Klinefelter described a group of phenotypically abnormal male human beings who had unusually small testes, were sterile, had certain female physical characteristics (such as enlarged breasts and wide hips), and in some cases had subnormal mental capabilities. About one of every thousand males exhibits these symptoms, which together are known as *Klinefelter syndrome*. Genetic studies of the chromosomes in men having this disorder show that all of them have an XXY karyotype (Figure 3-6).

Clearly, fertility and many other phenotypic characteristics are affected by the X chromo-

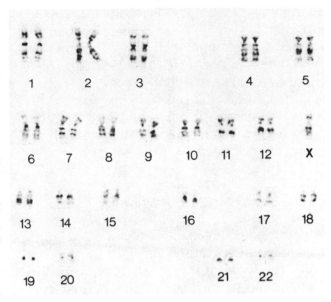

FIGURE 3-5 Karyotype analysis reveals the XO genotype; the diagnosis for people who have this genotype is Turner syndrome. (Courtesy of Patricia Jacobs, Director, Wessex Regional Cytogenetics Unit.)

some constitution. Infertility in people having Turner or Klinefelter syndrome results from the absence of functional reproductive organs (ovaries or testes) in these individuals because their cells contain one X chromosome too few or too many.

Hermaphroditism (a term derived from the names of the greek god Hermes and the goddess Aphrodite) is an exceedingly rare sexual abnormality in human beings. In the past seventy years, only about 350 examples of true human hermaphrodites have been reported in the world. These people have both ovarian and testicular tissue, but neither is functional in the sense that both sperm and eggs are produced. The tissues of true hermaphroditic individuals consist of a mixture of cells with either an XX or an XY karyotype.

XYY Males

As we have just seen, having one too few or one too many X chromosomes causes abnormal mental and physical characteristics in human beings. This does not seem to be the case for an extra Y chromosome. Although the Y chromosome is essential for male sexual differentiation, it is the only chromosome on which no other functional genes are known to be carried.

Karyotype studies of prison inmates in Scotland in 1965 revealed that 3.6 percent of the men examined had an XYY chromosome constitution; in other words, all their cells carried an extra Y chromosome. This study led some people to speculate that the criminal behavior of men might be caused by an extra Y chromosome. The press and other media picked up the extra Y chromosome idea and encouraged this speculation about criminal behavior. In 1968, when it was discovered that a man named Richard Speck had murdered several nurses in Chicago, media reports based on a faulty karyotype analysis identified Speck as an XYY male. Subsequent analysis revealed that Speck's sex

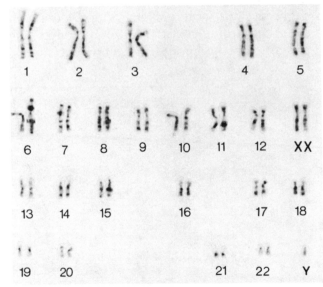

FIGURE 3-6 Karyotype analysis reveals the XXY genotype; the diagnosis for people who have this genotype is Klinefelter syndrome, which renders them sterile. (Courtesy of Patricia Jacobs, Director, Wessex Regional Cytogenetics Unit.)

chromosomes were normal—that is, he was XY—but the retraction received little attention.

We now know that about one of every thousand males (about 0.1 percent) in the general population has an XYY chromosome constitution, but that about 2 percent of the male inmates in penal and mental institutions have the XYY genotype. Despite their overrepresentation in prisons, however, only about 1 percent of all XYY males in the general population have a criminal record. Because ninety-nine of every hundred XYY males are living normal, socially acceptable lives, it is incorrect and misleading to conclude that XYY males are ''genetically'' prone to crime.

Despite these findings, in the 1970s a group of medical experts proposed a research study in which all newborn males delivered at a particular hospital would be genetically screened to determine which ones had an XYY chromosome constitution. These infants could then be followed for many years to see if they developed aggressive or criminal behaviors more frequently than males with XY chromosomes. Opponents of the proposed experiment argued that, if the parents of a newborn infant were told that he was XYY and hence at risk for aggressive, criminal behavior, they would raise and discipline their child with that possibility constantly in mind. Such an attitude might well lead to a self-fulfilling prophecy of behavioral problems and would certainly alter the parent-child relationship. The proposal was eventually discarded.

Down Syndrome

The most-common example of autosomal nondisjunction in human beings is *Down syndrome*, which results from *trisomy 21*—three copies of chromosome 21 instead of the normal two. About five thousand babies with Down syndrome are born in the United States each year. Down syndrome occurs in all human populations at all ages, but the risk that a woman will bear a Down child increases markedly past age thirty-five (Figure 3-7). For that reason, all physicians (because of possible legal action) now recommend amniocentesis for all pregnant women of age thirty-five or older. *Amniocentesis* is a prenatal diagnostic procedure by which Down syndrome can be detected *in utero* by karyotyping cells from the embryo. More than a hundred other genetic abnormalities can be detected *in utero* by amniocentesis, as will be described in Chapter 15.

With modern medical care, the life expectancy of a person having Down syndrome is now forty years or more. Because Down children ultimately exhaust both the financial and the emotional resources of their families, most of them are eventually placed in public institutions that are often crowded, understaffed, and

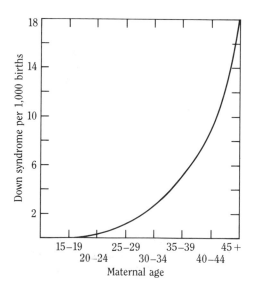

FIGURE 3-7 The frequency of Down-syndrome newborns in relation to the age of the mother.

underfunded. The plight of Down syndrome adults and other mentally deficient persons who have been committed to institutions is particularly tragic. However, with early physical therapy and continued encouragement by family and therapists, some of those having Down syndrome can achieve a measure of independence in living. Still, most couples who learn that the pregnant mother will bear a Down or other severely handicapped child elect to abort the pregnancy.

Other Chromosomal Abnormalities

Two other trisomies are relatively common among human beings, though much less so than trisomy 21. Both trisomy 13 (*Patau syndrome*) and trisomy 18 (*Edwards syndrome*) produce children with severe mental and physical abnormalities. These children usually do not survive for more than a few months after birth.

Many other chromosomal defects have been identified through the examination of aborted fetuses. Surprisingly, about 15 percent (some estimates are higher) of all human pregnancies terminate in spontaneous abortion. When cells from aborted fetuses are karyotyped, about half show some kind of chromosomal abnormality. The most common are **aneuploidy** (the lack or excess of one or several chromosomes), **polyploidy** (an excess of a complete set of chromosomes), and **translocations** (in which the same or different chromosomes have exchanged fragments with one another).

Although meiosis and mitosis function properly most of the time, abnormal gametes arise fairly frequently. The presence of such gametes means that abnormal zygotes (a **zygote** is the union of a sperm and an egg to form the first diploid cell of a new individual) also are produced frequently. Embryos that develop from genetically abnormal zygotes may abort spontaneously or may survive to term, producing a genetically handicapped individual (Box 3-2).

We have seen in this chapter that genes are located in chromosomes and that gross chromosomal abnormalities can cause hereditary diseases. Since these initial chromosomal discoveries, geneticists have shown that abnormalities in genes change the chemical reactions that take place in cells. And it is the altered cellular chemistry that ultimately changes the organism's phenotype. Therefore, before we can proceed to a discussion of normal and abnormal genes, we must explore the basic chemistry that determines the functions of cells. The next several chapters deal with the composition of cells and the chemistry of gene expression.

Summary

Animal chromosomes are of two general types: sex chromosomes (X and Y in human beings) and autosomes. The cells of normal human beings have two sex chromosomes and twenty-two pairs of autosomes. In a fundamental way, sex chromosomes determine a person's sex: XX individuals are female; XY are male.

Sex-linked traits are those encoded by genes located on an X chromosome (the Y chromosome carries only one known gene, that for male sex determination). The pattern of inheritance of sex-linked traits is non-Mendelian.

The nondisjunction of homologous chromosomes in meiosis results in gametes that have an abnormal number of chromosomes. The nondisjunction of X chromosomes affects the sex and fertility of animals of various species in different ways. In human beings, nondisjunction of the X chromosome in either parent results in children having either an XO (female) or

Inborn Errors of Metabolism

BOX 3-2

Archibald Garrod, an English physician who practiced medicine at the turn of this century, made a major contribution to our understanding of inherited human metabolic diseases. But, as happened with Mendel's experiments, the significance of Garrod's discoveries was generally unappreciated for more than thirty years.

In 1909, Garrod published a book in which he described four human diseases; *alkaptonuria*, *pentosuria*, *cysteinuria*, and *albinism*. He diagnosed the first three diseases on the basis of the presence of unusual chemical substances excreted in the person's urine. Alkaptonuria, in particular, is easily diagnosed because on exposure to air the urine turns black. Albinism is characterized by a defect in cellular pigmentation—the absence of a substance called *melanin*—that is especially noticeable in the skin, hair, and eyes (see photograph).

Garrod realized that albinism was similar to the other three diseases because people with albinism lacked skin pigments whose synthesis depends on enzymes. He deduced that the missing pigments must be synthesized by the action of enzymes that were also genetically determined. Garrod proposed three possible explanations for albinism: "We might suppose that the cells which usually contain pigment fail to take up melanins formed elsewhere; or that the albino has an unusual power of destroying these pigments; or again that he fails to form them." He went on to argue that the last explanation was correct and that the lack of melanins resulted from a missing enzyme.

Garrod was familiar with Mendel's discoveries concerning the patterns of inherited traits in peas. He recognized that the four human diseases that he had been studying were caused by inherited, recessive genes because the pattern of inheritance behaved according to Mendel's laws. With remarkable insight, he proposed that all four human diseases were the result of a missing or inactive enzyme, the absence or malfunction of which was due to a defective gene that had been inherited. Garrod proposed the general expression *inborn errors of metabolism* to describe this group of diseases. Today hundreds of inherited *metabolic diseases* (as they are usually now called) have been identified in human beings.

In 1982, physicians at the University of California Medical

Albinism in a Hopi child (center). [Courtesy of the Field Museum of Natural History (#118), Chicago.]

School in San Francisco were, for the first time, able to detect an inherited metabolic disease in a fetus *in utero* and begin treatment even before the infant was born. Using the technique of amniocentesis (discussed in Chapter 14), they determined that the fetus's cells were unable to synthesize biotin, an essential vitamin, because of mutant genes that had been inherited. Giving the mother large doses of biotin during her pregnancy ensured that the fetus received enough of the vitamin to develop properly while still in the womb. After the infant was born, its diet could be supplemented with the vitamin that it was unable to synthesize because of the inherited defect.

an XXY (male) chromosome constitution. The phenotypic effect of XO is Turner syndrome; that of XXY is Klinefelter syndrome; both result in sterile individuals. Other sex-chromosome abnormalities caused by nondisjunction include an XYY (male) complement and three or more X chromosomes (female).

Nondisjunction is not restricted to sex chromosomes. Down syndrome is caused by the nondisjunction of an autosome—chromosome 21. All the cells of a Down person have three copies of chromosome 21.

Many other chromosomal abnormalities have been identified through the examination of aborted fetuses. The most-common classes of chromosomal abnormalities are aneuploidy, polyploidy, and translocations.

Key Words

aneuploidy   The characteristic of having one or more chromosomes too many or too few than the number normally found in an organism.

autosomes   All chromosomes in eukaryotes, excluding the sex chromosomes.

Down syndrome   An inherited human disease caused by an extra chromosome number 21 in all of the person's cells.

hermaphrodite   A human being in whom both the male and female reproductive organs are present; however, neither organ is functional, and so hermaphrodites are sterile.

Klinefelter syndrome   A syndrome characteristic of men who have an XXY chromosome constitution.

mutant   An organism that carries a mutation that causes a phenotypic difference from a wild-type organism.

mutation   A heritable change in the genetic information.

nondisjunction   Failure of homologous chromosomes (or chromatids) to separate properly at meiosis, resulting in gametes with too few or too many chromosomes.

polyploidy   The characteristic of having a multiple of the normal diploid chromosome number.

sex chromosome   The X or Y chromosome in human beings.

sex-linked trait   A trait determined by a gene in an X chromosome.

syndrome   A group of symptoms used to characterize a disease.

translocation   The movement of a of chromosomal segment from one chromosomal location to another, either in the same chromosome or to a different one.

Turner syndrome   A syndrome characteristic of women who have an XO chromosome constitution.

wild-type   An organism normally found in nature, in contrast with a mutant organism.

zygote   A fertilized egg formed by the fusion of male and female gametes; the first cell of a new organism.

Additional Reading

Carlson, E.A. *The Gene: A Critical History*. Saunders, 1966.

Davis, B.D. "Frontiers of the Biological Sciences." *Science 81*, August, 1981.

8 Hooke, E.B. "Behavioral Implications of the Human XYY Genotype." *Science*, 179, 139–151 (1973).

9 McLeish, J., and B. Snoad. *Looking at Chromosomes*. Macmillan, 1958.

10 Pierik, R.L.M. *In Vitro Culture of Higher Plants*. Martinus Nijhoff, 1987.

11 Stauffer, R.C., ed. *Charles Darwin's Natural Selection*. Cambridge University Press, 1987.

Study Questions

1 How many autosomes are present in a human skin cell? In an egg? In sperm? In a human zygote?

2 What is meant by a reciprocal genetic cross?

3 Name three human genetic disorders that are caused by nondisjunction of chromosomes.

4 What is the chromosome constitution in cells taken from a hermaphrodite?

5 How would you determine whether a male had a XYY genotype?

6 Why do almost all women who are diagnosed as carrying a Down-syndrome fetus elect an abortion?

7 What is the difference between aneuploidy and polyploidy?

Essay Topics

1 Discuss why genetic crosses give different results if the mutant gene is on an autosome rather than on the X chromosome.

2 Discuss one or more genetic reasons why a woman may have a spontaneous abortion.

3 Explain how we know that genes are located in chromosomes.

The Chemistry of Cells

Molecules of Life

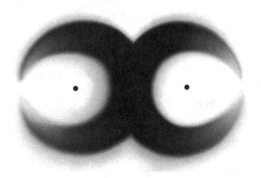

One of the most difficult issues in science is to decide
when a particular phenomenon is worth investigating.
HERMAN BONDI, *physicist*

ALL LIVING ORGANISMS are made up of cells. The simplest organisms, such as bacteria, are single cells; more-complex organisms, such as plants and animals, consist of millions, billions, even trillions of cells. The human brain alone is estimated to have 20 billion nerve cells that regulate mental processes and physiological functions.

Two essential properties of living cells are growth and reproduction. Bacteria growing in a liquid medium in a test tube duplicate their genetic information and make identical copies of themselves every half hour or so. Bacteria grow and reproduce exponentially: one cell produces two, two produce four, four produce eight, and so on. If the growth of bacteria in cultures were not limited by nutrients or space, within two days a single bacterium would produce so many cells that their combined weight would exceed the weight of the earth. Obviously, the growth and reproduction of all cells—not only bacteria—are limited by genetic and environmental factors.

People grow to a certain size and then stop growing; each organ in the human body also grows to a certain size and then stops. When tissues in plants or animals are damaged, the cells are stimulated to grow, the damaged tissue is restored, and the cells stop growing. In addition, cells in many tissues are constantly being replaced. What controls the growth, reproduction, and functioning of the various kinds of cells in plants and animals? The properties of all living cells, from bacteria to multicellular organisms, are controlled by genetic information. Therefore, an understanding of the chemical composition, physical structure, and biological functions of the genetic material is essential to an understanding of living organisms.

The properties of all living cells derive solely from the chemical and physical properties of the atoms and molecules of which they are composed. Atoms combine chemically to form simple molecules, which in turn may combine to produce more-complex molecules. Many thousands of different molecules contribute to the structure and functioning of cells, and aggregations of cells make up the tissues of living organisms.

We can think of an organism as something like a house. A house is a complicated structure of wood, metals, concrete, and other materials. It has plumbing to carry water, wiring for electricity, mechanical devices to heat or cool the rooms, paint and other materials to protect its surfaces, and devices to warn of fire or unwelcome intruders. If we were to see all the pieces of a house piled in random fashion—the boards, nails, wires, pipes, shingles, and so on—we would have no idea of precisely how the pieces should be put together. To organize all these materials, we would need a blueprint. In living cells, it is the genetic information that supplies the blueprint for the organization of all the components and for the synthesis of the protein and other molecules that together produce the wondrous property of life.

To understand how single cells and organisms grow, reproduce, evolve, and age, we must first understand the properties of the building materials of cells: atoms and molecules. This chapter describes the basic principles that govern the chemical and physical interactions of atoms and molecules and how the functions of cells derive from the organization and chemical reactions of the molecules in them.

Elements

All matter is composed of elements. An **element** is a substance that cannot be broken down further or transformed into another substance by ordinary chemical or physical means. For example, metals such as iron, copper, and gold do not yield simpler substances when they are subjected to chemical or physical treatment. In all the matter that exists on earth, and presumably elsewhere in the universe, only ninety-two elements occur naturally (though heavier elements have been created in the laboratory by physicists). Hydrogen and helium, the lightest of the elements, were created when the universe came into existence. All the heavier elements, such as carbon, nitrogen, sulfur, sodium, and phosphorus, and others that are necessary for life were created through the explosions of stars (novas and supernovas) billions of years ago (see Box 4-1).

Carbon, because of its unique chemical properties, plays a key role in cells. Every carbon atom can form chemical bonds with as many as four carbon atoms or with any other element or combination of elements. As a result, extremely complex organic molecules can be constructed from aggregates of carbon atoms. Organic molecules are molecules that contain one or more carbon atoms; inorganic molecules contain no carbon atoms. A simple organic molecule, such as methane, contains one carbon atom chemically bonded to four hydrogen atoms (CH_4); a molecule of ethyl alcohol contains two carbon atoms, six hydrogen atoms, and one oxygen atom, (CH_3CH_2OH); a molecule of one of the proteins in milk contains more than five thousand atoms ($C_{1864}H_{3012}N_{468}O_{576}S_2$).

The First Organic Molecules

All living cells synthesize countless different kinds of organic molecules. But the cells themselves are composed of millions of different organic molecules that can be synthesized only in cells. Which came first, molecules or cells?

As the earth and the rest of the solar system were forming, such simple inorganic molecules as minerals, salts, and gases were synthesized as a consequence of the high temperatures and pressures prevailing within the earth and on its surface. But what was the source of the complex organic molecules that led to the development of primitive cells and ultimately of complex living organisms? How did the small molecules essential to life, such as amino acids, nucleotides, and sugars, first arise? And what is the origin of such large molecules (macromolecules) as proteins and nucleic acids?

While the earth was still quite hot and relatively fluid, there was very little oxygen or liquid water in the atmosphere or on the surface. Gigantic electrical storms raged across the skies and enormous volcanoes spewed forth matter from the molten interior. Radiation from space and from radioactivity within the earth itself supplied the energy necessary for chemical reactions to take place among the elements on the earth's surface.

To simulate this early condition, scientists have exposed simple gas mixtures such as nitrogen (N_2), hydrogen (H_2), carbon dioxide (CO_2), and water vapor (H_2O) to electrical discharges or to high heat and pressure (Figure 4-1). Under such conditions, many of the complex organic molecules that serve as the building blocks of biologically active molecules form spontaneously. Presumably, such complex organic molecules could also have formed in the chemical and physical environment of the primitive earth.

Where Did the Elements Come From?

BOX 4-1

A quite convincing body of physical and astronomical evidence points to the fact that the universe and all matter that now exists was created from 10 to 20 billion years ago in an explosion of unimaginable intensity that is referred to as the "Big Bang." Before that event, there was no universe; neither time, nor matter, nor energy, nor even space itself existed before the Big Bang. We cannot even hazard a scientific guess about where the matter and energy contained in that primordial explosion came from. However, from the moment of that singular event, all the laws of physics as they are understood today have applied, and they can be used to predict how the universe expanded, cooled, and evolved to give rise to the atoms, stars, and galaxies that we now observe.

At the moment of the Big Bang, there was only energy; atoms could not have existed at the elevated temperature of the explosion. However, the enormous amount of energy in the fireball of the Big Bang instantly began to radiate outward, creating space and time. Within billionths of a second, the temperature of the fireball began to drop and particles of matter began to form. (Energy and matter are interconvertible, according to Einstein's famous law $E = mc^2$: energy is equal to mass multiplied by the square of the speed of light in a vacuum.) Within seconds after the explosion, energy was converted into protons, which are hydrogen atoms without an electron. Hydrogen, then, was the first and the simplest element. The energy in this dense soup of protons was enough to fuse some protons together, creating an isotope of helium, the next element. Protons are still being fused into helium in nuclear reactions deep within our sun; these reactions supply the energy that sustains all life on earth.

As the cloud of hydrogen nuclei expanded and cooled further, the force of gravity began to pull parts of it together to form local concentrations of denser material; the formation of stars and galaxies had begun. Billions of years ago, a galaxy known as the Milky Way was formed, and within that galaxy, about 5 billion years ago, a unique solar system consisting of an average-size star (our sun) and its nine planets was formed.

The many different elements that are now found on the earth and other planets in our solar system were—except for hydrogen and some of the helium—formed in a complex series of nuclear reactions in the hot interior of stars formed much earlier in the history of the universe. The heavier elements, such as carbon, oxygen, nitrogen, iron, and lead, were formed in nuclear reactions within stars and were spread through space after the stars exploded in spectacular explosions called *supernovas*. Even now some stars are being formed and some are disintegrating; this continuing process allows us to decipher the history and evolution of the universe.

The elements on other planets and elsewhere in the universe are identical with the elements on earth. We know this because each star gives off a characteristic pattern of light called a *spectrum*, which, when analyzed, tells us what elements are present in the star. Because our planet contains almost all of the elements, it could not have been formed during the early history of the universe, when only the lighter elements were available. (Even today most of the matter in the universe is still in the form of hydrogen.) Billions of years from now, our sun will eventually explode, again spewing its atoms and those of earth and the other planets back into space.

READING: Freeman Dyson, *Origins of Life*. Cambridge University Press, 1985.

Apparently, the geological changes that took place early in the history of the earth favored the formation of increasingly complex organic molecules. Had the earth been closer to or farther away from the sun, its evolution would have been quite different. The other planets of our solar system were formed at the same time as the earth and contain the same elements, but they evolved differently. And, as far as we know, no living organisms ever arose on them. Planets that are too far from the sun (Mars and Jupiter) are too cold and barren to allow for the development of cells, and planets that are too close to the sun (Mercury and Venus) are too hot. Indeed, if the earth's orbit had been just slightly different, it is likely that the molecules of life would never have formed and that the earth too would be without living things.

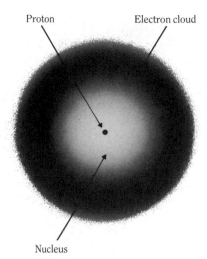

FIGURE 4-2 A model of the hydrogen atom. The nucleus of a hydrogen atom is a single proton with a positive (+1) charge. The proton is surrounded by the electron, which is represented as a cloud of negative (−1) charge. All elements normally contain equal numbers of protons and electrons so that they have no net electrical charge.

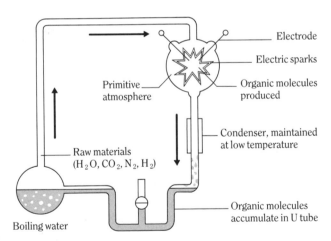

FIGURE 4-1 A diagram of the Miller-Urey experiment; the apparatus was designed by Harold Urey and Stanley Miller. A mixture of gases contained in the large glass enclosure was sparked and produced complex organic molecules such as amino acids and sugars. This experiment showed that the biological molecules essential to living *could* have been synthesized in the environment presumed to have existed on earth billions of years ago. Recent studies, however, raise some questions about the early earth's atmosphere and surface environment.

The Structure of Elements

All elements (atoms) are composed of fundamental subatomic particles called **protons**, **neutrons**, and **electrons**. A proton and a neutron are dense, heavy particles located at the center of an atom. Protons and neutrons are physically identical subatomic particles except that protons have a positive electrical charge (+1), whereas neutrons have no electrical charge and hence are neutral. An electron has a mass that is only a tiny fraction of the mass of a proton or a neutron, but it carries an equivalent unit of negative electrical charge (−1) that is able to neutralize the unit of positive charge on a proton. Electrons do not exist as discrete heavy particles like protons and neutrons, but as clouds of negative charge distributed in orbits (shells) around the nucleus of the atom (Figure 4-2). The chemical properties of an element are

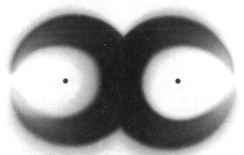

FIGURE 4-3 The distribution of the electron cloud around two hydrogen atoms bound together in hydrogen gas. A chemical (covalent) bond is formed because the electrons are shared by both protons in the hydrogen nuclei.

due solely to the number of electrons in the outer orbits around the nucleus.

Two atoms of hydrogen join together chemically through the equal sharing of their two electrons, as shown in Figure 4-3. Because the two electrons are no longer uniquely associated with the separate protons, but are shared by both, the hydrogen atoms become bonded together. For heavier elements, such as sodium and chlorine, electrons are usually shown as discrete particles for convenience in specific orbits around the nucleus of an atom as in Figure 4-4. Each orbit of an atom can accommodate only a certain number of electrons according to the rules of quantum mechanics worked out by physicists. The first orbit, or shell, can accommodate two electrons; the second, eight; the third, eight; the fourth, eighteen; and so forth.

Common table salt (NaCl) consists of an atom of sodium (Na) and an atom of chlorine (Cl) that are chemically joined together by the sharing of one electron (Figure 4-4). When sodium chloride salt dissolves in water, the atoms separate and are kept apart by the molecules of water that surround them.

The chlorine atom in solution retains the electron that it had been sharing and becomes a negatively charged chloride ion (Cl^-); the sodium atom, which has lost that electron, becomes a positively charged ion (Na^+). An **ion** is an atom in which the number of electrons is greater or fewer than the number of protons in its nucleus. In other words, the atom is no longer neutral with respect to electrical charge. Molecules also may be either neutral or positively or negatively charged. An atom or a molecule that has a net positive or a net negative electrical charge is said to be *ionized*.

The chemical properties of all atoms and molecules are due solely to the interactions and exchanges of electrons; the nuclei of atoms are never changed or affected in chemical reactions. To change the nucleus of an atom requires a nuclear reaction rather than a chemical one.

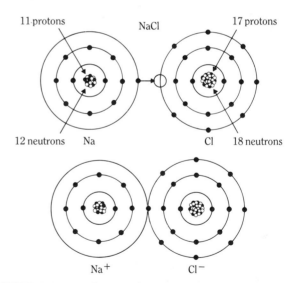

FIGURE 4-4 Sodium and chlorine atoms are joined by another kind of chemical bond called an ionic bond. Because the shared electron is more strongly attracted to the chlorine atom, it becomes negatively charged. Correspondingly, the sodium atom is positively charged because it has lost one of its electrons.

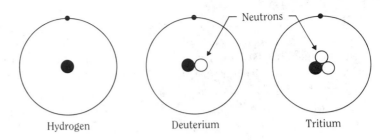

FIGURE 4-5 All three isotopes of hydrogen have one proton in their nucleus but differ in the number of neutrons. Deuterium is a stable isotope of hydrogen, but tritium is a radioisotope and emits radioactivity when the tritium nuclei decay.

Isotopes and Radioactivity

The weight of an atom is the sum of the weights of all the protons and neutrons in the nucleus (the contribution of electrons to weight is negligible). However, the atoms of an element may have different numbers of neutrons in their nuclei and therefore have different atomic weights. Different forms of an element whose atoms differ in weight are called **isotopes**.

Variations in the number of neutrons in the nucleus of an element do not affect the chemical properties of the element, but they do alter the element's atomic weight and sometimes the stability of the nucleus. Hydrogen, which always has only one proton in its nucleus, provides a simple example. Three isotopes of hydrogen occur in nature, with atomic weights of 1, 2, and 3 (Figure 4-5). These isotopes are hydrogen (one proton, no neutrons), deuterium (one proton, one neutron), and tritium (one proton, two neutrons). All three hydrogen isotopes have identical chemical properties, because all of them have one proton and one electron. They differ only in their atomic weights and in the stability of their nuclei. Deuterium is a stable isotope of hydrogen; when it chemically bonds to oxygen, it behaves chemically like hydrogen and forms molecules of "heavy" water (D_2O). Tritium is a radioactive (unstable) istotope of hydrogen, because its nucleus spontaneously "decays," or breaks apart. In the process, an electron is created from the conversion of energy into mass in the nucleus and is ejected from the tritium nucleus.

The emission of subatomic particles and radiation such as x-rays and gamma rays from unstable atomic nuclei is known as **radioactivity**; similarly, isotopes that emit subatomic particles and radiation are called **radioisotopes**. The nuclear decay of tritium converts a neutron into a proton and is equivalent to the acquisition of a positive electrical charge by the nucleus. The end result is that the tritium is converted into an isotope of helium, which has two protons in its nucleus.

Radioisotopes have a characteristic **half-life**, which is defined as the time that it takes for half of the nuclei of any given sample of a radioactive element to decay and be converted into some other element. Tritium has a half-life of 12.3 years; in that length of time half of all the tritium nuclei in a sample will have decayed. In the next 12.3 years, half of the remaining tritium nuclei will decay, and so on. Although the half-life of any radioactive isotope can be determined with great accuracy, it is impossible to predict which nuclei in a sample will decay and which will remain stable through any given period of time (although all the nuclei will decay eventually).

Most other elements also have stable isotopes and radioactive isotopes. Carbon, for example, has six naturally occurring isotopes. The most-prevalent isotope of carbon, ^{12}C (six protons and six neutrons), accounts for 98.6 percent of all carbon atoms in nature. Carbon-14 (^{14}C) is a radioactive isotope with six protons and eight neutrons. This isotope is rare in nature, but it is manufactured in large quantities for use in research and for medical purposes. It is also a by-product of atomic bomb explosions. Carbon-13 (^{13}C) is a stable isotope that is heavier than ^{12}C.

Researchers use both radioactive isotopes and heavy isotopes in conducting experiments in genetics, biochemistry, and medicine. Radioactive isotopes are often called tracer elements, because the emission of particles or energy from their molecules enables experimenters to locate them in cells or tissues and to observe their roles in the chemical reactions that take place within cells. Researchers incorporate heavy isotopes into molecules to make them more dense than the molecules synthesized from light isotopes (the most-common isotopes of an element). They can readily distinguish the heavier molecules from the millions of lighter molecules in cells and thus advance their understanding of cellular chemical reactions.

Isotopes became generally available for biological and medical research only after the Second World War when atomic reactors were developed that were capable of manufacturing various isotopes in large quantities at low cost. Today, thousands of compounds containing radioactive and heavy isotopes are available for research and for other applications.

Chemical Bonds

As mentioned earlier in this chapter, the chemical bonds that join atoms together in molecules are the result of interactions between the electrons in the outermost orbits of the atoms. When two electrons are shared equally by two atoms, a **covalent bond** is formed. A molecule of hydrogen gas, for example, is composed of two hydrogen atoms that share their two electrons equally (Figure 4-3), though it is impossible to determine which electron belongs to which atom. It is the chemical sharing of the negatively charged electrons that holds the hydrogen atoms together. Covalent bonds require more energy to separate the two atoms than is required to break other kinds of chemical bonds.

An **ionic bond** is the result of an *unequal* sharing of electrons by atoms. Ionic bonds are not as strong as covalent bonds. A good example of an ionic bond is table salt (NaCl), in which the atoms of sodium and chlorine are held together primarily by their respective positive and negative electrical charges. Sodium atoms have only a single electron in their outer shell, which means that the electron is easily removed. Chlorine atoms have seven electrons in their outer shell and need only one more to fill it. When atoms of sodium and chlorine join, the shared electron is more closely associated with the chlorine atom than with the sodium atom. Consequently, the sodium atom becomes positively charged (it becomes a positive ion) and the chlorine atom becomes negatively charged (a negative ion). Chemical bonds between most elements are usually a combination of covalent bonds and ionic bonds because most atoms are joined both by the sharing of electrons and by electrical charge.

Another kind of chemical bond, one that is particularly important in living cells, is the **hydrogen bond**. Hydrogen bonds are much weaker than either covalent bonds or ionic bonds. Even so, they are responsible for the remarkable stability of the genetic information in cells and for the three-dimensional shape of proteins. Consider, for example, a molecule of ordinary water (Figure 4-6), in which two hydrogen atoms are chemically bonded to an oxygen atom. The two hydrogens' electrons are strongly associated with the outer orbit of electrons around the oxygen atom. Because those electrons are more tightly bound to the oxygen atom than to the hydrogen atom, the water molecule is said to be **polar**—that is, there is a slight positive electrical charge around the hydrogen atoms at one end of the water molecule and a slight negative charge around the oxygen atom at the other end of the molecule. As a result of this polarity of charges, the hydrogen atoms of one water molecule are attracted to the oxygen atoms of other water molecules, as indicated by the dotted lines in Figure 4-7.

To break the hydrogen bonds that hold water molecules together, some form of energy is required. For example, heat energy breaks the bonds that hold water molecules together in ice, changing the ice to liquid water. And heat energy breaks the bonds that hold the molecules together in liquid water, changing the water to steam. The reason that water is such a good solvent of charged molecules like salt is that the water molecules surround the sodium and chloride ions and neutralize their charges. It so

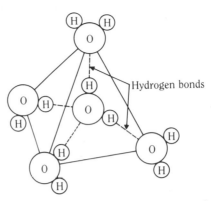

FIGURE 4-7 Hydrogen bonds hold water molecules together in particular orientations when the temperature is lowered and ice forms. Heating ice destroys the crystalline structure of ice and the water molecules become random in orientation. Further heating breaks the hydrogen bonds and allows water molecules to escape from the liquid as steam.

happens that most biological molecules are electrically charged, and the solvent property of water has a strong influence on how they interact and function in living cells.

Because the number of hydrogen bonds between molecules in certain substances is quite large, the overall energy required to break all the hydrogen bonds can be very great. The Lilliputians who tied Gulliver up were aware of this principle: they used thousands of thin threads that together were as strong as a few thick ropes. It is the energy of hydrogen bonds that maintains the three-dimensional structures of biological molecules such as proteins and nucleic acids (DNA and RNA). When foods are cooked, it is primarily the disruption of hydrogen bonds in the protein molecules that alters the texture of the food and makes it palatable. A common example of this is the change in the protein of a raw egg after it is boiled.

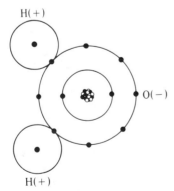

FIGURE 4-6 A molecule of water (H_2O) is formed when two hydrogen atoms become chemically joined to an oxygen atom by the sharing of electrons. Because of the positive charge associated with the two hydrogens and the negative charge associated with the oxygen, water exhibits an electrical plus-minus polarity. For that reason it is called a polar molecule.

Chemical Reactions

All chemical bonds—covalent, ionic and hydrogen—can be broken if the molecules are exposed to enough energy. That energy can also create new bonds and as a result produce new substances. That is precisely what happens in chemical reactions. **Chemical reactions** are defined as interactions between atoms and molecules that lead to the formation of new substances. Reactions in which carbon-containing molecules participate are *organic*; reactions in which carbon atoms are excluded are *inorganic*.

Chemical reactions are written as equations with the molecules that react with one another on one side and with the molecules produced by the reaction on the other side. For example, chlorine gas and methane gas react in the presence of light energy to produce hydrochloric acid and methylchloride:

$$Cl_2 + CH_4 \xrightarrow{\text{Light energy}}$$

Chlorine   Methane

$$HCl + CH_3Cl$$

Hydrochloric Acid   Methylchloride

The number of each kind of atom must always be the same on both sides of the equation. In this equation, for example, there are two chlorine atoms, four hydrogen atoms, and one carbon atom on each side.

Many reactions require more than one molecule of a substance, as in the manufacture of aspirin from salicylic acid and acetic anhydride:

$$2 C_7H_6O_3 + C_4H_6O_3 \rightarrow 2 C_9H_8O_4 + H_2O$$

Salicylic   Acetic   Aspirin   Water
acid   anhydride

Many reactions, like this one, proceed spontaneously when the substances are added to a solvent such as water. Other reactions require the assistance of an agent called a catalyst. A catalyst is defined as any substance that increases the rate of a reaction without itself being altered or used up in the reaction.

Many metallic compounds act as catalysts in both inorganic and organic reactions. For example, common hydrogen peroxide (H_2O_2) is a very stable compound. But if a pinch of magnesium oxide (MnO_2) is dropped into a container of hydrogen peroxide, oxygen gas will be released instantly causing the liquid to effervesce:

$$2 H_2O_2 \xrightarrow{MnO_2 \text{ (catalyst)}} 2 H_2O + O_2$$

Cells are able to grow and function because thousands of different chemical reactions are taking place within them every second. Some sort of catalyst is required for every chemical reaction that takes place in an animal, from the digestion of food to the production of nerve impulses in the brain. Those catalysts are **enzymes**, which are large protein molecules that help bring molecules together and accelerate the rates of reaction between them. And, in all organisms, genes encode the information that directs the synthesis of enzymes.

The property of "being alive" depends on the chemical reactions that occur moment by moment in the cells of an organism. Death occurs when a few crucial (and eventually all) chemical reactions cease. The question of whether "life" depends solely on the existence of chemical reactions or whether something more is required, especially in regard to human life, still cannot be answered to everyone's satisfaction (see Box 4-2). What is certain, however, is that many of the chemical reactions that are essential to life are identical or very similar in all living organisms from bacteria to human beings.

For example, all organisms require energy. All plants—grasses, flowers, grains, trees—convert the light energy from the sun into chemical

Does "Life" Result Only from the Chemistry of Molecules?

BOX 4-2

The question of what is unique to living organisms compared with nonliving objects like rocks and chairs has fascinated philosophers, scientists, and religious leaders for thousands of years. In the past, it was assumed that human beings, and even less-complex organisms, were endowed with some form of spirit or supernatural energy that made them alive. People, in particular, were (and are) believed to have souls.

With the advent of modern science and the enormous increase in our understanding of physics and chemistry, some scientists argued that eventually all the properties of life will be explained solely in terms of the physical and chemical properties of atoms and molecules. This philosophical idea is known as *reductionism*; that is, the idea that all aspects of living organisms, even human thought and behavior, can ultimately be explained by simple laws of nature.

The position of reductionist scientists can best be understood by analogy. For example, we know that a jet airliner is composed of thousands of individual parts—bolts, screws, motors, wheels—none of which can fly. Yet, when all the parts are assembled in just the right way, the plane is capable of flight, a property not possessed by any of the components. So it is with life. Millions of atoms and molecules, though not alive individually, can assemble in just the right arrangement to produce a cell that is alive.

Other scientists and philosophers challenge this reductionist viewpoint. They argue that life will never be explained even from a complete understanding of the structures and functions of molecules in cells. Their view (usually called *holism*) is that the ultimate property of things as complex as airplanes or cells is always greater than the sum of their parts. Moreover, the final property can never be predicted even from a complete understanding of the components. Thus, in reference to the airplane analogy, without knowing beforehand that

Dried tardigrade, as seen in a scanning electron microscope. (Courtesy of Robert Schuster, University of California, Davis.)

airplanes can fly, we could never make the prediction of flight by analyzing the parts separately. And, without having observed life, we could never predict it by studying molecules.

Certain experiments, however, do seem to support the view that life does derive solely from the structure and function of molecules in cells. At a temperature of absolute zero ($-273°C$), chemical reactions among atoms and molecules cease, because there is no heat energy to support them.

Yet many organisms—bacteria, seeds, and protozoa—can endure temperatures close to absolute zero and survive. Thus, the property that we call life does not necessarily disappear when all cellular functions and chemical reactions come to a temporary halt.

Another example is that a complex microscopic animal called a tardigrade (water bear) can be completely desiccated (dried out) and then, after months or years, can be restored to the living state. Although tardigrades are complex, multicellular animals, they can survive desiccation, which prevents most chemical reactions in cells. As a consequence of these experiments, some scientists support the idea that the property "life" derives solely from the way in which molecules are organized and function in cells. Others insist that life is a mysterious property that will never be explained by science.

energy by the process of **photosynthesis**. The result of the series of chemical reactions in photosynthesis is to combine carbon dioxide from the air with water in the plants to synthesize sugars and to release oxygen into the air. Thus, by means of special photosynthetic enzymes, plants convert light energy into chemical energy, which is stored as molecules of sugar. Animals that eat the plants have different enzymes in their cells that break down the sugar molecules, and they have other enzymes that synthesize new molecules from the chemical energy supplied by the sugars. The kinds of chemical reactions in any plant or animal are ultimately determined by the array of genes carried in its chromosomes and by the way in which those genes are regulated and expressed. Biochemists have shown that the chemical reactions and the molecules synthesized in the cells—proteins, carbohydrates, DNA—are fundamentally the same in all living organisms. This fact has profound implications for the origin, relatedness, and evolution of all living organisms on earth, which will be discussed in later chapters. For the moment, however, we turn our attention to the important biological molecules found in all cells.

Summary

The properties of all living cells derive solely from the chemical and physical properties of the atoms and molecules of which they are made and how they are organized. All matter is composed of elements—substances that cannot be broken down further or transformed into other substances by ordinary chemical or physical means. Carbon is a key element in organic molecules and, therefore, in living cells.

All elements, or atoms, are composed of subatomic particles called protons, neutrons, and electrons. Protons have a positive electrical charge and electrons have a negative electrical charge (neutrons are neutral). The number of electrons that an atom has equals the number of its protons. An atom that has a greater number of electrons than protons is a negatively charged ion, whereas an atom having a greater number of protons than electrons is a positively charged ion.

Protons and neutrons reside in the nucleus of an atom and their combined weights constitute the weight of the atom. However, atoms of an element can differ in the number of neutrons that they possess, which means that their atomic weights can differ. These different forms of an element are isotopes, which are either stable or unstable (that is, radioactive).

Chemical bonds between atoms in a molecule are determined by interactions of electrons. In a covalent bond, two electrons are shared equally by two atoms. In an ionic bond, the sharing of electrons is not equal. Covalent bonds are stronger (that is, breaking them requires more energy) than ionic bonds. A third type of bond, and the weakest of the three, is the hydrogen bond. Many biological molecules have a large number of hydrogen bonds, and so the overall energy required to disrupt them can be very great. Hydrogen bonds stabilize the three-dimensional structures of proteins and the long strands of nucleic acids (DNA and RNA).

Chemical reactions are interactions between atoms and molecules that lead to the formation of other substances. Some chemical reactions are spontaneous; others require a catalyst. Enzymes are biological catalysts. They are large proteins that help to bring atoms or molecules together to accelerate rates of reactions.

Key Words

catalyst Any substance that increases the rate of a chemical reaction without being used up itself in the reaction; enzymes are catalysts.

chemical reaction Interactions between atoms and molecules that lead to formation of other substances.

covalent bond A chemical bond created by the equal sharing of electrons by two atoms.

electron A negatively charged particle that orbits the nucleus of an atom.

element A substance that cannot be broken down further by ordinary chemical or physical means.

enzyme A protein that increases the rate of a chemical reaction in a cell.

half-life The time that it takes for one-half of the nuclei of a radioactive isotope to decay and be converted into some other element.

hydrogen bond The sharing of a hydrogen atom between two other atoms; weaker than a covalent or ionic bond.

inorganic Molecules or reactions in which carbon atoms do not participate.

ion A positively or negatively charged atom.

ionic bond A bond caused by the electrical attraction between two atoms; weaker than a covalent bond.

isotopes Different forms of an element; their chemical properties are identical, but their weights differ because of differences in the number of neutrons in the nucleus.

neutron A particle in the nucleus of an atom that is identical with a proton but does not carry an electrical charge.

organic Molecules or reactions that include carbon atoms.

photosynthesis The cellular process of converting light energy into chemical energy.

polar molecule One in which the charges are distributed unequally among the atoms such that some are positive and others negative; water is a polar molecule.

proton A positively charged particle in the nucleus of an atom.

radioactivity The release of energy from an atom's nucleus in the form of particles or radiation.

radioisotope An isotope whose nucleus is unstable and decays spontaneously, emitting energy.

Additional Reading

Boas, A.P. "What Made the Moon?" *The Geographical Magazine*, July, 1987.

Cloud, P. *Oasis in Space: Earth History from the Beginning*. Norton, 1987.

Dulbecco, R. *The Design of Life*. Yale University Press, 1987.

Dyson, F.J. *Origins of Life*. Cambridge University Press, 1985.

Monastersky, R. "The Plankton-Climate Connection." *Science News*, December 5, 1987.

Weinberg, S. *The First Three Minutes: A Modern View of the Origin of the Universe*. Bantam, 1979.

Study Questions

1   Define the essential chemical difference between an element and a molecule. Between alcohol and table salt.

2   What component of an atom determines its chemical properties?

3   Name three different kinds of chemical bonds. Which one is the strongest? The weakest?

4   How do isotopes differ from radioisotopes?

5   What chemical process is used by plants to obtain energy?

6   How were the heavier elements like iron created?

7   What molecules catalyze chemical reactions in cells?

Essay Topics

1   Discuss the significance of enzymes in determining the functions of different kinds of human cells.

2   Discuss what you think are the criteria necessary to say that something is "alive."

3   Find out about and describe the properties of some other chemical reactions that interest you.

5

Cells and Molecules

Living Organisms

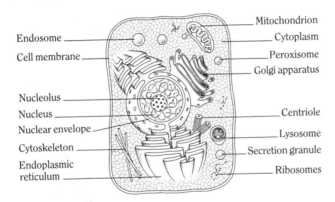

Endosome

Cell membrane

Nucleolus

Nucleus

Nuclear envelope

Cytoskeleton

Endoplasmic
reticulum

Mitochondrion

Cytoplasm

Peroxisome

Golgi apparatus

Centriole

Lysosome

Secretion granule

Ribosomes

Animal cell (eukaryote)

Every true scientist should undoubtedly muster sufficient courage and integrity to resist the temptation and the habit of conformity.

ANDREI SAKHAROV, *Soviet physicist*

Achievement in science, as in art and business, is accomplished by people willing to stand up for what they believe.

DAVID GOLDFARB, *Soviet biologist*

WHAT IMPRESSES people most when they first observe and study nature is the enormous diversity of living organisms. What, if anything, do all these millions of diverse kinds of organisms—from vanishingly small microorganisms to awesomely large whales and redwood trees—have in common? Are there any fundamental unifying principles that are shared by all organisms? Are the molecules in all organisms the same or different? As we will see, the most-extraordinary fact about living organisms is not that they are so different, but that they are so similar in genetic and biochemical organization.

One unifying principle of biology is that all living organisms, from the smallest and simplest to the largest and most complex, are composed of **cells**, which are the smallest living units that are able to grow and reproduce. Cells are highly organized ensembles of molecules and macromolecules in which chemical reactions occur that together produce the unique property we define as life. Cells grow and reproduce independently, as do bacterial or yeast cells, or they may grow cooperatively, as in the tissues of plants and animals.

Another unifying principle is that the overall chemical composition of cells, no matter what their function or type, is basically the same. The proteins and DNA found in the simplest bacteria are structurally and functionally quite similar to those found in a human cell. Of particular significance is that the chemical structure of DNA is the same, as is the genetic code (see Chapter 9). Proteins in all cells are synthesized from the same twenty different amino acids. Even the biochemical mechanisms by which proteins are synthesized are quite alike in all types of cells.

The observable differences between species are due primarily to differences in some genes and to differences in the way genes are expressed in development (discussed in Chapter 10). For example, the hemoglobin in the blood of a chicken and that in a dog are quite similar, as are many enzymes. However, certain proteins (and genes) in the two species are significantly different, as is the regulation of gene expression in development. Together, these differences give rise to different species.

The third unifying concept is that all organisms are related to one another and have evolved from some common ancestor. Simple cells presumably arose on earth once there were sufficient kinds and amounts of the appropriate molecules. More-complex cells and eventually multicellular organisms are presumed to have evolved from these simple cells by various mechanisms, including **natural selection**, which is defined as the differential reproduction of genetically different individual organisms as a result of their adaptation to particular environments. This important biological concept was proposed and documented by Charles Darwin. Many other scientists before him had similar ideas, but it was Darwin who presented the theory of natural selection in 1859 in his book *On the Origin of Species by Means of Natural Selection.* In Darwin's words:

As many more individuals of each species are born than can possibly survive, and, as a consequence there is a frequently recurring struggle for existence, it follows that any being, if it varies however slightly in any manner profitable to itself, under the complex and sometimes varying conditions of life, will have a better chance of surviving, and thus be naturally selected. From the standpoint of inheritance, any selected variety will tend to propagate its new and modified form.

Evolutionary studies since Darwin's time have lent strong support to the basic concept of natural selection and to the idea that all organisms evolved from simple cells that arose by chance from random chemical reactions billions of years ago (Box 5-1).

When Is Matter Alive?

Because all organisms are composed of cells, it is helpful to understand what it means to say that a cell or an organism is alive. Instinctively, each of us knows when a plant, animal, or person is alive—or dead. What properties or characteristics do we measure to determine whether an organism is, in fact, living? No single, all-encompassing definition of life is acceptable to everyone, but there are certain properties of living cells that distinguish them from nonliving matter.

1 **Metabolism**. Metabolism is the sum of all the chemical processes in a living organism. Cells are able to extract energy from the environment to fuel the chemical reactions that they need to grow and reproduce. Most bacteria utilize chemicals that are supplied by their environment, whether it be soil, water, or a human stomach. Some forms of bacteria (cyanobacteria—also referred to as blue-green algae), as well as all plants, are able to carry out photosynthesis, the process of converting the energy in sunlight into chemical energy that can be used for growth. The ultimate source of energy for all plants and animals is sunlight, which can be used directly by all plants and indirectly by animals, who obtain energy by eating plants or other animals.

2 **Growth**. As a result of the utilization of energy and synthesis of new molecules, cells increase in size and weight.

3 **Reproduction**. Cells and microorganisms reproduce, giving rise to two identical copies of themselves. Growing cells eventually reach a size at which they divide, giving rise to two progeny cells that also grow and reproduce.

4 **Mutation**. In the process of growth and reproduction, cells occasionally undergo a mutation, which is a permanent change in the genetic information in the DNA. Although each cell's DNA has a small probability of undergoing a mutation, among the billions of cells in a population the chance that one or a few cells may mutate is quite high. Mutations that occur in sperm and egg cells, giving rise to progeny with new characteristics, are the ultimate source of genetic differences between organisms and the origin of new species.

5 **Response**. Organisms and even individual cells respond and react to their environment. Stimuli change the chemical reactions in cells and change the behaviors of organisms.

6 **Evolution**. Because of mutations and other genetic mechanisms, the genetic information in a population of organisms changes through time. Some of these genetic changes in turn cause structural and functional changes in certain members of the population that make those members better able to survive and reproduce in their particular environments. This process of biological change and diversification through time is called biological evolution.

These six characteristics apply to all living cells; inanimate matter may undergo some but not all of these processes. Because viruses—particles composed of DNA (or RNA) and pro-

How Did Cells Arise?

BOX 5-1

Fossils of primitive bacteria and algae have been preserved in sedimentary rocks. Electron micrographs of these fossils show structures that are similar to those in modern microorganisms. Some of the rocks in which such bacterial and algal fossils are found are more than 3 billion years old. Thus, it seems likely that the first primitive cells arose shortly after the earth was formed, or at least within a billion years or so. For most of the earth's history, life consisted of single cells; all of the complex plants and animals, including those that have become extinct, arose within the past half billion years, according to the fossil records.

How did the first simple cells arise? Most scientists believe that life began with *protocells*—clusters of complex proteinoid molecules similar in many ways to primitive bacteria and algae but not yet "alive." Under laboratory conditions thought to resemble the conditions on the primitive earth, some simple proteinoid microspheres have been formed that have many of the properties of present-day bacteria (see the photograph below). They absorb dyes in the same manner as bacteria and often divide like bacteria. They can carry out a limited number of enzymelike chemical activities, and they have a well-defined structure, including membranes that regulate the entry and exit of small molecules and of electrically charged atoms. Although these proteinoid microspheres contain no genetic information and cannot regulate their growth, and thus are not alive, they do have a significant number of the important properties found in living cells today. It is enticing to speculate that such proteinoid microspheres are models of the first true cells.

The origin of cells on earth is believed to have occurred in a series of highly determinate events that may be unique to our planet. There is no evidence so far that similar chance events have occurred on any other planet in our solar system. However, the universe contains enormous numbers of galaxies and stars, though astronomers have yet to find any signs that these other stars possess planets on which the physical and chemical conditions would be hospitable to the evolution and survival of life forms as we know them. Many people still believe, however, that life does exist on some far-off planet.

READINGS: Steffi Weisburd, "The Microbes That Loved the Sun," *Science News*. February 15, 1986. Ben Patrusky, "Protolife's Clouded Beginnings," *Mosaic*, 15(3), 1984. Sidney W. Fox and Anwar K. Khoury, "The New Evolutionary Paradigm," *BioEssays*. December, 1988.

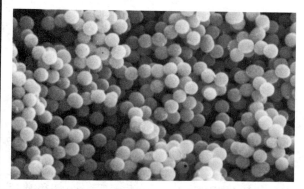

Proteinoid microspheres can be created in the laboratory and have some of the properties of living cells, including the ability to form buds that can separate from the original microspheres to form new ones. (Photograph from *The Emergence of Life*, Basic Books, Inc., 1988, and Steven Brooke Studios.)

A   A plant virus

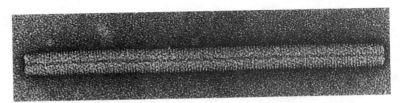

Tobacco mosaic virus

B   Human viruses

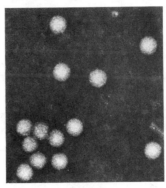

Polio virus

tein that infect, grow, and reproduce in living cells — fail to exhibit all of these properties, they fall somewhat ambiguously between living and nonliving matter (Figure 5-1). Outside living cells, viruses are not able to grow and reproduce. However, once they infect cells, they are able to take over the cellular machinery of their hosts in such a way as to enable them to reproduce themselves in great numbers — thereby qualifying them as being alive.

Genetic Information Governs Biochemical Processes

Many fundamental advances in genetics have emerged from studying bacteria because they are the simplest cells and are easily grown and manipulated in the laboratory (Box 5-2). One species in particular, *Escherichia coli* (Figure 5-2), a bacterium occurring naturally in the human gut, has been the subject of extensive research. Because *E. Coli* reproduces every half hour or so under laboratory conditions, and because the chemical environment in which *E. coli* cells are grown can be manipulated experimentally, this organism is used for all sorts of genetic and biochemical experiments. And, because there are donor and recipient strains of *E. coli*, genetically different bacteria can be mated to determine the location of genes in the bacterial DNA. (Bacterial mating is discussed in Chapter

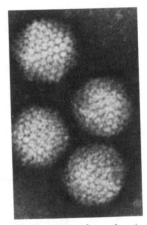

Cold virus (adenovirus)

FIGURE 5-1   Photographs of viruses observed in the electron microscope: (A) tobacco mosaic virus, a plant virus; (B) viruses that infect human cells (polio attacks nerve cells, and adenovirus causes colds). (Tobacco mosaic and polio viruses courtesy of R. Williams and H. Fisher, *An Electronmicrographic Atlas of Viruses*, Charles C. Thomas, Publishers; cold virus courtesy of M. Wurtz and E. Kellenberger.)

Scientific Discovery and the Prepared Mind

BOX 5-2

Nobel laureate and biochemist Albert Szent-Györgyi is often quoted as saying, "The creative scientist sees what everyone else has seen, but thinks what no one else has thought." A classic example of this kind of scientist is Alexander Fleming, best known for his discovery of the antibiotic pencillin.

Fleming noticed in 1929 that some Petri plates used to culture bacteria were contaminated with molds. He also noticed that bacterial growth was inhibited in the moldy areas. Now, this observation had undoubtedly been made before in many laboratories, but there is no evidence that any scientist before Fleming had ascribed any significance to the phenomenon or grasped its implications. Fleming, however, reasoned or knew intuitively that the molds growing on the plate were producing some chemical substance that could inhibit bacterial growth. He also realized the potential value of such a substance.

Fleming proceeded to isolate the inhibitory compound, which he named after *Penicillium*, the species of mold that produced it. For a number of reasons, the world was not prepared for this miracle drug. One reason was that sulfanilamides (sulfa drugs) were in widespread medical use at the time and were effective in treating many bacterial infections. Also, penicillin was chemically unstable and difficult to purify. As a result, penicillin did not come into widespread use until World War II, when it was used to treat soldiers for gonorrhea and other bacterial diseases. Today, penicillin and similar chemically synthesized derivatives are among the most-effective drugs for treating many kinds of bacterial infections.

What is less well known about Alexander Fleming is that his mind had been prepared for this

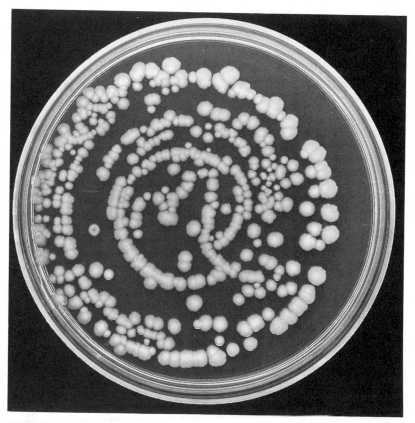

Each colony began with one bacterial cell on the surface of the medium in the Petri plate. Each colony consists of millions of genetically identical bacteria.

discovery by another important observation that he had made years earlier. In 1922, Fleming reported on "a remarkable bacteriolytic element found in tissues and secretions." While routinely culturing bacteria-containing nasal secretions from a patient, he noticed that something in the secretions seemed to "dissolve" the bacteria and prevent their growth. He followed up on this chance observation and found that there was a bacterial inhibitory substance in many human secretions, including tears and saliva, that would destroy almost any kind of bacteria. He called this substance *lysozyme*—an enzyme that disintegrates bacteria and other kinds of cells. It was Fleming's discovery of lysozyme some seven years before his discovery of penicillin that alerted him to be on the lookout for substances that inhibit bacterial growth.

Another recent example of a prepared mind is that of Michael Zasloff who, in 1987, discovered a whole new class of antibiotics. Zasloff works with the African toads (*Xenopus laevis*) and frequently operates on them to remove eggs that are used in developmental research (see Chapter 10). He noticed that the wounds in the toads never became infected even though they were swimming around in tanks of germs. By analyzing the skin, he discovered that it contains small proteins that act as antibiotics and protect the toads from infection. Zasloff named the new drugs *magainins* (from the Hebrew *magain*, meaning shield). The magainins are now being tested in animals to determine their effectiveness in curing infections.

READING: Gunther S. Stent, "Prematurity and Uniqueness in Scientific Discovery," *Scientific American*. December, 1972.

6). An *E. coli* bacterium contains about four thousand genes in its single DNA molecule. The precise location and functions of about one-quarter of the genes in *E. coli* are known, making it the best-understood organism from a genetic standpoint.

In some respects, bacteria are biochemically more efficient than human cells. When supplied with water, a few minerals, and the simple sugar glucose, bacteria grow and reproduce rapidly. Using only the carbon, oxygen, and hydrogen atoms in glucose and various essential salts such as sodium, potassium, and magnesium, bacteria can synthesize the many thousands of biological molecules and macromolecules that they need. Among the biological molecules syn-

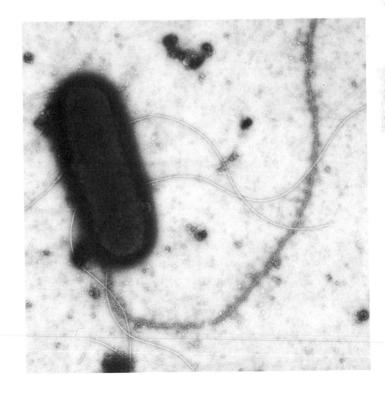

FIGURE 5-2   Electron micrograph of an *E. coli* bacterium showing its sex pilus (long appendage) and flagella (light fibers) that are used for swimming. (Courtesy of B. Eisenstein.)

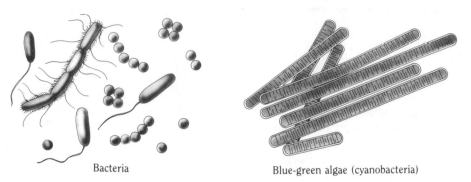

Bacteria   Blue-green algae (cyanobacteria)

FIGURE 5-3 Drawings of various prokaryotes, cells that occur singly or as long filaments. Cyanobacteria contain pigments that allow them to convert light energy into chemical energy by photosynthesis.

thesized by enzymes in *E. coli* are vitamins, amino acids, fats, proteins, RNA, and DNA. Human cells, on the other hand, are unable to synthesize certain vitamins and certain amino acids, which must be obtained from food.

How are bacteria able to direct and coordinate the synthesis of all the different biological molecules that they need to grow and reproduce, using only the sugar glucose and a few minerals? The information that directs the chemical reactions in *E. coli* (and in all other organisms) is contained in its **genome**, which is defined as the total amount of genetic information in an organism. In *E. coli* and other bacteria, the genome is a single molecule of DNA. In more-complex organisms, the genome is one complete set of the DNA contained in chromosomes (23 chromosomes for human beings). Structurally, the chromosome in bacteria is not equivalent to the more-complex chromosomes in plant and animal cells, because these chromosomes contain not only DNA but also proteins and other molecules that are associated with the DNA. Moreover, plant and animal chromosomes are in the nuclei of cells, whereas bacteria have no nuclei. However, the genetic information en-coded in all chromosomes is contained in DNA molecules.

Kinds of Cells

In the simplest life forms, organism and cell are identical. Bacteria and blue-green algae (also called cyanobacteria) are single cells that belong to the superkingdom of organisms called **prokaryotes** (Figure 5-3). Yeasts, molds, and protozoa are also single-celled organisms, but their cellular structures and functions are more complex than those of bacteria and they belong to the other superkingdom of organisms, the **eukaryotes** (Figure 5-4). All plants and animals are multicellular eukaryotes: their individual cells have the same basic properties that are found in yeasts and protozoa. A prokaryote has no nucleus and reproduces itself by simply dividing (Figure 5-5). A eukaryote has a nucleus and many other specialized cellular structures. It replicates by the sequential processes of mitosis and cytokinesis.

Biologists estimate that as many as 80 percent of all cells on earth are prokaryotes, and multicellular plants and animals probably would not survive without the biochemical assistance of these microorganisms. We generally are not aware of the enormous number of bacteria around and in us, however, because we do not see them. The human body contains more bacteria than human cells, yet they add only a tiny fraction to total body weight. The skin is covered with numerous different kinds of bacteria that are adapted to various environments. In moist underarm areas, there are more than 2 million bacteria per square centimeter of skin. The presence of normal skin bacteria often prevents invasion by harmful bacteria. Internally, bacteria contribute to normal body functions such as digestion; in fact, human stools consist mainly of billions of bacteria.

Cells are highly organized structures whose components function and interact in precise ways. The cellular components of prokaryotes such as bacteria coexist together in the **cytoplasm**, the fluidlike interior of a cell that is encased by a cell membrane and cell wall. (Most animal cells have cell membranes but no cell walls.) In bacteria, all components are free to move about in the cytoplasm, although larger structures such as DNA tend to be localized in certain sections of the cell (Figure 5-6).

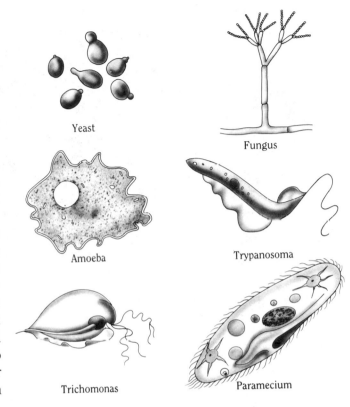

Yeast

Fungus

Amoeba

Trypanosoma

Trichomonas

Paramecium

FIGURE 5-4 Drawings of various simple eukaryotes. Protozoa such as amoeba and paramecia are single cells, as are yeasts. All multicellular plants, animals, and fungi (molds) are eukaryotes.

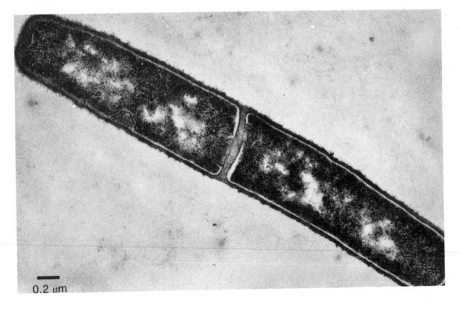

0.2 μm

FIGURE 5-5 A cross-section of dividing bacteria as seen in the electron microscope. The light areas inside the cell are DNA. The small dark particles are ribosomes, sites of protein synthesis in the cell. (Courtesy of E. Kellenberger and A. Benichou-Ryter.)

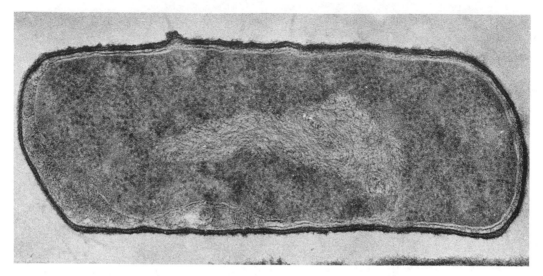

FIGURE 5-6   An electron micrograph of a bacterium, *Bacillus subtilis*. The large DNA molecule (center) is not surrounded by a nuclear envelope as in eukaryotic cells. (Courtesy of A. Ryter and F. Jacob, *Ann. Institute Pasteur*, 107 (1964), 384, and American Society for Microbiology.)

Eukaryotic cells contain numerous structures called organelles—mitrochondria, chloroplasts, Golgi bodies, lysosomes—that perform specialized functions in cells. Although mitochondria and chloroplasts contain small DNA molecules that contribute to the functions of those organelles, the hereditary information in all eukaryotic organisms is carried by the chromosomes in the nucleus. (Mitochondria are "inherited" only from the mother because they

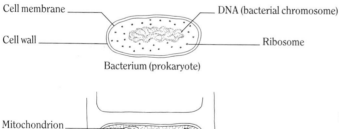

FIGURE 5-7   The various cellular structures that characterize prokaryotic and eukaryotic cells.

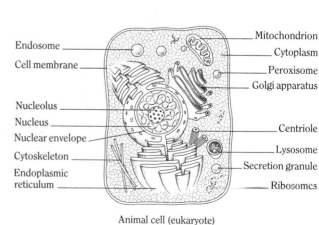

TABLE 5-1 Comparison of cell structures in prokaryotes and eukaryotes. In prokaryotes, all cellular components are in the cytoplasm. Eukaryotic cells have a defined nucleus and cytoplasm that are physically separated by the nuclear envelope. Other cellular organelles in eukaryotes are also encased by membranes.

Cell structure	Function	Present in	
		Prokaryotes	Eukaryotes
Cell wall	Structural	Yes	Yes (in plants, not in animals)
Cell (plasma) membrane	Regulation of substances moving into and out of cell	Yes	Yes
Ribosomes	Structures used for protein synthesis	Yes	Yes
DNA	Carries genetic information	Yes	Yes
Nucleus	Membrane-enclosed structure containing chromosomes	No	Yes
Nucleolus (pl. nucleoli)	Structure in the nucleus where ribosomes are assembled	No	Yes
Endoplasmic reticulum	Membranes used to transport proteins	No	Yes
Golgi bodies	Membranous sacs used to transport proteins	No	Yes
Lysosomes	Organelles used to degrade and dispose of worn-out cellular material	No	Yes
Vacuoles	Fluid-filled sacs	No	Yes
Mitochondrion (pl. mitochondria)	Organelle used to synthesize energy (ATP)	No	Yes
Chloroplasts	Organelles that carry out photosynthesis	No	Yes (plants only)
Mitosis	Process of separation of replicated chromosomes into new cells	No	Yes
Meiosis	Process of chromosome segregation into gametes	No	Yes

are present in large numbers in eggs, but sperm do not contribute any mitochondria.) Table 5-1 lists the major cellular components that are shared by both prokaryotes and eukaryotes, as well as those found only in eukaryotes. The most-significant thing about all organisms—prokaryotes and eukaryotes—is that cellular functions are highly organized and obey the laws of chemistry and physics.

Bacterial cells are extremely simple in their molecular organization and differ from eukary-otic cells in that the functions of DNA and ribosomes are not compartmentalized into or-ganelles but are present throughout the cyto-plasm (Figure 5-7). In fact, chloroplasts and mitochondria in eukaryotic cells are thought to have evolved from primitive bacteria that some-how merged with primitive eukaryotic cells bil-lions of years ago. The evidence supporting this idea derives from the fact that the structure of the ribosomes (described in Chapter 8) in mito-chondria more closely resemble those in bacteria than they do the ribosomes located in the cytoplasm of plant and animal cells.

Eukaryotic cells are unquestionably more complex than prokaryotic cells. Nevertheless, important similarities also exist between them, which suggests that eukaryotic cells evolved from prokaryotic ancestors. That this is so becomes more evident when the biochemical processes of the two types of cells are compared. The structures of DNA, RNA, and protein molecules are often identical in both kinds of cells. Both kinds of cells use similar mechanisms in the synthesis of proteins, and the metabolic pathways of eukaryotic cells are closely related to those of prokaryotes.

Cells Adapt to Environments

Bacteria regulate the synthesis of important macromolecules such as DNA, RNA, and protein according to the chemical composition of the medium (the environment) in which they are grown. **Generation time** is the interval in which the number of bacteria doubles. It is the rate of growth, irrespective of the chemical environment, that dictates the overall macromolecular composition of the bacterial cells. In different environments, bacteria will express different genes, synthesize different enzymes, and exhibit different phenotypes.

Bacterial adaptation to the chemical environment is accomplished by regulating the expression of genes so as to produce only those enzymes that are required for optimal growth in a particular environment. If bacteria are supplied with all of the twenty different amino acids in addition to glucose, they are able to grow much more rapidly because they do not need to produce the almost one hundred enzymes required to synthesize these amino acids, and so the genes for synthesizing the amino acids remain

unexpressed. The energy saved by the bacteria is redirected into increased synthesis of other molecules that enable the bacteria to grow twice as rapidly—that is, half the former generation time (Figure 5-8). (Regulation is described in Chapter 10.)

The alteration of the size and macromolecular composition of bacteria in response to a nutritional change is a simple example of the way in which a cell's phenotype is determined by the interaction of its genotype with the environment. In bacteria growing in a medium supplemented with amino acids, a number of genes are "switched off," but they can be switched back on again if the amino acids in the medium are used up or removed. Specific genes in animal cells are usually switched on or off for good, because specialized cells generally perform the same functions throughout the animal's life. Heart cells and nerve cells are not required to change their functions—indeed, for them to do so might lead to an animal's death. And heart cells do not synthesize the digestive enzymes that are synthesized in stomach cells even though both kinds of cells have identical sets of chromosomes.

Cells that express different genes and, as a result, have different functions are said to be differentiated from one another. The mechanisms of **cellular differentiation** are defined as permanently different patterns of gene expression in eukaryotic cells (prokaryotes do not differentiate). The mechanisms of cellular differentiation are still poorly understood. However, this is a very active research area of genetics and biology because differentiation is crucial to the correct development of organisms. Many chemical substances such as alcohol and the tranquilizer thalidomide can adversely affect gene expression in developing human embryos. High levels of alcohol in a pregnant woman's blood can cause mental retardation in the in-

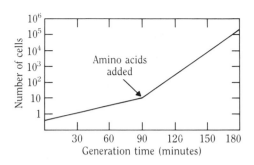

FIGURE 5-8　Effect of the chemical composition of the medium on the rate of growth of bacteria. Addition of amino acids to a simple salts-sugar medium increases the rate of growth (shortens the generation time) because the bacteria divide more often.

fant. Taken early in pregnancy, the drug thalidomide can prevent normal development of arms. In both of these examples (discussed in Chapter 14), there is nothing genetically abnormal with the affected persons and they, themselves, can produce normal progeny. Substances that produce developmental abnormalities in animals are called **teratogens** (from the Greek *teras*, monster, and *gen*, born). More will be said about these substances in Chapter 14.

The environment in which any organism develops and lives plays a crucial role in determining its phenotype. For example, bacteria having identical genotypes will produce colonies of different colors, depending on the kind of chemicals added to the growth medium. Plants that are genetically identical will grow to different heights or yield different amounts of seeds, depending on the environmental conditions of soil, water, and light. Similarly, a person's weight, height, strength, and intelligence are strongly affected by the environments to which he or she is subjected both before and after birth.

Molecules and Macromolecules

Molecules are formed by combining relatively few atoms; macromolecules are considerably larger and range from thousands to millions of atoms—all chemically bonded to one another.

One general class of biologically important macromolecules is **proteins** (from the Greek *proteios*, meaning of the first rank), which consist of long chains of smaller molecules called **amino acids**. As mentioned earlier, only twenty different amino acids are used by all organisms in the synthesis of proteins (Figure 5-9). Proteins have essential structural roles in the architecture of cells; however, their most-significant role is to catalyze (speed up the rate of) chemical reactions. Proteins that catalyze chemical reactions are called enzymes, and every cell contains thousands of different enzymes. Enzymes can increase the rate at which a chemical reaction occurs by as much as a million billion (10^{15}) times. Without enzymes, biochemical reactions would proceed much too slowly to sustain life at temperatures that the cell could tolerate.

Enzymes are chains of amino acids joined together by a particular kind of covalent bond that is called a **peptide bond**; chains of amino acids, correspondingly, are called **polypeptides** (Figure 5-10). If you examine the chemical structures in Figure 5-9, it is apparent that all twenty amino acids have the same chemical structure except for different chemical groups (R-groups) shown outlined on the top side of each amino acid. It is the R-groups that make each amino acid functionally different. Figure 5-10A shows how two amino acids are joined by chemically bonding the amino (NH_3^+) end of

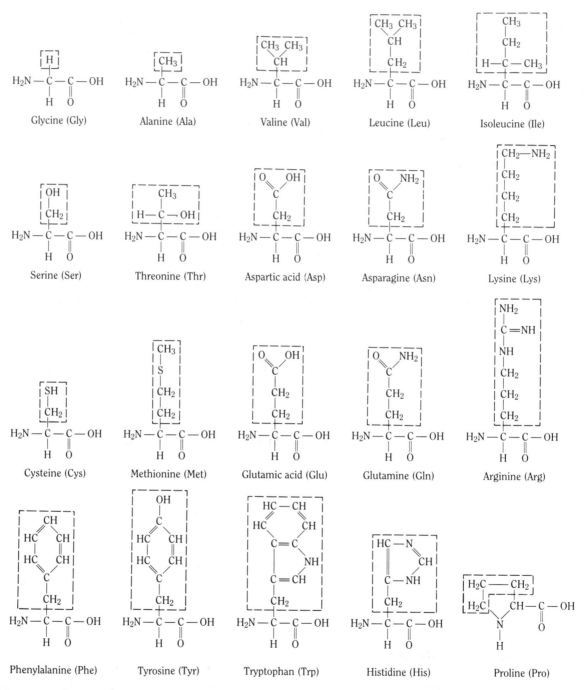

FIGURE 5-9 Chemical structures of the twenty different amino acids found in the proteins of all organisms. Notice that all amino acids have the same basic structure and differ only in their side chains.

one amino acid with the carboxyl (COO⁻) end of another amino acid. When the peptide bond is formed, a molecule of water (H_2O) is eliminated. Figure 5-10B shows four amino acids joined together to form a polypeptide. How the particular amino acids are chosen and joined together is a complicated series of processes that is discussed in Chapter 8. However, keep in mind that, although proteins have different functions in bacteria and human cells (that is, they have different amino acid sequences), all proteins are synthesized from the identical set of twenty amino acids by mechanisms that are essentially the same in all organisms. (Some amino acids may be chemically modified in cells *after* the protein is synthesized.)

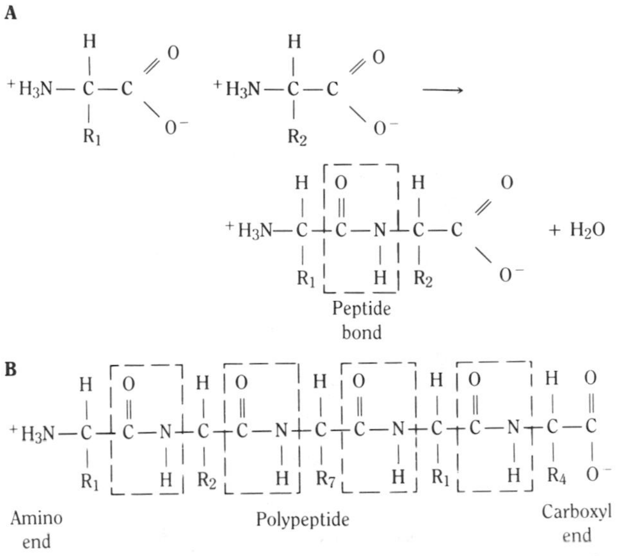

Structures of Proteins

If a protein consists of only a single polypeptide chain, then the terms protein and polypeptide mean the same thing. However, most proteins consist of two or more polypeptide chains folded together in a precise manner. For example, human **hemoglobin**, the protein in red blood cells that binds and transports oxygen, consists of four polypeptide chains. However, these four chains are not identical. Rather, there are two pairs of polypeptides called alpha (α) and beta (β) chains. The identical α-chains have 141 amino acids, whereas the β-chains have 146 amino acids. Every human red blood cell contains from 200 million to 300 million hemoglobin molecules (represented as $\alpha_2\beta_2$).

Proteins perform many other functions in addition to catalyzing chemical reactions. They transport nutrients and other kinds of small molecules into and out of cells. Hemoglobin, for example, transports oxygen to body cells and removes carbon dioxide. Proteins are required for all physical movement—muscles are made of proteins. Skin is made of protein. Bone,

FIGURE 5-10 The peptide bond. Any two amino acids can be joined together by enzymes.
A. The amino end ($-NH_3^+$) of one amino acid is covalently linked to the carboxyl end ($-COO^-$) of another amino acid by formation of a peptide bond and the release of a molecule of water in the enzymatically catalyzed reaction.
B. When peptide bonds link many amino acids together, a polypeptide results. The side groups (R_1, R_2, R_7, and so on) represent different amino acids. For example, the amino acid with R_1 might be methionine, the one with R_2 might be leucine, and so forth.

which provides the body's mechanical support, is made up of proteins and mineral salts. Bacteria swim by means of a special taillike structure—a *flagellum*—that is composed of protein. The regulation of genetic expression in DNA is accomplished by proteins, and the growth and differentiation of plant and animal cells are regulated by proteins.

The unique and virtually unlimited variety of shapes that proteins can assume gives enormous functional versatility. Protein structure is characterized by four levels of complexity called primary, secondary, tertiary, and quarternary structure (Figure 5-11). The **primary structure** of a protein is the sequence of amino acids carried in its polypeptide chains. The amino acid

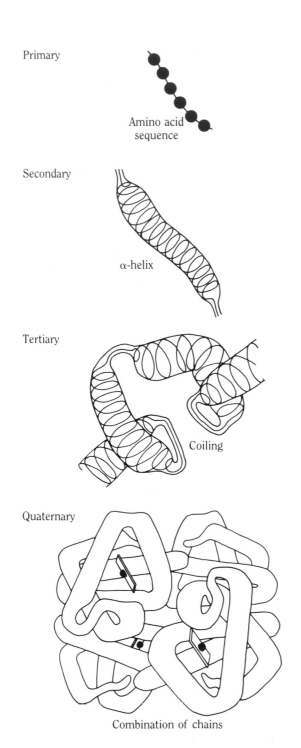

Primary

Amino acid
sequence

Secondary

α-helix

Tertiary

Coiling

Quaternary

Combination of chains

sequence also determines the protein's **secondary structure**, which refers to the parts of the polypeptide chain that are folded or helical in configuration. The particular kind of secondary structure is determined by the hydrogen bonds that form between various amino acids in the polypeptide chain. The **tertiary structure** is the overall three-dimensional configuration of the polypeptide chain—that is, the way in which it is twisted or coiled around itself. For example, the three-dimensional structure of a polypeptide chain that contains cysteine amino acids is partly determined by chemical bonds called **disulfide bridges**, which form between the sulfur atoms of pairs of cysteine amino acids. If you reexamine the structure of cysteine (Figure 5-9), you will notice that this amino acid has a sulfur atom that is free to form a covalent bond with another sulfur atom. To appreciate the importance of cysteine and disulfide bridges, consider what people are actually paying for when they go to a hairdresser for a permanent. They are paying for the destruction and rearrangement of disulfide bonds in the proteins of their hair fibers (See Box 5-3). Finally, the **quaternary structure** refers to the way in which two or more polypeptide chains interact and fold around one another to produce a functional protein molecule.

DNA and RNA

Two extremely important classes of macromolecules are DNA (deoxyribonucleic acid) and

FIGURE 5-11 The structure of proteins. The primary structure is determined by the sequence of amino acids in a polypeptide chain. This sequence also determines how many folded or helical regions (the secondary structure) the polypeptide will have. The tertiary structure refers to the specific three-dimensional shape of the polypeptide. The quaternary structure refers to the shape of a protein such as hemoglobin that contains more than one polypeptide chain.

Hair Styles
Burning Your Disulfide Bridges

BOX 5-3

People spend vast amounts of time and money trying to improve the appearance of their hair—an effort that is actually an applied exercise in biochemistry. Human hair, like the hair of other animals, consists of structural proteins called *keratins*—long protein fibers twisted into helices or other regular structures that are linked to one another by sulfur–sulfur bonds (—S—S—) known as disulfide bridges. The stiffness of the keratin protein matrix is determined by the number of disulfide bridges that connect the proteins. The many disulfide bridges in the keratins in animal horns, hooves, and claws make these protein structures extremely hard and inflexible; hair has fewer disulfide bridges; making it stiff yet flexible; and the low-sulfur proteins in skin make it quite soft and pliable.

A human hair is assembled from protein fibers that are wound together in a regular helical arrangement. When hair is brushed or combed, the helix in each fibril is stretched, temporarily breaking some of the hydrogen bonds holding the helix in position. When the tension is released, the protein helices reform, and the disulfide bridges that connect the various strands reorient the hair fibers into the normal hair pattern.

A permanent wave breaks the disulfide bridges with heat and chemicals and then allows them to reform in different arrangements after the hair is shaped into the desired style by the hairdresser. As long as these new disulfide bridges remain, the hair keeps its new curl. As new hair grows out, normal hair appearance returns as the disulfide bridges in the new hair assume their natural configuration. An average human hair grows about 6 inches per year. This might not seem like very much, but it means that each hair follicle cell must spin out about ten turns of a helical keratin protein every second.

The proteins that make up hair are synthesized in special cells of the skin and scalp. Thus, adding proteins to hair cannot affect the "health" of hair; the quality of skin and hair proteins is determined by the cells that synthesize these structural proteins. In short, the claims of cosmetics and shampoo manufacturers that their products can improve skin or hair are actually intended to improve their profits.

READINGS: Tom Conry, ed., *Consumer's Guide to Cosmetics*. Anchor Press/Doubleday, 1980. T.J. Laughlin and T.M. Ferrell, "Biotechnology in the Cosmetics Industry," *Bio/Technology*. October, 1987.

RNA (ribonucleic acid). DNA is the molecule that carries the genetic information in all cells; viruses carry genetic information in either DNA or RNA molecules. Cells also use RNA molecules, but primarily for directing the synthesis of proteins (discussed in Chapter 8). DNA and RNA are **nucleic acids**—DNA was originally isolated from the nuclei of cells and behaves chemically like an acid (hence the term, nucleic acid). Both are polynucleotide chains whose uniqueness is determined by the particular sequence of simple molecules called bases, even as the uniqueness of proteins is determined by the particular sequence of amino acids.

Bases

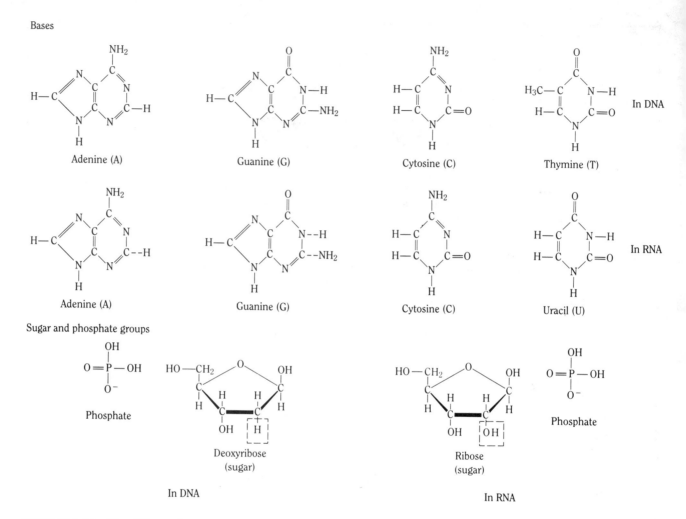

FIGURE 5-12   The different bases in RNA and DNA, and the sugar and phosphate groups in DNA and RNA.

Four different **bases** called adenine (A), guanine (G), cytosine (C), and thymine (T) are present in all DNA molecules (Figure 5-12A). The bases A, G, C, and uracil (U) are found in RNA. Thus, the bases in RNA and DNA are the same except that thymine (T) is found only in DNA and uracil (U) is present only in RNA. (There are rare exceptions to these rules that need not concern us here.) Deoxyribose sugars are present in DNA and ribose sugars in RNA (Figure 5-12B); phosphates are found in both molecules.

The synthesis of DNA and RNA macromolecules in cells requires the chemical joining of bases, sugars, and phosphates (Figure 5-13). Enzymes join these components together in the cytoplasm of cells and thus supply organisms with all of the different **nucleotides** (base–

sugar–phosphate) that are required for the synthesis of RNA and DNA (discussed in Chapters 6 and 7). The uniqueness of each RNA and DNA polynucleotide is determined by the particular

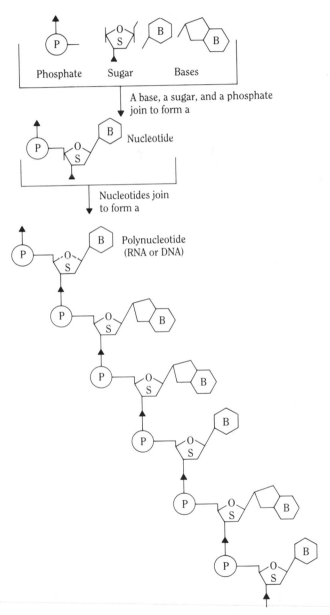

FIGURE 5-13 A schematic diagram showing how bases, sugars, and phosphates are joined together to produce a polynucleotide chain — either RNA or DNA.

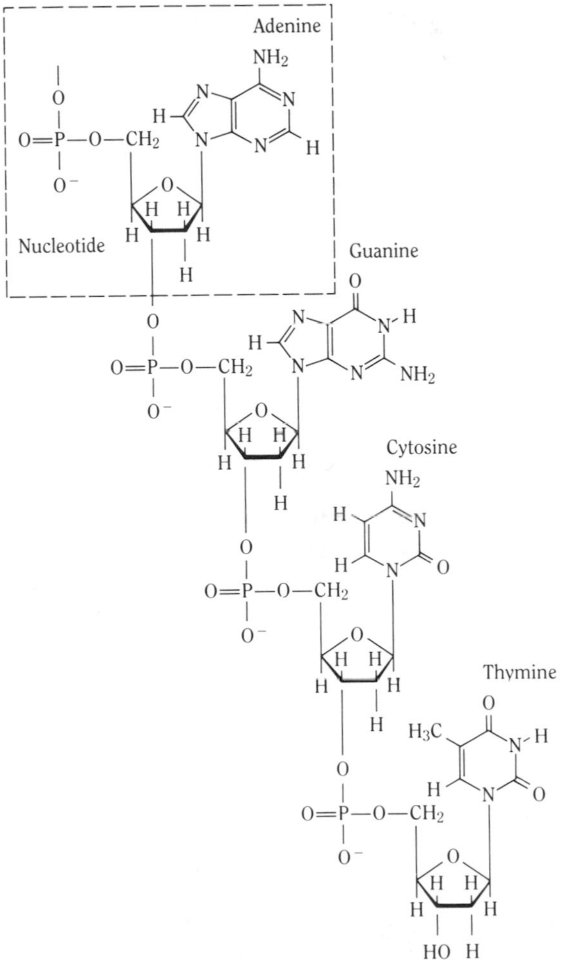

FIGURE 5-14 The chemical structure of a chain of DNA showing the four different bases found in all DNA molecules in all organisms.

sequence of bases of the molecule (for example, AAATCGGCAAC . . . for a DNA molecule or UU-CAAAUUGGCAU . . . for an RNA molecule).

All DNA has deoxyribose sugars and all RNA has ribose sugars; these sugars differ in an OH and H group at one position in the molecule as shown in Figure 5-12B. The chemistry of a short polynucleotide showing all of the atoms and the covalent bonds joining nucleotides together is shown in Figure 5-14.

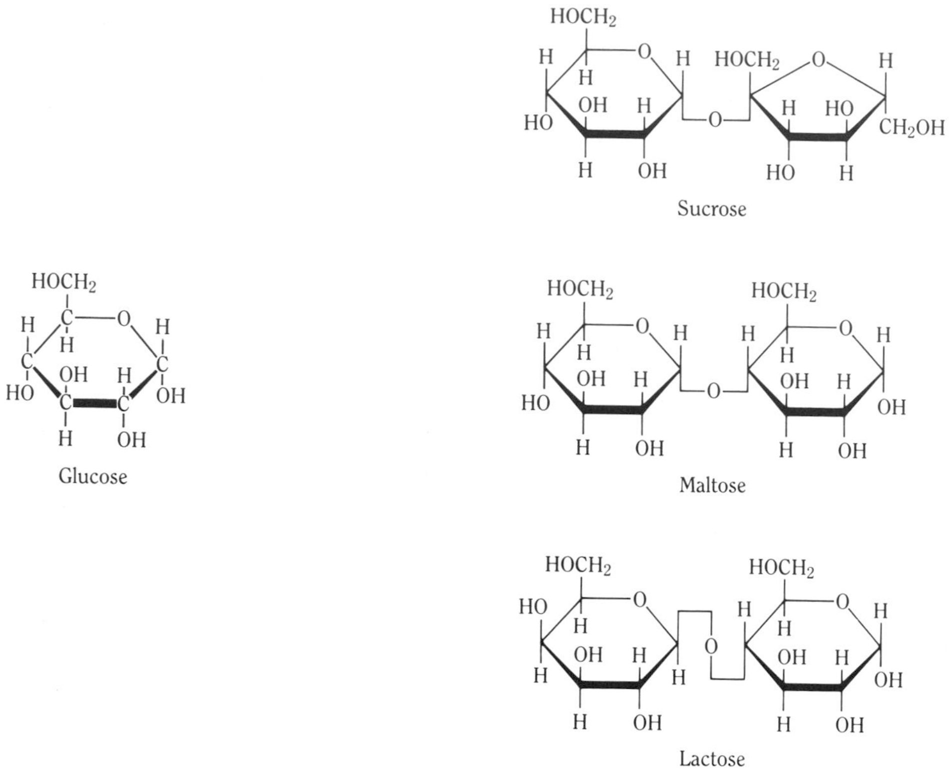

HOCH₂ ... Sucrose

Glucose

Maltose

Lactose

FIGURE 5-15 Glucose is the sugar molecule that is metabolized by all cells to produce energy. A single molecule of glucose and an alternate form of it called fructose bond together to produce a sucrose (common table sugar), a disaccharide that comes from cane or beet. Maltose, which consists of two molecules of glucose, is a common sugar in grains; lactose, a combination of galactose and glucose, is found primarily in milk. Longer chains are called polysaccharides, or carbohydrates.

The nucleic acids contain the information that directs the synthesis of enzymes and other proteins. The addition of each atom to a molecule is catalyzed by an enzyme that generally is specific for that chemical reaction and no other. Because many thousands of different chemical reactions occur in a cell every second, thousands of different enzymes are required—and the information for their synthesis is carried in DNA. Much of what follows in the next few chapters elaborates on how nucleic acids and proteins are synthesized and interact to produce the enormous diversity of organisms in nature.

Carbohydrates and Sugars

Another important class of macromolecules are **carbohydrates**, which consist of long chains of sugars (also called **polysaccharides**). The most-important sugar in all organisms is glu-

Cellulose

FIGURE 5-16 Cellulose is the most-abundant carbohydrate (polysaccharide) in plants and consists of enormously long chains of glucose molecules.

cose, a **monosaccharide**; plants synthesize glucose from water and carbon dioxide by photosynthesis, whereas animals obtain glucose from eating plants or animals. Glucose molecules can be joined together by enzymes to yield various kinds of **disaccharides**, which are pairs of sugar molecules (Figure 5-15). The general chemical formula for all sugars and carbohydrates is $C_nH_{2n}O_n$ in which n can be any number ranging from one to many thousands.

Plants and animals construct extremely long chains of sugar molecules in their cells as a means of storing energy and for structural purposes. The most-abundant carbohydrate in the world is **cellulose**, which is the major component of wood, paper, cotton, and other fibers (Figure 5-16). Another common carbohydrate in plants is **starch**, which consists of glucose molecules joined together in a chemically different way from those in cellulose. Animal cells use sugar for energy and for other metabolic needs; however, excess sugar is stored in the

form of **glycogen**, which consists of branched chains of glucose containing more than a million atoms (Figure 5-17).

A great deal of glycogen in animals is stored in the liver. In periods of nutritional deprivation, the liver's glycogen is broken down and the glucose provides the animals with energy. In recent years, a number of hereditary human diseases have been identified that are characterized as glycogen-storage diseases (Table 5-2). Each of these diseases is named for the biochemist who worked out the chemistry of the particular form of glycogen abnormality.

Lipids

Lipids, sometimes referred to as fatty acids, play important roles in the chemistry of cells. However, lipids and fatty acids do not mean the same thing; fatty acids are components of some lipids. Like other kinds of oil and grease, lipids

FIGURE 5-17 Glycogen is a carbohydrate found in animal cells and consists of more than a million glucose molecules chemically bonded together. Glycogen is used to store energy in the body and is broken down during exercise.

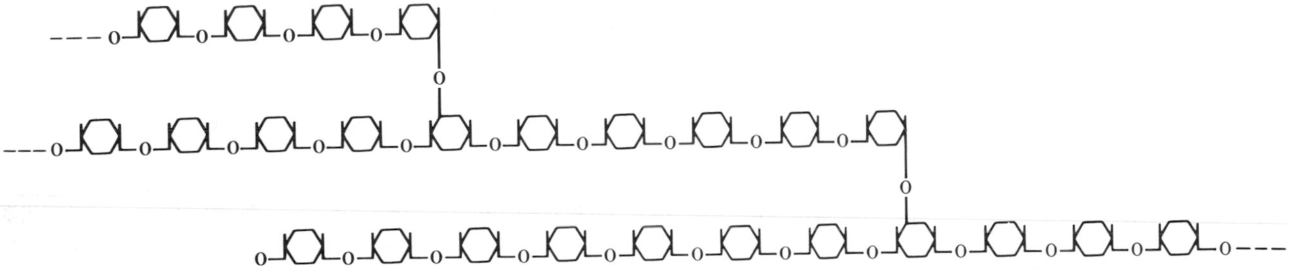

Glycogen

TABLE 5-2   Human hereditary glycogen storage diseases.

Type	Inheritance	Organ affected	Symptoms
Von Gierke's disease	Autosomal recessive	Liver and kidney	Enlargement of the liver. Failure to thrive.
Pompe's disease	Autosomal recessive	All organs	Cardiorespiratory failure causes death, usually before age two.
Cori's disease	Autosomal recessive	Muscle and liver	Like Von Gierke's disease but less severe.
Andersen's disease	Autosomal recessive	Liver and spleen	Progressive cirrhosis of the liver. Causes death, usually before age two.
McArdle's disease	Autosomal recessive	Muscle	Painful muscle cramps. Otherwise patient is normal and well developed.
Hers's disease	Autosomal recessive	Liver	Like Von Gierke's disease but less severe.

are not soluble in water. Their insolubility is due to their chemical composition—they consist of long chains of carbon and hydrogen atoms with a few oxygen and phosphorus atoms (Figure 5-18). **Lipids** are essential components of cell membranes and regulate the flow of substances into and out of cells. Lipids and membranes are essential because they help to compartmentalize various cellular components. Lipids also par-

ticipate in the synthesis of vitamins and hormones. Certain fatty acids (as well as vitamins and amino acids) are essential in the human diet because human cells are unable to synthesize them.

The key to all chemical reactions and molecules synthesized in cells is the array of enzymes that they contain. Genes carry the information

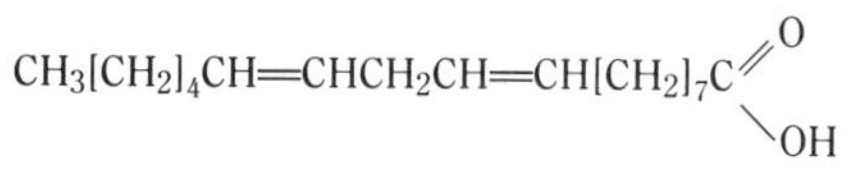

Linoleic Acid

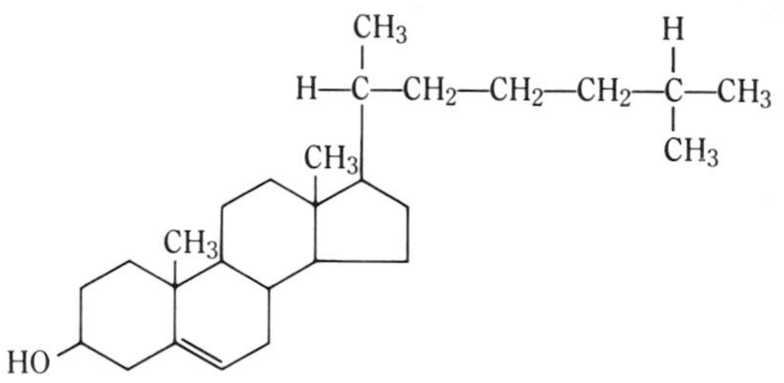

Cholesterol

FIGURE 5-18   Lipids (fats) are components of plant and animal cells, primarily in membranes. Linoleic acid is the most-abundant fatty acid in animal cells. It is an unsaturated fatty acid because of the double bonds between certain carbon atoms in the chain. When a fatty acid is saturated, it means that hydrogen atoms have been attached to those carbons and the double bonds are eliminated. Cholesterol is found in blood and membranes. Too much cholesterol in the circulation contributes to heart and artery diseases.

that directs the synthesis of enzymes, and so we now turn our attention to the structure and function of genetic information and how it is expressed in the form of enzymes and other proteins.

Summary

A cell, which is composed of organized ensembles of molecules, is the smallest living unit that can grow, change, and reproduce. Living cells are distinguished from nonliving matter by the following properties: metabolism, growth, reproduction, mutation, responsiveness to stimuli, and evolution. Nonliving matter (crystals) may have some of these properties but not all.

Cells are classified as either prokaryotic or eukaryotic. A prokaryotic cell has a simple molecular organization. It lacks a nucleus, and it reproduces by simply dividing. A eukaryotic cell has a complex molecular organization, the components of which function and interact with precision. It has a nucleus, and it reproduces by the sequential processes of mitosis and cytokinesis (Chapter 2).

A molecule is a combination of relatively few atoms compared with the thousands or even millions of atoms that make up macromolecules. Proteins are biologically important macromolecules that not only function structurally in cells, but also catalyze chemical reactions. There are four levels of complexity in a protein's structure: primary (the sequence of amino acids in its polypeptide chains); secondary (the folded or helical configuration of its polypeptide chains); tertiary (the three-dimensional configuration of its polypeptide chains); and quaternary (the interaction and folding of two or more of its polypeptide chains around one another).

Nucleic acids (DNA and RNA) are another important class of macromolecules. They are chains of nucleotides; a nucleotide consists of a base, a sugar, and phosphate. DNA carries the genetic information in all cells and in many (but not all) viruses. The primary function of RNA is to direct the synthesis of proteins.

Carbohydrates (long chains of simple sugars) and lipids (long chains of CH_2 groups) are macromolecules that serve essential functions in cells. Sugar and carbohydrates are primarily sources of energy; lipids are components of membranes. These macromolecules are used by cells in other ways as well.

Key Words

amino acids   The twenty different small molecules that are found in all proteins.

bases   The chemical subunits in DNA and RNA whose sequence encodes the genetic information.

carbohydrate   A large molecule consisting of chains of sugars; cellulose and starch are carbohydrates.

cells   The fundamental units of living organisms; all cells are capable of reproducing themselves.

cellulose   A long chain of sugar (glucose) molecules found in wood, paper, cotton, and other fibers.

cytoplasm   everything in a cell exclusive of the nucleus.

differentiation   The process by which cells become increasingly (and irreversibly) specialized in their functions in tissues and organisms during development.

disaccharide   Two sugar molecules joined together.

disulfide bridge   A covalent bond between two sulfur atoms; a disulfide bridge binds two cysteine amino acids in a protein together.

eukaryotes   The superkingdom of all organisms other than bacteria; all eukaryotes have a true nucleus and undergo mitosis and meiosis.

evolution   Process of biological change and diversification through time; also, changes in gene frequencies in populations that accumulate through time.

generation time   The time required to double the number of cells in a population.

genome   The total amount of genetic information (haploid number of chromosomes) in an organism.

glycogen   A long chain of sugar (glucose) molecules found in animals.

hemoglobin   The oxygen-carrying molecule in red blood cells.

lipids   Large molecules, some of which contain fatty acids, that are insoluble in water.

metabolism   The sum of all the chemical processes in an organism.

monosaccharide   A single sugar molecule.

nucleic acid   A chain of nucleotides; RNA and DNA.

nucleotide   A small molecule consisting of a base, a sugar, and phosphate.

peptide bond   The specific covalent bond that joins amino acids together in polypeptide chains.

polypeptide   A chain of amino acids; one chain of a protein that has several chains, such as hemoglobin.

polysaccharide   A chain of sugar molecules; another name for carbohydrate.

primary structure   The sequence of amino acids in a polypeptide chain.

prokaryotes   The superkingdom of all forms of bacteria.

protein   One chain or several chains of amino acids joined together by peptide bonds; proteins serve catalytic or structural functions in cells.

quaternary structure   The folding together of two or more polypeptide chains to produce a functional protein.

secondary structure   The helical configuration of a polypeptide chain.

starch   A carbohydrate in plants.

teratogen   Any agent that causes defects in a developing embryo.

tertiary structure   The overall three-dimensional configuration of a polypeptide chain.

thalidomide   A tranquilizer drug and a teratogen that affects limb development in human embryos.

Additional Reading

Block, I. "The Worlds Within You." *Science Digest*, September-October, 1980.

Fox, S.W. "New Missing Links." *The Sciences*, January, 1980.

Kutter, G.S. *The Universe and Life*. Jones and Bartlett Publishers, Inc., 1987.

Lovelock, J.E. *Gaia: A New Look at Life on Earth*. Oxford University Press, 1979.

Porter, K.R., and J.B. Tucker. "The Ground Substance of the Living Cell." *Scientific American*, March, 1980.

Weinberg, R.A. "The Molecules of Life." *Scientific American*, October, 1985.

Study Questions

1   What are the three unifying principles of biology?

2   About how many different kinds of proteins can be synthesized in a typical bacterium?

3   List five ways in which prokaryotes differ from eukaryotes.

4   What structures or organelles in human cells resemble bacteria?

5   What kind of substances can cause birth defects?

6   What are three chemical properties that distinguish a protein from a nucleic acid?

7   What are some kinds of energy-producing molecules in cells?

Essay Topics

1   Discuss how you think each of the four forces of nature affects cells.

2   Describe the functions of several classes of macromolecules in cells.

3   Discuss how *you* think cells arose on earth.

6

DNA
Structure and Replication

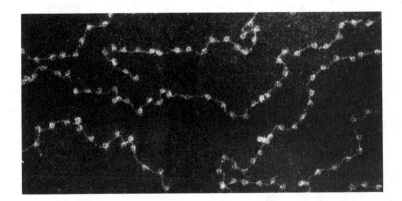

"'I can't believe that,' said Alice. 'Can't you?' the Queen said in a pity tone. 'Try again: draw a long breath, and shut your eyes.'

Alice laughed. 'There's no use trying,' she said, 'one can't believe impossible things.' 'I dare say you haven't had much practice,' said the Queen."

LEWIS CARROLL, *author*

A ONE-PAGE ARTICLE titled "Molecular Structure of Nucleic Acids" that appeared in the British journal *Nature* in April, 1953, initiated a revolution in biology, the consequences of which are still being explored. Francis Crick, an English physicist, and James Watson, an American biologist, began their paper with this modest introduction: "We wish to suggest a structure for the salt of deoxyribose nucleic acid (D.N.A.). This structure has novel features which are of considerable biological interest." The biological interest was indeed considerable, and the scientific implications were revolutionary, because, for the first time, the structure and functions of the genetic information in all living organisms became comprehensible according to the laws of chemistry and physics.

The structure of the DNA molecule proposed by Watson and Crick has survived the test of time in all of its essential details. Implicit in their proposed DNA structure was the chemical basis for the three basic functions of the genetic material: replication, recombination, and mutation. Watson and Crick concluded their now well-known article with this understatement: "It has not escaped our notice that the specific pairing we have postulated immediately suggests a possible copying mechanism for the genetic material."

With the structure of the DNA molecule before them, scientists could readily understand how genetic information could be encoded in DNA and how the hereditary information could be accurately duplicated. Within a few years of the discovery of DNA's structure, the chemical nature of genes and their regulation also became understood. Biologists can now understand how genetic information is encoded, deciphered, and used to direct the synthesis of proteins in all living cells. Before considering these topics, however, we will examine the key experiments that led to and followed the discovery of the structure of DNA.

DNA Carries Genetic Information

As we know, the genetic information of cells resides in their DNA molecules. How was this crucial fact, nowadays taken for granted, discovered? As with several other fundamental concepts of genetics, including Gregor Mendel's discoveries of the patterns of inheritance and Charles Darwin's idea of natural selection, DNA was not readily accepted as the carrier of genetic information until many years after its initial discovery. A key observation pointing to DNA as the genetic material was made in England in 1928 by a microbiologist, Frederick Griffith. He had noticed that two major strains of the bacterium *Diplococcus pneumoniae* could be isolated from patients with symptoms of pneumonia. He called these strains type R pneumococci (*rough* colony phenotype) and type S pneumococci (*smooth* colony phenotype).

When Griffith inoculated mice with type S bacteria, they contracted pneumonia and died. When he injected mice with type R bacteria, however, they did not get the disease. Knowing that bacteria are killed by heating (which is why surgical instruments are sterilized by flaming or boiling), Griffith next heated the type S bacteria before injecting them into the mice. This time no disease was produced, proving that the pneumonia was indeed *caused* by live, but not dead, type S bacteria (Figure 6-1).

Then Griffith made a remarkable discovery. When he took type R bacteria that were unable

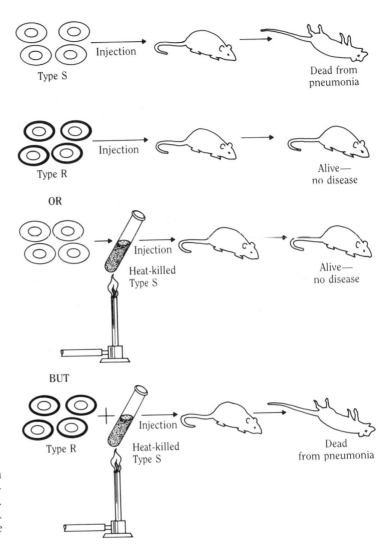

FIGURE 6-1 Griffith's experiments in which mice were injected with type S or type R pneumococci. Type S caused pneumonia and the mice died. Type R was harmless, as were heat-killed type S. However, a mixture of type R and heat-killed type S caused pneumonia and, again, the mice died.

to cause disease and mixed them with heat-killed type S bacteria that were also unable to cause disease and then injected the mixture into mice, the mice contracted pneumonia and died. Moreover, Griffith was able to isolate millions of live type S bacteria from the dead mice. The puzzling question was: How did the live type S bacteria arise in the mice? Griffith was unable to answer this question.

DNA Transforms Bacteria

Oswald T. Avery and his collaborators at Rockefeller University pursued the question that Griffith's experiments raised for many years. Eventually, they became convinced that pieces of DNA from the heat-killed type S bacteria could somehow enter the type R bacteria in the infected mice and genetically transform type R

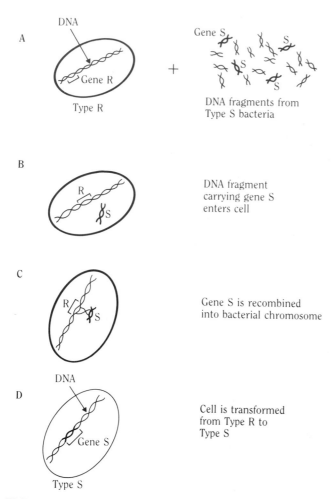

A

DNA

Gene R

Type R

+

Gene S

DNA fragments from
Type S bacteria

B

R

S

DNA fragment
carrying gene S
enters cell

C

R

S

Gene S is recombined
into bacterial chromosome

D

DNA

Gene S

Type S

Cell is transformed
from Type R to
Type S

FIGURE 6-2 Type R bacteria are transformed into type S: (A) fragments of DNA from type S bacteria are added to type R bacteria; (B) some type R cells receive DNA carrying gene S and (C) recombine that fragment of DNA into their chromosome; (D) the expression of gene S (the fragment of DNA containing gene R is lost) changes type R into type S bacteria. This process is called bacterial transformation.

into type S pneumococci (Figure 6-2). They assumed that this genetic change was heritable—that the bacteria would continue to grow and reproduce as type S pneumococci and thus be able to cause disease. This process of changing both the genotype and phenotype of bacteria by

transferring DNA from one type of cell into another is called **transformation**. As we shall see in Chapter 12, bacterial transformation using pieces of DNA is one of the key tools in genetic engineering, which permits genes from virtually any organism to be inserted into microorganisms.

In a paper published in 1944, Avery stated that "among microorganisms the most striking example of heritable and specific alterations in cell structure and function that can be experimentally induced and are reproducible in well defined and adequately controlled conditions is the transformation of specific types of pneumococcus." In later reports, Avery and his colleagues continued to claim that the transformation of bacteria was due to DNA and that DNA carried the genetic information. Yet most biologists at the time did not accept their view.

In the 1940s, proteins were favored as the molecular carriers of genetic information. After all, proteins were known to be extremely complex molecules and to vary tremendously in size, structure, and activity. And were not the traits of organisms also complex and enormously variable? The complexity of proteins and the complexity of organisms was too obvious a connection for most scientists to ignore.

Moreover, DNA appeared to be a chemically uninteresting molecule. At that time in history, DNA was thought to consist of a monotonous, repeating structure consisting of only four different bases. Thus, it was not possible in the 1940s for scientists to envisage how such a molecule could contain the enormity of genetic information needed by cells. As a final argument against DNA being the genetic material, many scientists held that the experiments done in Avery's laboratory were technically flawed. Even the best chemical techniques available at the time produced preparations of bacterial DNA used to transform one type of pneumococcus into the other that were contaminated with about 1 percent of protein. Many critics claimed that this small amount of protein, not the DNA, was actually responsible for bacterial transformation.

However, Avery and his collaborators were correct in their conclusions, and history now accords the work of Griffith, Avery, and others the recognition that they did not receive at the time. DNA does indeed contain the genetic information as was proved by research in which the experimental organisms were viruses (Box 6-1).

DNA Structure

The combined insights of Watson and Crick (Figure 6-3) enabled them to assemble an array of observations made by a number of scientists into a model that accurately represented the DNA molecule. They were aware of two crucial facts concerning the four bases in DNA—adenine (A), thymine (T), guanine (G), and cytosine (C), whose chemical sequence encodes the genetic information. First, Watson and Crick knew that the order of the four bases in DNA is random; that is, any one of the four bases is free to follow any other base in the chain of DNA. Second, from the work of the biochemist Erwin Chargaff, they knew that the number of adenine bases in DNA always equals the number of thymine bases and that the number of guanine bases always equals the number of cystosine bases. The fact that A = T and G = C is known as the **equivalence rule**. This rule played a crucial role in the Watson-Crick model by suggesting to them that A always pairs with T and G always pairs with C in the DNA double helix. However, the equivalence rule does not determine the relative amounts of the base pairs in any DNA molecule, which varies widely in different organisms, particularly among species of bacteria.

FIGURE 6-3 James Watson (left) and Francis Crick (right) in 1962 at the time that they were awarded the Nobel Prize. (United Press International.)

By 1950, the idea that molecules could assume a helical configuration was well known. A helix is like a coil spring or the threads on a screw. Linus Pauling, a chemist at the California Institute of Technology, had demonstrated in the late 1940s that protein molecules, made of long chains of amino acids, are twisted into helical structures. Several years later, two English scientists, Maurice Wilkins and Rosalind Franklin, obtained x-ray photographs of purified crystals of DNA. If the pattern of spots on these x-ray films (Figure 6-4) could be interpreted, the physical structure of the DNA molecule could be deciphered. Watson and Crick examined the x-ray photographs obtained by Wilkins and Franklin and drew the following conclusions:

1 DNA is composed of two helical chains of nucleotides twisted together to form a double helix.
2 The width of the double helix is about 20 angstroms (1 angstrom = 10^{-10} meter).
3 There are ten base pairs per turn of the helix, and the helix makes one complete turn every

The Hershey-Chase Experiment
Final Proof That DNA Carries Genetic Information

To understand the proof that DNA carries genetic information, it is necessary to know what bacteriophages (usually called *phages*, from the Greek *phagein*, "to eat") are and how they grow and reproduce in the cells that they infect. A phage is a virus that infects bacteria—a particle consisting of a DNA molecule (some viruses have RNA instead of DNA) that is packaged inside a head made up of proteins (see Figure 5-1). Viruses grow and reproduce only inside living cells; although most viruses survive outside cells, they are biologically inactive until they infect cells. The capsular proteins of viruses determine what kinds of bacteria, plants, or animals that they can infect, because it is the proteins that recognize and attach to specific receptor proteins on the surfaces of cells. That is why *tobacco mosaic virus* attacks and infects only tobacco leaves and why the *AIDS virus* attaches to and infects only certain human cells.

A classic experiment using phages was performed in 1952 by Alfred Hershey and Martha Chase at the Cold Spring Harbor Laboratories on Long Island, New York. It was known at that time that phages carried the genetic information for directing the growth and reproduction of

BOX 6-1

new phages upon infecting *E. coli* bacteria. Because chemical analyses showed phages to be composed only of DNA and protein, the crucial question is: Which part of the phage carries the genetic information—the DNA or the protein molecules?

Proteins in the tail fibers of T-phages recognize specific proteins on the surface of the bacteria, allowing the phages to attach to the bacterial cell wall and inject their DNA. From this point on, the infection is irreversible, and the bacteria are destined to burst open about 30 minutes after being infected. However, for about 15 minutes after infection, no phage particles can be detected inside the infected bacteria, even if the cells are broken open and the contents analyzed for phages. New phage particles begin to appear in the bacteria after 15 minutes; within an hour, most of the bacteria have burst open releasing several hundred new phages.

Hershey and Chase described the results of the experiment as follows:

1 Most of the phage DNA remains with the bacterial cells.

2 Most of phage protein is found in the leftover fluid.
3 Most of the infected bacteria remain competent to produce phage.
4 If the mechanical stirring is omitted, both protein and DNA remain attached to the bacteria.
5 The phage protein removed from the cells by stirring consists of more or less intact, empty phage coats, which may be therefore thought of as passive vehicles for the transport of DNA from cell to cell, and which, having performed that task, play no further role in phage growth.

Hershey and Chase concluded that phage DNA molecules contain the genetic information for directing the synthesis of new phages when the DNA is injected into bacteria. Surprisingly, the actual numbers obtained in the Hershey-Chase phage experiment were less conclusive than the numbers obtained in the bacterial transformation experiments. By 1952, however, the scientific community was more ready to embrace DNA as the genetic material. And with the announcement of the structure of DNA by Watson and Crick the following year, DNA became universally accepted as the molecule carrying the genetic information.

FIGURE 6-4   A photograph of a DNA diffraction pattern obtained by shining a beam of x-rays through a partly crystallized sample of DNA. The structure of the DNA can be deduced from the regular pattern and positions of the dark spots on the film. (Courtesy of R. Langridge, University of California, San Francisco.)

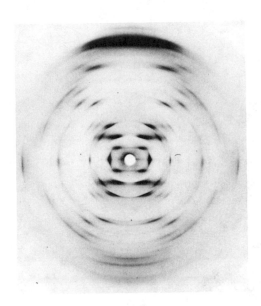

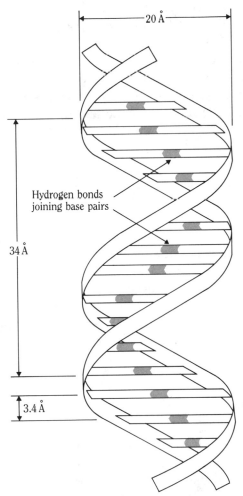

FIGURE 6-5   The Watson and Crick model for DNA. From x-ray photographs, they were able to deduce that the molecule is a double helix with the dimensions indicated. The complementary base pairs are separated by 3.4 angstroms along the chains, and there is one complete helical turn every 34 angstroms (1 angstrom, Å, equals 10^{-8} centimeters).

34 angstroms along the length of the DNA chain (Figure 6-5).

4   In the two chains of the helix, adenine must always appear opposite thymine and guanine opposite cytosine to satisfy the equivalence rule. This makes the chains **complementary**; that is, the information (sequence of bases) in one chain dictates the sequence of bases in the other chain.

5   The stability of the double helix is accounted for by the enormous number of hydrogen bonds that connect the complementary bases along the entire length of the double helix (Figure 6-6).

It is the particular order of the bases in the two DNA chains that determines the genetic information in the DNA molecule; how this information is encoded and decoded is discussed in Chapters 8 and 9. However, some other important structural features of the DNA molecule should be mentioned at this point. The backbone of each helical chain consists of repeating units of sugar molecules (deoxyribose)

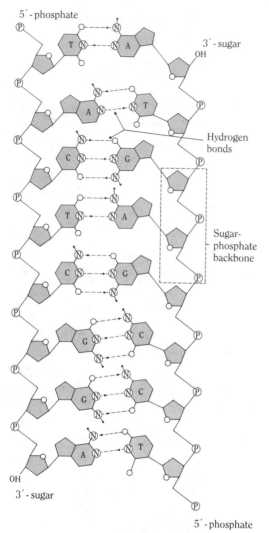

FIGURE 6-6 DNA strands are joined by hydrogen bonds (dashed lines) between complementary base pairs. The strands are said to be complementary because, if you know the sequence of bases in one strand, you can write down the bases in the other (antiparallel) strand. Guanine-cytosine pairs are joined by three hydrogen bonds; adenine-thymine pairs by only two. DNA molecules with a higher G-C content are more stable than those with a higher A-T content.

and phosphates (PO_4^- groups), which are what makes it an acid. Also, each chain of the double helix has a defined chemical orientation; at one end of each chain, called the 3-prime (3') end, is a deoxyribose sugar, and, at the other end of the chain, the 5-prime (5') end, is a phosphate group. The two chains of the helix are chemically *antiparallel* (oriented in opposite directions), as shown in Figure 6-6.

Not only were Watson and Crick able to figure out the structure of DNA, but they also pointed out how the genetic information could be accurately duplicated generation after generation in all organisms. However, how DNA first came to be the carrier of genetic information in all organisms remains a mystery (see Box 6-2).

DNA Replication

Each single chain of the DNA double helix contains all of the genetic information because the bases in one chain determine what the bases in the other chain must be. Replication of the genetic information is accomplished by synthesizing two new chains of DNA in which bases pair up with their complementary bases in the existing chains. In this way, replication gives rise to two double-helical DNA molecules that are informationally identical because they contain the same sequence of bases as did the original DNA molecule.

Examination of the Watson-Crick model (Figures 6-5 and 6-6) leads to the prediction that DNA molecules should replicate in a **semiconservative** manner; that is, after each chromosome (DNA) duplication, newly replicated DNA molecules should contain one preexisting chain and one newly synthesized chain. Several years after the Watson-Crick structure of DNA was announced, Matthew Meselson and Franklin Stahl at the California Institute of Technol-

DNA from Space
A Science Fantasy?

BOX 6-2

How did life really begin on earth? Nobody knows, but one indisputable fact is that DNA molecules carry the genetic information in every organism on earth. Did today's DNA molecules evolve from simpler molecules that formed billions of years ago or did they arise elsewhere in the universe? Was the earth "seeded" from space with cells or organisms? If it was, then perhaps there are other intelligent life forms elsewhere in the universe. And, if they exist, perhaps we can communicate with them.

One logical way to communicate over interstellar distances would be to transmit information on one particular wavelength, much as we do here on earth for radio, television, and shortwave communication. According to many experts, the wavelength of choice to transmit signals into space or to receive them is at the wavelength of the excited hydrogen atom, because hydrogen is the most-abundant element in the universe. We assume that any civilization as scientifically developed as our own would have discovered the spectrum of hydrogen light emitted by the stars. Many scientists believe that if we hope to detect meaningful signals—signals that would, of necessity, have traveled thousands or even millions of years—we should search the heavens with antennas tuned to the wavelength of the hydrogen atom.

Even searching for a message from space at just one wavelength, however, would be an enormously expensive and time-consuming task.

A few years ago, some of the world's most renowned physicists, astronomers, engineers, and biologists gathered at an international meeting to consider alternative means of communication. Are there ways to communicate other than with electromagnetic signals traveling at the speed of light? What would be the most-desirable properties for an information-containing signal that has to cover enormous distances of space and that may take millions of years to reach a receiver? Clearly, what is needed is a signal that contains as much information in the smallest and stablest form possible so that its information content will not diminish over time or distance. Ideally, the information in the message should be able to reproduce itself if it is to be disseminated as widely as possible. And, like the advertising messages that appear on radio or TV, interstellar messages should be repeated over and over to reach as many receivers as possible in the universe. Some kind of booster mechanism should also be employed—here on earth, signals that are transmitted over thousands of miles are electronically boosted so that they arrive at their destination with their informational content and signal strength intact.

Some scientists believe that DNA molecules have properties that make them ideally suited for transmitting information through eons of time and light-years of space. A molecule of DNA could wander forever through the void of space essentially unchanged except for an occasional mutation. If, by chance, a DNA molecule landed in a suitable environment, its message might begin to be reproduced in enormous numbers. Recently, Sidney W. Fox of the University of Miami proposed that the formation of amino acids and other molecules on the primitive earth might have afforded just such an environment for the reproduction of nucleic acids such as DNA. Could a DNA molecule have arrived from space billions of years ago?

Suppose that a molecule of DNA containing a message from space did arrive on earth billions of years ago and became encapsulated within a protocell, began replicating, and eventually evolved into human beings. Although this idea makes for a good science-fiction story, it ultimately pushes the question of how life began further out into space.

READING: Francis Crick, *Life Itself: Its Origins and Nature.* Simon & Schuster, 1981.

ogy devised an experiment to test whether replication is indeed semiconservative. In theory, DNA could replicate in a conservative manner; that is, after replication, the two original chains could remain hydrogen-bonded together and the two newly synthesized chains could become partners. However, conservative replication does not occur, as the Meselson-Stahl experiment demonstrated.

When a solution of cesium chloride (CsCl)—a salt similar to sodium chloride but heavier—is centrifuged for a long time (24 hours or more), a density gradient is established in the tube, because of the counteracting forces of sedimentation and diffusion on the salt molecules. If DNA is added to the CsCl solution before centrifugation, it, too, will sink in the salt solution, eventually settling and forming a narrow band at the position in the gradient where its density precisely matches the density of the salt solution. (The principle is the same as that governing the level at which your body floats. In fresh water, you usually sink below the water because the weight of water that you displace is less than the weight of your body. But, if there is salt in the water, as there is in sea water, part of your body floats above water because salt is heavier than water. And the saltier the water, the higher you float, because the submerged part of your body displaces an amount of salt water equal to the weight of your entire body.)

The Meselson-Stahl experiment was designed to produce "heavy" DNA by growing *E. coli* bacteria in a medium containing a stable, heavy isotope of nitrogen (^{15}N). This is referred to as "density labeling" the DNA. The more-abundant isotope of nitrogen (^{14}N) is "light" because it contains one neutron less than does ^{15}N. Thus, it follows that any molecules synthesized with ^{15}N atoms will be heavier than those synthesized with ^{14}N. After many generations of growth in a medium containing $^{15}NH_4Cl$

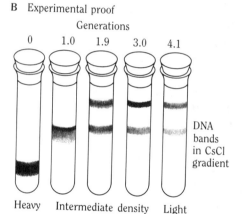

FIGURE 6-7 The Meselson-Stahl experiment confirmed the prediction of the Watson-Crick model that DNA replicates by a semiconservative mechanism. After each generation, one old chain of DNA is joined to a newly synthesized chain. When the bacterial DNA was analyzed by centrifugation, the heavy, intermediate, and light DNA molecules could be visualized.

(heavy ammonium chloride), all the bacteria will have DNA heavier than ordinary ^{14}N-containing DNA. The bacteria are then collected by centrifugation and the ^{15}N medium is removed. Growth of the bacteria is then continued in ^{14}N medium.

If DNA replicates in a semiconservative manner, after one generation of growth in the light

TABLE 6-1 Comparison of the number of base pairs in DNA molecules of different organisms.

Organism	Number of base pairs (× 1,000)	Length in micrometers (.000001 meter)
Tumor virus	5.1	1.7
Phage φX174	5.4	1.8
Phage T4	116	55
Smallpox virus	190	63
E. coli	4,000	1,360
Yeast	13,500	4,600
Drosophila	165,000	56,000
Human being	3,000,000	990,000

Note: A tumor-causing virus has about 5,000 base pairs, whereas the haploid set of 23 human chromosomes has 3.0 billion base pairs. Assuming an average of 1,000 base pairs per gene, the tumor virus contains only 5 genes, whereas the human genome could contain 3.0 million genes. In reality, most sequences in human DNA are not used or are not functional (so-called junk DNA). Calculations indicate that there are probably no more than 100,000 genes in human beings. (Assuming that each gene, on average, has 1,000 base pairs, calculate the fraction of the total DNA in human beings that carries useful genetic information).

medium, all of the DNA molecules in the bacteria should be of hybrid, or intermediate, density because all should consist of one heavy (old) chain and one light (new) chain (Figure 6-7A). After two generations of growth, half of the DNA molecules should be of hybrid density and half should be fully light. The results of the Meselson-Stahl experiment confirmed these predictions (Figure 6-7B). Bacterial DNA does indeed replicate semiconservatively as Watson and Crick suggested. Because of complementary base-pairing one preexisting chain is joined by hydrogen bonds to the newly synthesized chain. Since this experiment was performed, it has been shown that DNA is replicated in the same way in cells of plants and animals. We now know that the Watson-Crick model for the structure of DNA and semiconservative DNA replication is valid for all organisms irrespective of the size of the DNA (Table 6-1).

Errors can occur in base-pairing during replication even though the enzymes participating in DNA replication (discussed later in this chapter) have evolved to be extremely accurate. In fact, it is estimated that the rate of error is less than one per billion base pairs. (To appreciate this degree of accuracy, consider the number of errors that occur in an automated automobile assembly line.) However, a change in even a single base pair in DNA in a human sperm or egg can cause an inherited disease.

Plasmid DNA

DNA molecules are found not only in the chromosomes in the nucleus, but also in the cytoplasm of cells. Human cells contain hundreds of mitochondria, each of which contains a few or even many small, circular DNA molecules. Each of these DNAs carries about thirteen genes.

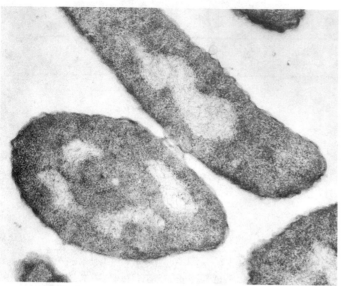

FIGURE 6-8 A donor bacterium containing an F-plasmid (rod-shaped) is conjugating with a recipient bacterium. The two bacteria are physically joined and DNA is transferred from donor to recipient at the point of contact. (Courtesy of David P. Allison, Oak Ridge National Laboratory.)

Plant cells contain chloroplasts (the organelles that carry out photosynthesis), which also have small, circular DNA molecules. Bacteria, too, contain physically closed, circular DNA molecules called **plasmids**. They are visualized under the electron microscope as circles, some of which may be twisted into loops.

Bacteria may contain only a few plasmids per cell or they may have hundreds of copies. The plasmids may be quite small (a few genes) or very large (one hundred genes or more). Because plasmids contain DNA and genes, they have greatly assisted scientists who study the structure and functions of DNA. Unlike the large DNA molecule that functions as a chromosome in bacteria and contains thousands of essential genes, bacterial plasmids contain genetic information that is *not* essential for the growth of cells. Thus, bacteria grow and repro-

duce equally well with or without plasmids. However, plasmids confer special properties to bacteria—for example, the ability to conjugate with other bacteria. **Conjugation** refers to the act of mating that bacteria undergo with members of their own species and even with bacteria of different species (Figure 6-8).

One of the best-characterized bacterial plasmids is the F (fertility)-plasmid in *E. coli*. This large plasmid (it contains about one hundred genes) is found in donor bacteria and allows them to transfer both bacterial plasmid and chromosomal DNA to recipient cells (this process is discussed in the next chapter). Plasmids provide an important mechanism by which bacteria can exchange genes. Presumably, these mechanisms for genetic exchange benefit bacterial survival in changing environments and have played important roles in their evolution.

Other plasmids found in many bacterial species are the R (resistance)-plasmids. These plasmids carry genes that code for enzymes that make the bacteria carrying them resistant to antibiotics such as penicillin and tetracycline. (R-plasmids also play an important role in genetic engineering, which is discussed in Chapter 12.) R-plasmids can be transferred by conjugation from one species to another—for example, from *E. coli* to *Gonorrhea nisseria*, a bacterium that causes a sexually transmitted disease. R-plasmids that make disease-causing bacteria antibiotic resistant also contribute to human health problems (see Box 6-3).

In the past twenty-five years, thousands of tons of antibiotics have been added to animal feed in the United States because the producers of beef, pork, chicken, lamb, and other meats became convinced (mostly by manufacturers of antibiotics) that antibiotics were needed to prevent diseases and to promote faster weight gain in their animals. However, we now realize that such widespread use of antibiotics creates a

Human Diseases and Plasmids

BOX 6-3

Travelers to foreign countries—especially in tropical climates in which warm, moist environments enable bacteria to thrive—frequently suffer from diarrhea, which can cause not only temporary discomfort but also severe illness. Severe diarrhea results in the depletion of body fluids and the loss of essential salts. If the fluids and salts are not replaced, the person may die.

Recently, it has been shown that many diarrhea victims carry in their digestive systems strains of *E. coli* that contain plasmids, some of whose genes direct the synthesis of the toxins that cause diarrhea. People born and raised in areas where the toxin-producing bacterial strains exist probably develop an immunity to the toxins as a result of repeated exposures.

But visitors to these areas have no immunity.

Another, more-severe disease is *hemolytic anemia*, which is caused by *E. coli* that enter the human circulatory system. When these bacteria are in the blood, they destroy red blood cells, resulting in severe anemia and death. These particular strains of bacteria have another kind of plasmid containing a gene that codes for the synthesis of *hemolysin*, a protein that destroys red blood cells.

If the bacteria that cause diarrhea or severe anemia are also resistant to several antibiotics, then it becomes more difficult to treat these diseases. In some cases, by the time the antibiotic resistance is discovered, the infection has progressed to the point at which treatment is ineffective and the patient does not recover.

It is becoming increasingly evident that even the development of antibiotic "wonder" drugs may not always have wonderful consequences. Because plasmids provide bacteria with the genetic information that they need to survive in antibiotic-containing environments, the wholesale, indiscriminate use of antibiotics must be reduced so that, when these drugs are really needed, they will be able to kill disease-causing bacteria.

READING: Richard P. Novick, "Plasmids," *Scientific American*. December, 1980.

serious health problem for both animals and human beings because many disease-causing bacteria have become resistant to antibiotics.

In 1987, scientists showed that an outbreak of food poisoning in the Midwest was caused by meat contaminated with *Salmonella* bacteria carrying a particular R-plasmid. They argued that the plasmid-bearing bacteria had been selected for in these animals because their feed contained antibiotics. This was the first convincing demonstration that antibiotic-resistant bacteria that arise in animals can also end up causing disease in people.

Origin and Direction of DNA Replication

Both circular DNA molecules (plasmids, mitochondria, and bacterial chromosomes) and linear ones (those in human chromosomes, for example) replicate by initiation at a specific location called the **origin of replication**. Plasmids and the bacterial chromosome have just one origin; however, the DNA molecules in plant and animal chromosomes have hundreds or thousands of origins of replication (Figure 6-9). If this were not so, the DNA simply could not be replicated in the time allotted for the S

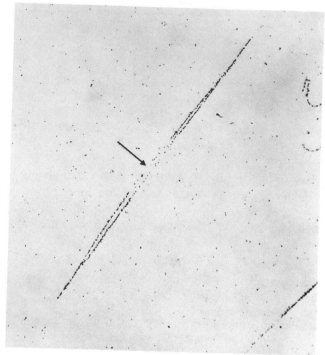

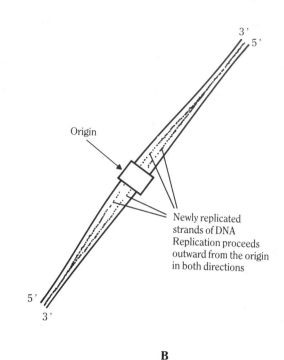

FIGURE 6-9   A. Autoradiograph of a region of replicating DNA. A small amount of radioactive thymine is added for a very short time (a pulse) to cells in which DNA is replicating. Then the level of radioactivity is increased. The DNA strands on either side of the origin are faintly labelled; further away the strands are darker because the film has been more exposed. (Courtesy of D. Prescott, University of Colorado, Boulder.)
B. Schematic diagram of what is taking place in part A.

phase of the cell cycle and the overall cycle would take much longer (see Figure 2-2). Also, except for plasmids and other very small DNA molecules, the replication of DNA is bidirectional—that is, synthesis proceeds in both directions from each replication origin. Although this doubles the rate at which a DNA molecule replicates, it creates difficulties in the untwisting of loops of DNA and in the unwinding of the double helix because strands are unwinding in opposite directions.

The enormously long DNA molecules in the forty-six human chromosomes (which together add up to more than a meter in length) present the cell with a considerable problem in packaging its DNA into a nucleus that is only about one-millionth of a meter in diameter. The packaging is accomplished by the tight wrapping of DNA around small molecules called **histones**. The structures formed by DNA wound about histones are called **nucleosomes**, which make chromosomes look like beads on a string when visualized by electron microscopy (Figure 6-10).

DNA Replication Enzymes

Accurate replication of a DNA molecule as long as those contained in human chromosomes is a far more complex task than the construction of an automobile. The Watson-Crick model of

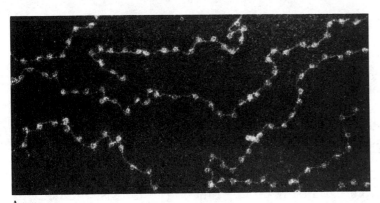
A

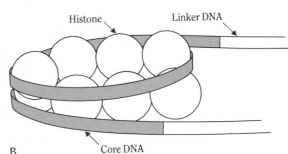
Histone Linker DNA
B Core DNA

FIGURE 6-10 Nucleosome structure of chromosomes: (A) an electron micrograph of a eukaryotic chromosome showing its beads-on-a-string appearance; (B) a schematic diagram of a nucleosome (one bead). (Electron micrograph courtesy of Ada Olins.)

DNA makes replication seem deceptively simple—merely a matter of inserting the correct complementary bases according to the preexisting single-strand templates. But even in a bacterium there are more than 3 *million* base pairs and in a human egg or sperm about 3 *billion*. To appreciate the immensity of the task of synthesizing a DNA molecule in *E. coli*, realize that you would have to correctly line up 3 million base pairs in 30 minutes or less. And this does not take into account that, elsewhere in the cell, all the millions of bases and sugar molecules are being synthesized and must arrive at the right place at the right moment.

The mechanism of DNA replication for some phages, plasmids, and for *E. coli* bacteria has been worked out in considerable detail, but how the DNA in the forty-six human chromosomes is synchronously replicated is still only partly known. About a hundred different enzymes are needed to synthesize the four bases (A, T, G, and C) and to join these bases to deoxyribose sugars and phosphates before they are inserted into a new strand of DNA. Other enzymes are needed to join the complementary bases together in the

new DNA strand, to untwist the helix so that the replication can proceed, to repair gaps and breaks in the DNA, and to perform other functions essential for accurate replication. A few of the well-characterized enzymes required for DNA replication in bacteria are listed in Table 6-2.

TABLE 6-2 Functions of some bacterial DNA replication enzymes.

Enzyme	Function
RNA primase	Synthesizes a short piece of RNA that is necessary to begin synthesis of each new DNA strand.
DNA gyrase	Twists circular DNA molecules into tight coils and untwists them.
DNA helicase	Unwinds strands of the double helix so that replication can proceed.
DNA replicase	Joins nucleotides together.
DNA polymerase I	Fills in gaps in newly replicated DNA strands and corrects errors in base pairing.
DNA ligase	Joins the sugar to the phosphate to make the sugar-phosphate backbone of each strand of DNA continuous.

DNA replication begins with one double-stranded helical molecule, which becomes two double-stranded helical molecules identical with each other and with the parent molecule from which they were replicated. If the replication process has copied the DNA accurately, the daughter molecules will contain identical genetic information. However, an error in the insertion of even a single base will produce a change in the genetic information carried by the replicated DNA. To minimize replication errors, cells contain enzymes whose functions are to monitor the accuracy of the replication process and to correct mistakes or repair damage. We now turn our attention to processes that oppose the faithful replication of DNA—mutation and recombination—and will learn why, without such mechanisms, life could not have diversified and evolved.

Summary

Deoxyribonucleic acid (DNA) is the molecule in cells that contains genetic information. The configuration of the molecule is a double helix, and it consists of two antiparallel strands of nucleotides, each of which contains a base, a sugar (deoxyribose), and a phosphate. The bases in DNA are adenine, cytosine, guanine, and thymine. An adenine molecule in one strand of DNA pairs with a thymine molecule in the other strand; similarly, cytosine pairs with guanine. This pairing makes the strands complementary. The sequence of bases in DNA determines the genetic information that it carries.

DNA is present not only in chromosomes (in the nuclei of eukaryotes), but also in other cellular structures. Small, circular DNA molecules are found in the mitochondria of animal cells and in the chloroplasts of plant cells. In microorganisms, these small, circular molecules of DNA are called plasmids, which are extremely useful to scientists in both basic and applied research.

DNA is replicated by the synthesis of two new strands; in this synthesis, new bases pair with their complementary bases in the existing strands. Newly replicated DNA thus consists of one preexisting strand and one new strand; such replication is semiconservative.

Many enzymes are required in the replication of DNA. They function to untwist the helix to allow replication to proceed; they catalyze the synthesis of bases and unite them with sugars and phosphates; they repair breaks in the DNA strands; and they join complementary base pairs in the replication of the two antiparallel strands.

DNA replication begins and ends at specific sites in the molecule that are determined by specific base sequences. The region in DNA where synthesis begins is the origin; the replication fork is the region of active synthesis of new strands of DNA.

Key Words

complementary   Refers to the base-pairing rules in which A pairs with T (or U) and G pairs with C.

conjugation   Mating between a donor bacterium and a recipient one such that DNA is transferred from the donor and stably maintained in the recipient.

equivalence rule   The amount of adenine in DNA equals the amount of thymine; the amount of guanine equals the amount of cytosine.

F-plasmid   Small, circular DNA molecule in *E. coli* bacteria that causes cell–cell conjugation. The F-plasmid can be transferred from a donor to a recipient bacterium or it can facilitate the transfer of chromosomal genes.

histone   A small DNA-binding protein thought to regulate gene expression and chromosomal structure in eukaryotic cells.

ligase   An enzyme that joins a sugar to a phosphate in a DNA strand.

lysis   The bursting open and death of a cell.

nucleosome   A beadlike structure observed in eukaryotic chromosomes that consists of DNA wrapped around histones.

origin of replication   A site in a chromosome where replication of a new DNA molecule is initiated.

phage (bacteriophage)   A virus containing either RNA or DNA that infects and destroys (lyses) bacteria.

plasmid   An extrachromosomal, circular DNA molecule found in many kinds of bacteria. Plasmids are self-replicating and may exist in many copies.

polymerase   An enzyme that replicates a single strand of either DNA or RNA.

primase   A special RNA polymerase that initiates replication of a DNA strand.

replicase   An enzyme that replicates DNA by adding complementary bases.

replication fork   A region in DNA where duplication of both strands is taking place.

semiconservative replication   Refers to the replication of a double-stranded DNA molecule consisting of one old polynucleotide strand and one newly synthesized strand.

terminus of replication   A site in a chromosome where replication of DNA is terminated.

topoisomerases   Enzymes that can twist and untwist DNA helices and circles.

transformation   The stable insertion of a fragment of DNA into a cell, causing its genotype and phenotype to be changed.

Additional Reading

Block, I. "How Life Works: The World Within You." *Science Digest*. September-October, 1980.

Cech, T.R. "RNA as an Enzyme." *Scientific American*. November, 1986.

Check, W. "Plasmids: More Than Genetic Debris." *Mosaic*, March-April, 1982.

Day, M. "The Biology of Plasmids." *Scientific Progress*, 71, 203 (1987).

Grivell, L.A. "Mitochondrial DNA." *Scientific American*, March, 1983.

Kornberg, R.D., and A. Klug. "The Nucleosome." *Scientific American*, February, 1981.

Lin, E.C.C., R. Goldstein, and M. Syvanen. *Bacteria, Plasmids, and Phages: An Introduction to Molecular Biology*. Harvard University Press, 1984.

Watson, J.D. *The Double Helix: A Personal Account of the Discovery of DNA*. Atheneum, 1968.

Yates. G.T. "How Microorganisms Move Through Water." *American Scientist*, July-August, 1986.

Study Questions

1   What process changes harmless bacteria into pathogenic (disease-causing) bacteria?

2   Why did scientists think for many years that proteins carried genetic information?

3   If 30 percent of a DNA molecule consists of adenine bases, what percentages are cytosine, guanine, and thymine?

4   If DNA replication were conservative instead of semiconservative, what fraction of the DNA molecules would be half-heavy, half-light after one generation in "light" medium in the Meselson-Stahl experiment?

5   What is the meaning of the equivalence rule for DNA? Express your answer in words.

6   What kind of chemical bond gives DNA its stability in keeping the two strands together?

7   What two experiments proved that DNA is the genetic material?

Essay Topics

1   Discuss the significance of plasmids in causing human diseases.

2   Explain the key steps in the phage experiment that proved that DNA was the genetic material.

3   Discuss the various steps of DNA replication at a replication fork.

7

DNA
Recombination and Mutation

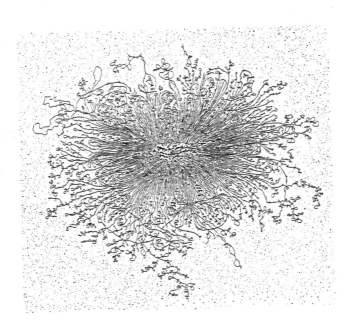

In one way or another, we are all genetic lemons.
FRED BERGMAN, *biologist*

THE GENETIC PROPERTIES of DNA such as replication or recombination could be studied, in principle, in any organism. However, some organisms are more suitable for the study of particular genetic properties than are others. We have already seen how Mendel devoted himself to studying peas. He also tried to breed and study the inheritance of traits in bees, but these experiments did not pan out. The fruit fly *Drosophila* has been studied by geneticists for almost a hundred years because it has a short generation time (a couple of weeks compared with about twenty years for human beings) and produces enormous numbers of offspring. If a geneticist wants to study how an animal's sex influences the inheritance of traits, he or she clearly will choose to work with flies rather than peas.

Beginning in the 1940s, geneticists became more interested in molecular aspects of the genetic information. How do genes direct the synthesis of protein molecules, what kind of molecules regulate the expression of genetic information, and so forth? For these kinds of questions, they realized that peas and flies are not suitable organisms; they are too complex. The molecular geneticists—as they came to be called—decided to study very simple organisms such as bacteria or even smaller bacterial viruses. They hoped that the fundamental properties of the genetic information carried in the DNA of these organisms would turn out to be more or less universal. That hope, fortunately, has turned out to be true time and time again. For example, the genetic code is amost universal, as we will see in Chapter 8. The molecular mechanisms of DNA replication are very similar except for minor details both in a bacterium and in a human cell. So are the molecular mecha-

nisms for recombining fragments of DNA molecules in cells or for changing the arrangement of bases in a DNA molecule by mutation.

The enormous variety of organisms that are alive today and the even greater number of species that have become extinct suggest that DNA sequences must continually change if organisms are to adapt and survive. New genes and combinations of genes are necessary if organisms are to flourish in new environments and new species are to evolve. On the one hand, DNA molecules must conserve and reproduce their genetic information with utmost fidelity; otherwise, organisms already adapted to an environment may not survive and reproduce. Therefore, the mechanism of DNA duplication must be exceptionally accurate; any alterations or mistakes must be kept to a minimum. On the other hand, mutations and genetic diversity are essential to the survival of individual organisms under new conditions and ultimately to the continued survival of populations. The genetic information in DNA molecules must be susceptible to change. The need for genetic information to be preserved and the necessity for it to change constitutes a fundamental biological paradox.

This chapter explains how the unique structure of DNA resolves this paradox. It begins with a description of experiments showing that genetically recombinant individuals do, in fact, arise in populations of flies and how these experiments can be used to precisely determine the locations of genes in chromosomes. Then, to understand the mechanisms of DNA recombination and mutation at the molecular level, we will turn to experiments in which bacteria and viruses are the organisms under study.

Linkage of Genes

Once DNA was discovered to be an unbroken, enormously long molecule in the 1960s, it became obvious that genes on the same chromosome (DNA) are physically linked together. However, the concept of **linkage** had been deduced many years earlier by fruit-fly geneticists who performed crosses with genetically different flies. The significance of linkage is that it allows geneticists to predict the outcome of genetic crosses; it also facilitates the mapping of human genes and the diagnosis of human hereditary disorders. To simplify the description of linkage somewhat, we will begin by assuming that two genetic loci (pairs of alleles) are located very close together on a fly's chromosome. *Drosophila* have a haploid set of three autosomes and one sex chromosome (N = 4). The mutant flies differ from wild-type flies in a number of traits. One kind of mutant has very short wings and can hardly fly—this trait is called vestigal wing and the gene symbol is *vg*. Another kind of mutant has oddly shaped, curved wings and the gene symbol is *cr*.

Certain agreed-upon written conventions are used to show the location of each allele on one or the other of the two homologous chromosomes (Figure 7-1). For example, $vg\ cr^+/vg\ cr^+$ indicates that both chromosomes carry identical alleles at each of two loci, whereas $vg\ cr/vg^+\ cr^+$ indicates that the alleles at each locus are different. The plus sign (vg^+) signifies the wild-type (normal) allele. Absence of a sign means that the allele (*vg*) is mutant. It turns out that it is important to know the positions of the alleles on a pair of homologous chromosomes to correctly interpret the results of a genetic cross. However, the assignment of each allele can often be inferred from the frequencies of the different classes of recombinant progeny produced in a cross.

A Homozygous flies (mutant)

Vestigial wing

$$\frac{vg \qquad cr^+}{vg \qquad cr^+}$$

or

$(vg\ cr^+/vg\ cr^+)$

or

$$\frac{vg \quad cr^+}{vg \quad cr^+}$$

Curved wing

$$\frac{vg^+ \qquad cr}{vg^+ \qquad cr}$$

or

$(vg^+\ cr/vg^+\ cr)$

or

$$\frac{vg^+ \quad cr}{vg^+ \quad cr}$$

B Heterozygous flies (wild-type)

$$\frac{vg \qquad cr^+}{vg^+ \qquad cr}$$

or

$(vg\ cr^+/vg^+\ cr)$

or

$$\frac{vg \quad cr^+}{vg^+ \quad cr}$$

$$\frac{vg \qquad cr}{vg^+ \qquad cr^+}$$

or

$(vg\ cr/vg^+\ cr^+)$

or

$$\frac{vg \quad cr}{vg^+ \quad cr^+}$$

FIGURE 7-1 Conventions for showing the arrangement of linked genes. Each genetic locus is located at the same site on each of two homologous chromosomes, but the relative positions of different alleles on homologous chromosomes are important in analyzing genetic crosses.
A. If the alleles for a genetic locus are identical, the individual organism is homozygous for that locus. Two ways of writing the positions of the alleles on chromosomes are shown.
B. If the alleles for a genetic locus are different, the individual organism is heterozygous for each locus at which the alleles differ.

A Genetic Cross

The purpose of most genetic crosses is to first determine whether genes that code for two different traits are on the same chromosome or on different ones. (Mendel showed that genes on different chromosomes segregate independently from one another.) If the genes are on the same chromosome and relatively close to one another, then they show genetic linkage. Once linkage is

Genetic testcross

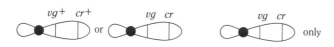

(female parent) ♀   ×   ♂ (male parent)

Wild type phenotype   Mutant phenotype

Gametes

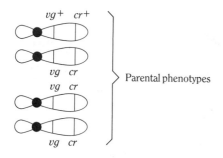

Progeny classes

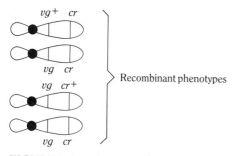

Parental phenotypes

Recombinant phenotypes

FIGURE 7-2 Diagram of a genetic testcross between male and female fruit flies. The two loci (*vg* and *cr*) are linked on the same autosome.

established, the distance between the genetic loci on the chromosome can be determined by the frequency of crossing-over of alleles during meiosis, which is revealed in the frequency of recombinant progeny.

Figure 7-2 shows a simple genetic testcross used to measure the distance separating two genetic loci. The genotypes of the male and female flies used in this testcross are set by preceding crosses to produce strains of mutant and wild-type flies in much the same manner that Mendel constructed his true-breeding strains of pea plants by repeated crosses. Note that, in the cross in Figure 7-2, the male fly carries all recessive alleles (it is homozygous recessive at both loci). This means that all progeny flies receive the identical pair of recessive

TABLE 7-1 Calculation of map units between loci from the frequency of parental and recombinant flies produced in a cross such as that shown in Figure 7-2.

Cross: $\dfrac{vg^+ \ cr^+}{vg \ \ cr}$ (Females) × $\dfrac{vg \ \ cr}{vg \ \ cr}$ (Males)

	Phenotype	Genotype	Number of progeny
Parental type	Wild-wild	$\dfrac{vg^+ \ cr^+}{vg \ \ cr}$	450
	Vestigal-curved	$\dfrac{vg \ \ cr}{vg \ \ cr}$	470
Recombinant type	Wild-curved	$\dfrac{vg^+ \ cr}{vg \ \ cr}$	43
	Vestigal-wild	$\dfrac{vg \ \ cr^+}{vg \ \ cr}$	37

Total 1,000

Frequency of recombinants $= \dfrac{43 + 37}{1,000} = .08$
$= 8$ percent

Definition: 1 percent recombinants = one map unit

vg   cr

|← 8 →| Chromosome

Map units

alleles from the male parent and the recombinant chromosomes in the progeny can come only from the female parent.

In the example in Figure 7-2, one locus determines whether the wing is of normal size (vg^+) or is very small (vg); the other determines whether the wing is straight (cr^+) or curved (cr). These genes are on an autosome, unlike the situation described in Chapter 3 in which the genes are on the X chromosome. The progeny flies from this cross show four phenotypes—two parental and two recombinant classes. The recombinants arise because some of the female gametes carry recombinant chromosomes as a result of crossing-over during meiosis (refer to Figure 2-7). The chance that a recombination will occur somewhere in the DNA between the two genes of interest is rather small, but the probability is directly related to how far apart the genes are on the chromosome.

The distance between two linked genes can be calculated from this cross as shown in Table 7-1. One thousand progeny flies are examined and, on the basis of the size and shape of the wing, each fly can be assigned to one of the four possible genotypes. The progeny types are easily distinguished: parental types are the most frequent; that is, most progeny have the same genotypes and phenotypes as the parents because of the way the cross was designed. However, a small number of progeny have different phenotypes—these are the recombinants. Their frequency can be calculated by dividing the total number of recombinant types by the total number of flies examined.

In Table 7-1, this turns out to be 8 percent recombinants. Geneticists agreed early on to define 1 percent recombinants as one **map unit** on the genetic maps that they began to construct for flies and other organisms. Figure 7-3 shows the map for one of the four chromosomes of *Drosophila*. A chromosome can have more

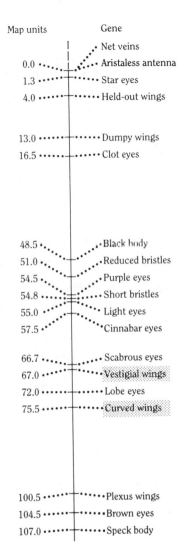

Map units	Gene
	Net veins
0.0	Aristaless antenna
1.3	Star eyes
4.0	Held-out wings
13.0	Dumpy wings
16.5	Clot eyes
48.5	Black body
51.0	Reduced bristles
54.5	Purple eyes
54.8	Short bristles
55.0	Light eyes
57.5	Cinnabar eyes
66.7	Scabrous eyes
67.0	Vestigial wings
72.0	Lobe eyes
75.5	Curved wings
100.5	Plexus wings
104.5	Brown eyes
107.0	Speck body

FIGURE 7-3 A genetic map of chromosome 2 of *Drosophila* showing the linkage of different genes.

than 100 map units (more than 100 percent) because distances are accurate and additive only over short segments. For long distances on a chromosome, the frequencies and map units are not accurate for complex reasons that are primarily of interest to geneticists. Until recently, detailed genetic maps were available for only a few organisms, such as flies, corn, tomatoes,

Sequencing the Human Genome

BOX 7-1

In the 1970s, chemical techniques were devised that permitted scientists to determine the sequence of the four bases (A, T, G, and C) in fragments of DNA. The techniques were laborious and time consuming. Yet within a few years the complete base sequence—all 5,375 bases—of a small virus was announced. By the mid-1980s, the complete sequence of a large human virus had been deciphered—the 172,282 base pairs in the Epstein-Barr (E-B) virus. Even with these achievements, the sequencing of all the DNA in the chromosome of *E. coli* still seemed like an impossible task because it has about 3 million base pairs.

However, DNA sequencing techniques improved so rapidly that by 1987 scientists were talking, not about sequencing the *E. coli* chromosome, which had become a relatively easy task, but of sequencing the entire human genome—all 3 *billion* base pairs. Automated DNA sequencing machines are now available that can "read" about 10,000 bases a day from a fragment of DNA with a rate of error of less than one mistake in a hundred. (So it requires several readings to be sure that a sequence is correct, but machines do not complain when asked to repeat a task.) Moreover, further improvements in sequencing machines may increase both the speed and the accuracy of reading the bases in DNA molecules.

The flood of new sequence information has already outstripped computer capacity—more personnel, better programs, and bigger computers are needed to enter, analyze, and retrieve the millions, and eventually billions, of bits of information. In 1983, Gene Bank was created at the U.S. National Laboratories in Los Alamos, New Mexico, to store all the DNA sequences that had been published. (A similar facility was established in Europe.) By the end of 1986, some 12 million bases had been recorded, and there was still a two-year back-log. Meanwhile, the U.S. Department of Energy and the National Institutes of Health have announced plans to sequence an entire human genome—a project in which hundreds of scientists will take part for decades at an estimated cost of billions of dollars.

Despite arguments over the scientific worth of such a grand sequencing project, it is already clear that the entire human genome will be sequenced. Like a mountain waiting to be climbed, it will be sequenced, if for no other reason than because it is there. But, as Robert A. Weinberg of the Whitehead Institute for Biomedical Research put it, "Those who fret about the sanctity of the human genome needn't worry—we may understand less about ourselves at the end of this project than when we began."

READINGS: Eugene F. Mallove, "Sequencing the Human Genome," *Computers in Science*. Premier Issue, 1987. Leslie Roberts, "Who Owns the Human Genome?" *Science*, 237, 358 (1987).

and bacteria. Traditional genetic techniques are not very helpful in mapping genes on human chromosomes. However, with new techniques, it soon will be possible not only to construct a detailed map of the twenty-three human chromosomes, but to obtain a complete sequence of the approximately 3 billion bases (see Box 7-1).

The genetic cross described in this section is a very simple one. The analyses of crosses dealing with more than two genes are more complicated. In many respects, recombination and genetic maps are easier to describe in bacteria because they have only a single chromosome (one DNA molecule). For this reason, bacterial genetics is often called haploid genetics in contrast with the diploid genetics of flies and human beings.

Recombination in Bacteria

Although it had been known for many years that plant and animal chromosomes could recombine during meiosis, nobody had thought of bacteria as sexual organisms until 1946, when Joshua Lederberg and Edward Tatum showed that genetically different strains of *E. coli* could mate with each other through conjugation. Strictly speaking, bacteria are not sexual organisms, which refers to male and female plants and animals that mate. However, because bacteria can exchange DNA with one another during a physical union, conjugation is analogous to a sexual mating. (Also, strictly speaking, the bacterial DNA is not a chromosome because it does not have the complexities of eukaryotic chromosomes, but it is convenient to refer to bacterial DNA as a chromosome.) It was fortunate for Lederberg and Tatum that they happened to choose strains of *E. coli* that are able to exchange DNA molecules upon cell–cell contact, because not all strains of *E. coli* and not all species of bacteria are able to conjugate with one another and thus exchange pieces of DNA.

In the original experimental work of Lederberg and Tatum, each of the two bacterial strains had different nutritional requirements that resulted from mutations in different genes (Figure 7-4). When the strains were grown separately, the minimal medium, which contained only mineral salts and the sugar glucose, had to be supplemented with the amino acid methionine and the vitamin biotin (for strain A) or the amino acids leucine and threonine (for strain B) or the bacteria would not grow. When the two strains were mixed together, the medium had to be supplemented with all four nutrients in addition to the required minerals and glucose.

When both strains were grown in the mixed culture, some of the bacteria exchanged genetic information, producing a few wild-type bacteria having no nutritional requirements. These relatively rare recombinant bacteria could be detected by spreading samples onto Petri plates that did not contain any of the four nutritional supplements. Under this condition, neither of the original mutant bacterial strains could grow, but wild-type recombinant bacteria did grow, eventually producing visible bacterial colonies on the plate. By the exchange and recombination of DNA molecules, these bacteria no longer carried the original mutations but had become wild-type for all genes and could grow on the minimal medium.

Where did the wild-type bacteria come from? One strain of *E. coli* (the donor) occasionally transfers some of its DNA to the other bacterial strain (the recipient) through conjugation (Figure 7-5). Once the DNA from the donor enters the recipient bacterium, the two DNA molecules may recombine, eventually giving rise to a bacterium that contains all wild-type genes. Thus, in bacteria, conjugation and recombination are mechanisms for creating genetic diversity. In a population of hundreds of millions of *E. coli* cells, some of which may be mutant, many new gene combinations can be produced by conjugation and recombination.

Plasmids Transfer Genes

Soon after the initial discovery that bacteria can conjugate and recombine their DNA molecules, a variety of related genetic and biochemical

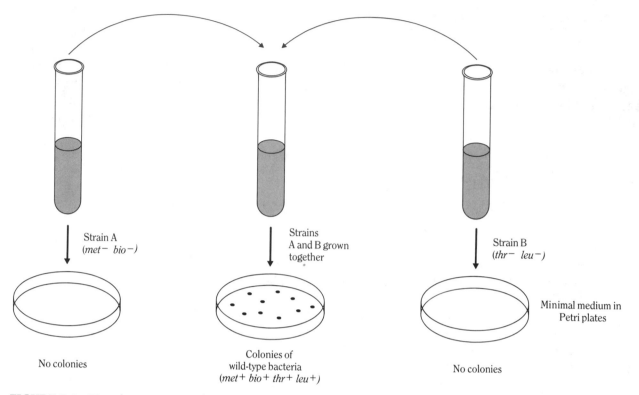

Strain A
(*met⁻ bio⁻*)

Strains
A and B grown
together

Strain B
(*thr⁻ leu⁻*)

Minimal medium in
Petri plates

No colonies

Colonies of
wild-type bacteria
(*met⁺ bio⁺ thr⁺ leu⁺*)

No colonies

FIGURE 7-4 First demonstration that bacteria can exchange genetic information by DNA transfer and recombination. Strain A has two mutant genes; strain B has two different mutant genes. When strains A and B are mixed together, DNA is transferred from one to another and wild-type bacteria that carry none of the mutant genes arise by recombination.

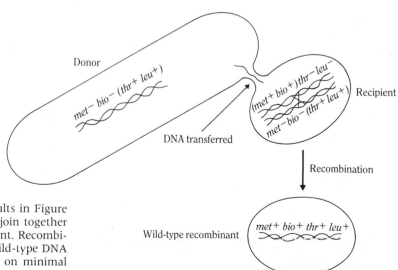

Donor

met⁻ bio⁻ (thr⁺ leu⁺)

DNA transferred

(*met⁺ bio⁺) thr⁻ leu⁻*
met⁻ bio⁻ (thr⁺ leu⁺)   Recipient

Recombination

Wild-type recombinant

met⁺ bio⁺ thr⁺ leu⁺

FIGURE 7-5 Conjugation explains the results in Figure 7-4. Donor and recipient bacteria physically join together and DNA is transferred from donor to recipient. Recombination in the recipient bacteria produces a wild-type DNA molecule and a wild-type cell able to grow on minimal medium.

experiments led to the discovery that what distinguishes donor from recipient bacteria is the presence in the donor of a small, circular DNA molecule called the **fertility factor**, or F-plasmid (discussed in Chapter 6). Any bacterium containing this F-plasmid has the donor phenotype and can transfer DNA; any bacterium without the F-plasmid has the recipient phenotype and can only accept DNA. The F-plasmid contributes nothing essential to the growth and survival of the individual bacterium that contains it; donor and recipient bacteria grow equally well. It is the F-plasmid, however, that enables the donor cell to attach to and conjugate with the recipient cell and then transfer DNA, which may include chromosomal genes, as well as the genes contained in the F-plasmid.

The F-plasmid DNA contains about one hundred genes; the functions of all but twenty or so are still unknown. The F-plasmid genes that *have* been identified participate in plasmid replication and transfer, including synthesis of the organ of attachment—a specialized appendage called a **sex pilus** (pl. pili). They are also responsible for the replication and physical transfer of one strand of newly replicated DNA from the donor to the recipient cell where it is copied again and converted back into a double helix. The recipient cell now contains the F-plasmid and itself becomes a donor.

Three Kinds of Donor Bacteria

It is possible to distinguish three genetically different kinds of donor *E. coli* bacteria. One type, called **F$^+$**, transfers only the F-plasmid to a recipient (**F$^-$**) bacterium. Normally, no chromosomal genes are transferred during conjugation between F$^+$ and F$^-$ cells. In about one out of every million cells, an F-plasmid becomes integrated into the chromosomal DNA, producing another type of donor called **Hfr** (high frequency of recombination). When Hfr bacteria conjugate with recipient bacteria, part or all of the donor's DNA is transferred along with the F-plasmid. The designation "Hfr" comes from the fact that recombination occurs frequently between the donated DNA fragment and the corresponding segment in the recipient bacterial chromosome.

Finally, the F-plasmid and the chromosome in an Hfr bacterium can become disintegrated. Again, with low probability, some adjacent bacterial genes may remain attached to the F-plasmid DNA when it exits from the bacterial chromosome. The bacteria formed in this process are called **F-prime** (F$'$) strains. They carry one or several chromosomal genes on the F$'$-plasmid as shown in Figure 7-6C. When such bacteria conjugate, the bacterial genes carried on the F$'$-plasmid are transferred to the recipient each time the plasmid itself is transferred.

The *E. coli* Genetic Map

When an Hfr bacterium conjugates with a recipient bacterium, each gene is transferred at a particular time after the mating process begins because the DNA is transferred in a linear fashion. The elapsed time can be measured because conjugation can be interrupted at any stage by agitating the bacteria in a kitchen blender (Figure 7-7). Only those genes that have already been transferred by the time that the bacteria are physically separated can produce recombinant bacteria. Such recombinant strains are detected by spreading a sample of the conjugated bacteria onto supplemented medium where only particular recombinant types are able to form colonies.

It requires approximately 100 minutes for the entire bacterial chromosome to be transferred, and so, by analogy, imagine travelling a high-

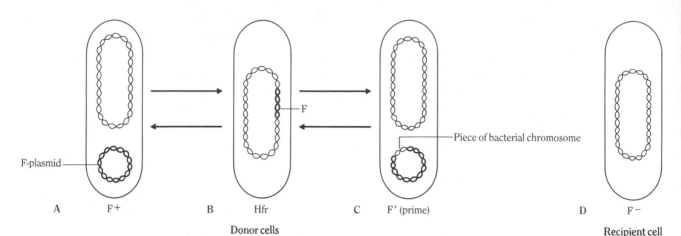

A   F+   B   Hfr   C   F' (prime)   D   F−

Donor cells   Recipient cell

FIGURE 7-6   Presence of an F-plasmid determines three types of donor cells; its absence produces a recipient cell: (A) an F$^+$ cell carries an independent, circular F-plasmid in the cytoplasm; (B) if the F-plasmid integrates into the chromosome, an Hfr cell is produced; (C) if the F-plasmid is excised (cut out) from the chromosome and carries with it some chromosomal DNA, the cell becomes F'; (D) a recipient cell without any F-plasmids is an F$^−$ cell. Hfr and F' cells transfer chromosomal genes; the F$^+$ cell transfers only plasmid genes.

way 100 miles long and mapping the towns on it according to the time it takes to get from town to town. Bacteria are normally able to grow and divide every 30 minutes under optimal laboratory conditions; yet, as we have just seen, they take at least 100 minutes to conjugate and transfer an entire chromosome. This observation led geneticist William Hayes, one of the discoverers of Hfr bacteria, to make the humorous observation that *E. coli* is the only organism in nature that engages in the sex act three times as long as its normal lifespan.

As more and more *E. coli* mutants became isolated and their genes positioned on a genetic map, it was eventually realized that every bacterial gene is linked to every other gene. Consequently, the *E. coli* genetic map appears to have no ends (Figure 7-8). This discovery led some geneticists to suggest that chromosomes in bacteria are physically circular DNA molecules like the smaller plasmids. The fact that the bacterial chromosome, like the F-plasmid, is a physically closed loop of DNA was later confirmed by elec-

tron micrographs of DNA molecules (Figure 7-9). Chloroplasts (organelles in plant cells) and mitochondria (organelles in both plant and animal cells) also contain physically closed circles of DNA (see Box 7-2). Human chromosomes, however, contain very long, linear, unbroken threads of DNA.

Mutations Change Genetic Information

A change in even a single base pair in the hundreds of thousands of bases in a DNA molecule can affect—sometimes fatally—a bacterium, plant, or animal. An examination of the Watson-Crick model of DNA helps us to see how, in the normal process of DNA replication, mutations may arise. A mutation is the result of a mistake in the replication mechanism. This mechanism normally ensures the pairing of A with T and G with C. However, no process,

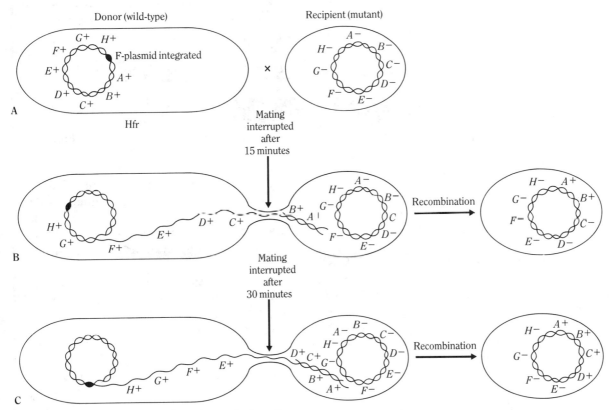

FIGURE 7-7   The interrupted-mating technique for constructing the genetic map of *E. coli*.
A. Wild-type Hfr bacteria are mixed with recipient mutant cells.
B. The Hfr chromosome is replicated and transferred. The mating can be interrupted at various times, and timing of gene transfer can be measured by detecting recombinant bacterial colonies on Petri plates.
C. As mating continues, more of the donor chromosome is transferred and the variety of recombinants that can be detected becomes greater. The *E. coli* genetic map is calibrated in minutes (Figure 7-8).

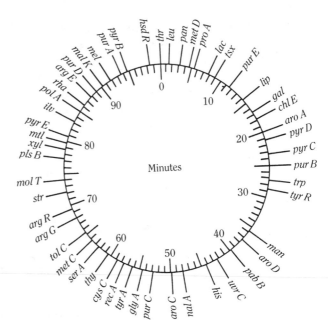

FIGURE 7-8   The genetic map of *E. coli*. Genes are indicated by three-letter abbreviations. If genes are part of a biochemical pathway, a capital letter is added (*met D* is a gene in the biosynthetic pathway for methionine). The *E. coli* genetic map is divided into units of 1 minute; the entire *E. coli* genetic map is set at 100 minutes, which is the time required to transfer an entire chromosome. About one thousand genes (one-fourth of the total) have been mapped in *E. coli*; only a few are shown here.

Mitochondria
Keys to Human Diseases and Evolution

All eukaryotes (fungi, plants, animals) have tiny organelles called *mitochondria* in the cytoplasm of their cells. These functionally specialized structures produce most of a cell's chemical energy in the form of *adenosine triphosphate* (*ATP*). Molecules of ATP are as essential to the activities of the cell as gasoline is to an automobile: they are the source of energy that sustains other chemical reactions in the cell. Cells that require more than normal amounts of energy, such as heart and muscle cells, have correspondingly more mitochondria and produce more ATP.

A small DNA molecule in each mitochondrion provides some of the necessary genetic information that enables it to perform its functions, and genes in the chromosomes in the nucleus supply the rest. Human mitochondrial DNA is physically circular like plasmid DNA and contains only 16,569 base pairs, the smallest number detected in mitochondria in any organism. In contrast, mitochondrial DNA in the fruit fly (*Drosophila*) is slightly larger—about 19,500 base pairs. And bread yeast (*Saccharomyces cerevisiae*) has the largest mitochondrial DNA—about 78,000 base pairs—even though it is a microorganism.

BOX 7-2

Mitochondrial Diseases

A human egg contains thousands of mitochondrial DNA molecules and hundreds of mitochondria because each mitochondrion contains many copies of its DNA. When a sperm fertilizes an egg, only its nucleus penetrates the egg and, consequently, both male and female offspring inherit mitochondrial DNA *only* from their mothers. The sperm's nucleus has no mitochondria. Thus, any defect in mitochondrial DNA is inherited by maternal lineage rather than by Mendelian inheritance of chromosomes from both parents. In recent years, a number of human diseases, particularly those of the optic nerves, skeletal muscles, heart, liver, and kidney, have been traced to defects in mitochondrial genes.

Each human mitochondrial DNA contains genes coding for thirteen enzymes, plus genes for ribosomal and transfer RNAs. If one of these thirteen enzymes is altered as a result of a mutation in mitochondrial DNA, certain tissues of the body may be affected, particularly nerve cells and skeletal muscles, which are especially dependent on mitochondrial enzymes. Some mitochondrial inherited diseases are rather mild, but others cause blindness or epilepsy. Mutations in mitochondrial genes can be distinguished from mutations in chromosomal genes because the patterns of inheritance for several generations are different.

Mitochondrial Evolution

Because the bases in mitochondrial DNA change through time as a result of mutations, the particular sequence of the 16,569 bases in different human populations provides a unique biochemical record of human migrations and human evolution. Populations that are closely related should have similar sequences; those that are more distantly related should have more-divergent sequences. By examining the patterns of base sequences, scientists have shown that the mitochondrial DNAs of native American Indians closely resemble those of Asian populations. This supports the view that human populations migrated from Asia to North America thousands of years ago, probably crossing a land bridge from Russia to Alaska.

Other studies have examined the mitochondrial DNAs from

people living in Africa, Asia, Europe, Australia, and New Guinea. Analyses of their mitochondrial DNAs led to the conclusion that probably all people alive today could be traced to a single female ancestor who lived in Africa as recently as 200,000 years ago. Not surprisingly, this hypothetical woman has been given the name Eve. Possibly, when enough different human Y chromosomes become available for sequence analysis, we may also discover "Adam," given that Y chromosomes have a paternal lineage.

Although there is still much controversy over the meaning of differences in base sequences in mitochondrial DNAs, any differences between mitochondrial DNAs can be used as a powerful tool for unravelling biological history on earth—especially human history.

READINGS: Rebecca L. Conn, "In Search of Eve," *The Sciences*. September-October, 1987. Douglas C. Wallace, "Mitochondrial Genes and Disease," *Hospital Practice*. October 15, 1986.

including DNA replication, is 100 percent accurate. Mistakes in the pairing of bases during DNA replication are extremely rare; in bacteria, for example, mistakes in base-pairing (**mutation rates**) are usually of the order of 1 in 10 million generations or less for any gene.

Mutation rates in human beings are more difficult to estimate because of the long human generation time and small human population size relative to bacteria. However, mutation rates have been measured for some human genes (Table 7-2). X-rays, viruses, and many environmental chemicals can increase the likelihood of damage to DNA, causing a change in one or more base pairs that may lead to cancer (discussed in Chapter 11). Also, errors in recombination during meiosis can produce mutations that result in hereditary diseases.

FIGURE 7-9 Electron micrograph of an *E. coli* DNA molecule. The molecule is an enormous circle of DNA that is somehow condensed and packaged into a bacterial cell. [Courtesy of Ruth Kavenoff. Bluegenes #1. © 1983. All rights reserved by Designergenes Posters Ltd. Posters and shirts available from Carolina Biological Supply, 2700 York Road, Burlington, North Carolina 27215. Telephone: (800) 334-5551.]

Mutations Are Spontaneous

Are mutations random and spontaneous in cells undergoing mitosis or meiosis or are they caused by specific interactions between an or-

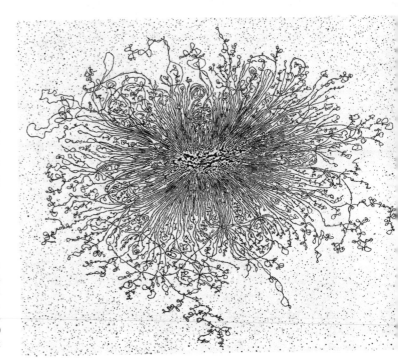

TABLE 7-2 Estimated mutation rates for a few genes responsible for human heriditary diseases.

Disease	Mutations (average per million genes per generation)
Hemophilia A—severe bleeding due to lack of factor VIII in blood	44
Hemophilia B—mild bleeding due to lack of factor IX in blood	2–3
Achondroplasia—dwarfism	10
Retinoblastoma—cancer of the retina of one or both eyes	8
Duchenne muscular dystrophy—progressive degeneration of muscle tissue	67

ganism and something in its environment? Scientists debated for centuries whether traits acquired by an individual organism can be passed on to its progeny, a phenomenon referred to as **Lamarckianism** (see Box 7-3). To some people, it seems obvious that their learned talents and acquired tastes will be passed on to their sons and daughters—that learned skills and behaviors can be inherited by succeeding generations. Many nineteenth-century scientists observed that the children of musicians tended to be musicians, the sons of athletes were usually athletic, and the children of wealthy parents generally became wealthy themselves. Thus, some scientists argued that acquired traits could be inherited. How various people have answered the question at different times has had serious consequences in human history—some of which are discussed in later chapters.

If the environment can influence the kind of mutations that occur in gametes, then inheritance of acquired traits is possible. However, if mutations simply occur at random in DNA, then the environment can play no role except by increasing or decreasing the overall frequency of random mutations.

That mutations do occur spontaneously and randomly in the DNA of bacteria was demonstrated in 1943 by Nobel laureates Max Delbrück and Salvador Luria. However, as their experimental design relied on statistical analysis (recall that Mendel also used statistics), not everyone was convinced by their experiment.

In 1952, Esther and Joshua Lederberg developed a simple **replica plating technique** (outlined in Figure 7-10) that they used to prove that bacterial mutants arise by chance. In this technique, several hundred bacteria are spread on the surface of agar medium in a Petri plate and are incubated overnight. By the next day, each bacterium has grown into a visible colony containing more than a million bacteria. Samples of the bacteria in each colony are then transferred by the use of a velvet-covered wooden block to several Petri plates that contain the same agar medium as the original Petri plate plus millions of T-viruses. (These plates are called replica plates because they are used to replicate the bacterial colonies.) Some bacteria from each colony stick to the threads of the velvet and can thus be transferred to the surface of each of the replica plates. Almost all of the bacteria on plates containing the T-viruses are killed. However, one or a few bacterial colonies do appear on these plates and grow in the same positions on each one; these positions correspond to those of bacterial colonies present on the original plate because the orientation of the original bacteria was maintained in their transfer.

The only conclusion that can be drawn is that virus-resistant bacteria already existed in the original colony. These were picked up by the velvet and transferred to the replica plates, where they subsequently grew to form new colonies. The mutation to virus resistance *must* have arisen in that bacterial colony during

How New Species and Organisms Arise

BOX 7-3

Why do certain species of plants and animals become extinct? How could complex animals have evolved from simpler ones? How did the enormous variety of life forms on earth arise? These questions have intrigued philosophers and scientists ever since they first began to systematically study nature.

In the early 1800s, a renowned French zoologist, Jean Baptiste Lamarck, put forth the idea that animals become better adapted to their environments with each generation, eventually giving rise to new types and new species. Lamarck's basic idea is summed up in what he called the law of inheritance of acquired characteristics: "All that has been acquired or altered in the organization of individuals during their life is preserved and transmitted to new individuals who proceed from those who have undergone these changes."

Lamarck used this idea to explain in ingenious ways the origin of animal characteristics. For example, he suggested that webbed feet in aquatic birds arose because they spread their toes in order to swim faster to catch fish. Gradually, the skin between the toes stretched to form a web, and this useful acquired character was passed on to progeny. Snakes supposedly arose from animals that were accustomed to crawling under bushes and along the ground in order to hide themselves from predators. Eventually, their legs became useless, and subsequent generations were born without legs. Moles gradually lost their eyes because eyes were of no use in their underground environment.

The classic Lamarckian example of an acquired trait is the giraffe's long neck. Lamarck reasoned that, because giraffes lived in arid regions where there was often little grass to graze, they were forced to eat the leaves of trees. This resulted in their having to stretch their necks to reach higher and higher leaves. The higher the animals had to stretch to obtain leaves, the longer their necks became, and elongated necks were then passed on to progeny until finally the giraffes' necks became long enough to reach leaves even in the tallest trees. Although these examples may seem foolish today, Lamarck was a very astute scientist and must be credited with one of the first rational attempts to account for the evolution of organisms.

It was Charles Darwin, born in 1809, the same year in which Lamarck's book was published, who eventually conceived of the idea of natural selection and "survival of the fittest." Neither Lamarck nor Darwin knew anything about the mechanisms of inheritance. Yet from years of careful observations of the characteristics of plants and animals, Darwin was able to deduce the basic mechanism of random genetic variation as the driving force in evolution. Experiments have shown that Darwin was right and Lamarck was wrong. The fundamental ideas of natural selection cannot be expressed more clearly than in Darwin's own words:

As many more individuals of each species are born than possibly can survive, and, as a consequence, there is a frequently recurring struggle for existence, it follows that any being, if it varies however slightly in any manner profitable to itself, under the complex and sometimes varying conditions of life, will have a better chance of surviving, and thus be naturally selected. From the standpoint of inheritance, any selected variety will tend to propagate its new and modified form.

READING: R. Lewin, "Lamarck Will Not Lie Down," *Science*, 213 (1981).

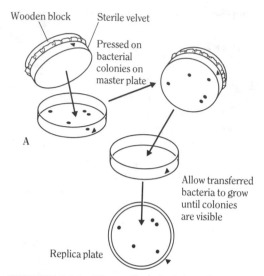

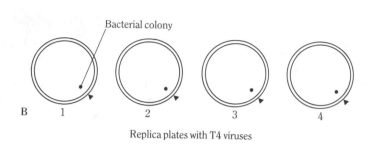

FIGURE 7-10   The replica plating techinque can be used to show that mutations occur spontaneously before cells are exposed to selective agents. A velvet pad is used to transfer bacteria from colonies on the original Petri plate to another plate, as shown in part A, or to several other plates. If the orientation of the wooden block is fixed, the pattern of colonies growing on the replica plates will be identical with that on the original plate if the medium is the same. If T4 phages are added (plates 1–4 in part B), all bacteria are killed unless mutant T4-resistant bacteria were present in one or more of the original colonies. Cells from those colonies (one is shown) are able to grow on the replica plates. This shows that the mutation to resistance occurred *before* the bacteria were exposed to the T4 phages.

growth of cells on the original Petri plate *before* the bacteria were exposed to the T-viruses.

It might still be argued that the mutation occurs by chance on the replica plates *after* the bacteria are exposed to the viruses. But then one must ask why the phage-resistant colonies always appear on several replica plates in *identical positions*, corresponding to that of one particular colony on the original plate. The only reasonable conclusion is that, as the bacteria grow on the original plate, a mutation occasionally arises in the DNA of one bacterium that makes all of its descendants in the colony resistant to infection. When these resistant bacteria are subsequently exposed to the T-viruses on the replica plates, they are the only ones able to grow and reproduce.

Why is this experiment so important? Partly because it provides an explanation for some modern ecological disasters, such as the emergence of DDT-resistant mosquitoes and other insects. It also explains why insects eventually become resistant to any pesticide, why weeds become resistant to herbicides, and why rats become resistant to rat poisons. Mutant organisms continually arise by chance in natural populations. In the right environment, the mutant types will flourish and the original nonmutant organisms will die out. The proof that mutant organisms arise spontaneously in populations also provides the genetic mechanism needed for natural selection to operate as proposed by Darwin. It is a tribute to Darwin's genius that he was able to convincingly document his idea of natural selection even though he knew nothing about chromosomes, genes, or mutations in DNA.

The demonstration of spontaneous mutations and of natural selection as alternatives to Lamarck's ideas has been accepted widely. However, the idea that acquired traits and not simply chance mutations may be inherited by future generations still attracts the attention of some scientists. In 1981, Edward Steele, an immunologist, provoked a scientific controversy by claiming that a particular kind of acquired immunity in mice (the immune system is discussed in Chapter 17) could be inherited by progeny mice in a Lamarckian fashion. Steele's results could not be repeated by other scientists, but the last word on the subject of Lamarckian inheritance has not yet been heard. Recent findings of "jumping genes" and other kinds of movable genetic elements (discussed in Chapter 12) still keep alive the possibility that some molecular mechanism may someday be discovered by which an acquired trait could become encoded in a piece of RNA or DNA that might subsequently be incorporated into the chromosomes of sperm or eggs. Such a finding would not negate the fact that most evolutionary changes result from random mutations and natural selection; it would simply demonstrate another biological mechanism by which organisms can diversify and evolve.

Summary

For organisms to continue to survive and reproduce, the DNA must be replicated with exceptional accuracy. However, mutations and genetic diversity are essential to their survival under new conditions and for evolution (emergence of new species).

DNA is an unbroken, very long molecule in which genes are physically linked together. Genes located relatively close to one another on the same chromosome are said to be linked. It is useful to know whether genes are linked in carrying out genetic crosses and in diagnosing human genetic disorders. In meiosis, fragments of DNA in homologous chromosomes may recombine through crossing-over. Early studies of recombination in *Drosophila* established the concept of linkage and led to the development of genetic maps of chromosomes. Studies of particular strains of bacteria led to the discovery that genetic diversity in bacteria occurs through recombination and conjugation. In conjugation, a donor bacterium, which contains an F-plasmid, mates with a recipient bacterium, which lacks the F-plasmid; newly replicated plasmid DNA and, in some cases, chromosomal DNA are transferred to the recipient. Donors are classified as F^+ (no chromosomal DNA is transferred); Hfr (high frequency of recombination in which all or part of the chromosomal DNA is transferred); and F' (in which plasmid and chromosomal DNA have become associated so that one or more chromosomal genes are transferred with the plasmid).

A mutation in a molecule of DNA changes the genetic information that it carries. It is the result of a mistake in the replication mechanism. Normally, in DNA that is undergoing replication, the pairing of bases (adenine with thymine and cytosine with guanine) is highly accurate. However, should a base pair with an incorrect base, the newly replicated DNA will contain a mutation. Mutations occur randomly and spontaneously in cells undergoing DNA replication and division.

Key Words

adenosine triphosphate (ATP) The energy-yielding molecule in cells that is used to drive chemical reactions.

F^+ *E. coli* Bacteria that contain an autonomous F-plasmid in the cytoplasm.

F' *E. coli* Bacteria that contain an autonomous F-plasmid that carries chromosomal genes in addition to plasmid genes.

fertility factor   A plasmid such as the F-plasmid that allows bacteria to conjugate and exchange DNA molecules.

Hfr *E. coli*   Bacteria that contain the F-plasmid integrated into the bacterial chromosome.

Lamarckianism   The inheritance of acquired traits; an idea named after the French scientist Jean Baptiste Lamarck.

linkage   The joint inheritance of two or more nonallelic genes because they are located close together (linked) on a chromosome.

map unit   A measure of linkage between genetic loci; 1 percent recombinants equals one map unit; human genetic maps are calibrated in map units called centimorgans.

mitochondrion (pl. mitochondria)   A self-reproducing organelle in all eukaryotic cells. The primary site for synthesis of ATP, which supplies energy to the cell.

mutation frequency   The number of mutant organisms in a population.

mutation rate   The number of mutations in a gene per cell generation.

replication plating   A technique in which bacteria are transferred to a series of Petri plates from an original plate by means of a velvet pad.

sex pilus   The physical appendage on bacteria that carry an F-plasmid.

testcross   The mating of male and female organisms of known genotypes; used to determine linkage and distance between loci.

Additional Reading

Brownlee, S. "The Lords of the Flies." *Discover*. April, 1987.

Conn, R.L. "In Search of Eve." *The Sciences*, September-October, 1987.

Drake, J.W., B.W. Glickman, and L.S. Ripley. "Updating the Theory of Mutation." *American Scientist*, November-December, 1983.

Graf, L.H. "Gene Transformation." *American Scientist*, September-October, 1982.

Stahl, F.W. "Recombination." *Scientific American*, February, 1987.

Tucker, J.B. "Gene Machines: The Second Wave." *High Technology*, March, 1984.

Wallace, D.C. "Mitochondrial Genes and Disease." *Hospital Practice*, October 15, 1986.

Study Questions

1   What are the three basic genetic functions of DNA?

2   What is meant when we refer to one map unit on a genetic map?

3   What are two different mechanisms for generating genetic diversity in a population of bacteria? In a human population?

4   What is the central idea of Lamarckianism?

5   Roughly how many different mutations can occur in a gene consisting of a thousand base-pairs?

6   If 10,000 bases a day can be sequenced reliably, about how long will it take to sequence the human genome assuming that each chromosome has, on average, one hundred million base-pairs?

7   If a mutation occurs in the DNA of mitochondria in a man's sperm, will this mutation be passed on to his progeny?

Essay Topics

1   Give your views as to whether or not the human genome should be sequenced.

2   Discuss why bacteria are more useful to geneticists than fruit flies for many kinds of experiments.

3   Explain how Mendel's results would have been affected if the traits that he studied had been determined by linked genes.

Transcription and Translation

Expression of Genetic Information

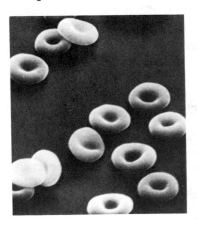 

Within the human body the RNA slides through the walls of the cell's nucleus, through infinitesimal tubes in the structure, and finds the pearllike ribosome bodies in the cytoplasm. The bodies MOVE *across* the long threadlike molecules of RNA and create the substances of the cell.

MICHAEL McCLURE, *poet*

DNA MOLECULES contain all the information that determines the growth and reproduction of cells, in addition to directing their specialized functions within an organism. DNA is responsible, directly or indirectly, for the synthesis of all other molecules in cells. In particular, it supplies the information for the synthesis of the two other major classes of information-containing molecules, ribonucleic acids (RNA) and proteins. Together, DNA, RNA, and proteins are the informational macromolecules of cells (Table 8-1).

To understand how genetic information is extracted from DNA and utilized by the cell, it is necessary to understand how RNA and protein molecules are synthesized and what functions they perform. For genetic information to be used by cells, the information carried by particular genes must be expressed. Therefore, genes must direct the synthesis of other kinds of molecules that function in accord with the genetic instructions.

Genes are expressed in two steps. The first step is **transcription**, in which many different RNA molecules are synthesized from genes in DNA. Transcription is the process by which the genetic information contained in sequences of bases (genes) in DNA is transferred into a complementary sequence of bases in RNA. The information in some of the RNA molecules is further deciphered in the next step of gene expression, **translation**, which converts the genetic information carried in the sequence of bases in RNA into a sequence of amino acids that determines a particular protein's structure and activity in the cell.

The Central Dogma

A few simple rules govern the exchange of information between DNA, RNA, and protein

TABLE 8-1   The three classes of informational macromolecules in cells.

Informational molecules	Properties
DNA A C G T T C C T G C A A G G	Double-stranded, helical nucleic acid. Information is contained in the sequence of bases A, T, G, C.
RNA A U U C G C U U G A G C	Single-stranded nucleic acid. May have some helical secitons. Information is contained in the sequence of bases A, U, G, C.
Protein Trp Phe Arg Gly Gly Ser Pro Ala Leu Ala Ala Ser Pro Trp Arg Leu Gly Ala Pro	Chain of amino acids. Information is contained in the sequence of the twenty different amino acids; this sequence determines the structure and function of a particular protein.

macromolecules in the cells of all organisms, as shown in a simple diagram devised by Francis Crick. The rules laid down in this diagram hold true for every cell and organism, and together they are known as the **central dogma** of molecular biology:

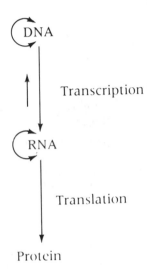

The long arrows in the diagram signify that information flows, in a two-step process, from DNA to RNA (transcription) and from RNA to protein (translation). The short arrow pointing from RNA to DNA indicates that, in a few exceptional circumstances, information can flow backward from an RNA molecule into DNA. This occurs only with certain RNA viruses, particularly those that are able to cause tumors in animals (discussed in Chapter 10). However, with this single exception, genetic information in cells always flows from DNA to RNA. The absence of a reverse arrow between protein and RNA is significant—information is never transferred from protein back into RNA or from protein into DNA. (You might consider how this particular aspect of the central dogma bears on Lamarck's idea of the inheritance of acquired traits, discussed in Box 7-3).

Finally, the diagram shows the transfer of information that can take place between identical kinds of molecules. Information in DNA is readily transferred to other DNA molecules in the course of replication, as indicated by the curved arrow around DNA. Similarly, the curved arrow around RNA indicates that in virus-infected cells information may be transferred from one viral RNA molecule to another. In uninfected cells this kind of exchange between RNA molecules never occurs.

These are strong rules. Perhaps you can now understand why molecular biologists are more impressed by the great fundamental similarities among cells than by their more easily observed differences. Furthermore, analyzing the informational molecules in different kinds of organisms provides important clues to the origin and evolutionary relatedness of cells and organisms (see Box 8-1).

Transcription

Transcription is the process by which RNA molecules are synthesized (transcribed) from DNA in cells. The molecular details of both transcription and translation are understood most thoroughly in bacteria, and so the description that follows is derived from bacterial experiments. The important similarities and differences between these processes in bacterial and animal cells are discussed later.

Types of RNA

Although all RNA molecules are composed of the same sugar-phosphate backbone and four bases (Figure 8-1), three functionally different types of RNA molecules—all transcribed from DNA—are found in all cells. **Messenger RNA** (mRNA) molecules carry information from

Archaebacteria
Are They the "Missing Link" Between Prokaryotes and Eukaryotes?

BOX 8-1

To the casual observer of nature, it is the differences between organisms that are most noteworthy. To the biochemist, however, it is the similarities between the molecules—especially the informational macromolecules, DNA, RNA, and protein—that are most significant. The processes of transcription and translation are remarkably similar in all cells, implying that these processes arose in primitive cells billions of years ago and stayed relatively unchanged as organisms evolved. Once the genetic code had become established (Chapter 9), any change in the basic processes of information transfer among molecules would probably have resulted in lethal changes for organisms.

New techniques of biochemical analysis have enabled biochemists to determine the base sequences of RNA and DNA molecules. Because ribosomes play such a crucial role in protein synthesis, sequence analysis of ribosomal RNA (rRNA) molecules provides a particularly reliable indicator of the relatedness between different species of organisms. That is, the more closely related species are, the more nearly identical are the sequences of bases in their rRNAs. The rRNAs are identical in the cells of all human beings, for example, as are the rRNAs in different strains of *E. coli*. Mutations that did not affect ribosome functions may have introduced a few base changes, but long stretches of the rRNAs from closely related organisms prove to be almost identical when analyzed. As would be expected, however, the base sequences for rRNA molecules in bacteria are quite different from the sequences for rRNA molecules in plant or animal cells, which presumably evolved from bacteria billions of years ago.

But a puzzling pattern emerges when the base sequences of rRNAs from certain unusual kinds of bacteria are examined. These organisms include certain anaerobic bacteria that survive only deep within rotting vegetation or inside a cow's stomach (they cannot tolerate exposure to oxygen), thermophilic bacteria ("heat lovers") that grow in hot springs where temperatures are high enough to kill any other kind of microorganism, and halophilic bacteria ("salt lovers") that are obliged to live in concentrated brine, where few other organisms can survive.

When biochemists determined the base sequences of the rRNAs from these unusual bacteria, they made a surprising discovery: the sequences were more similar to those in the rRNAs of plant and animal cells than to those in the rRNAs of other bacterial species. This finding suggests that these unusual bacteria, formerly believed to be close relatives of other bacteria, may in fact be the distant ancestors of eukaryotic cells. Biochemist Carl Woese has proposed the name *archaebacteria* for these organisms and has suggested further that they are the ancestors of both other prokaryotes and the eukaryotes.

Some of the answers to the mystery of the origin of life may emerge from further analysis of other informational macromolecules in archaebacteria. As DNA and RNA molecules from various organisms are sequenced, it may be possible to reconstruct the genealogy of cells. Woese believes that "biology is now on the threshold of a quieter revolution, one in which man will come to understand the roots of all life and thereby gain a deeper understanding of the evolutionary process."

Whether Woese's hypothesis is the correct interpretation of the sequence differences between rRNAs from various organisms, the base sequences of RNA and DNA do provide a glimpse into biological history.

READINGS: Carl Woese, "Archaebacteria," *Scientific American*. June, 1981. L. Margulis, *Early Life*. Jones and Bartlett Publishers, Inc., 1982.

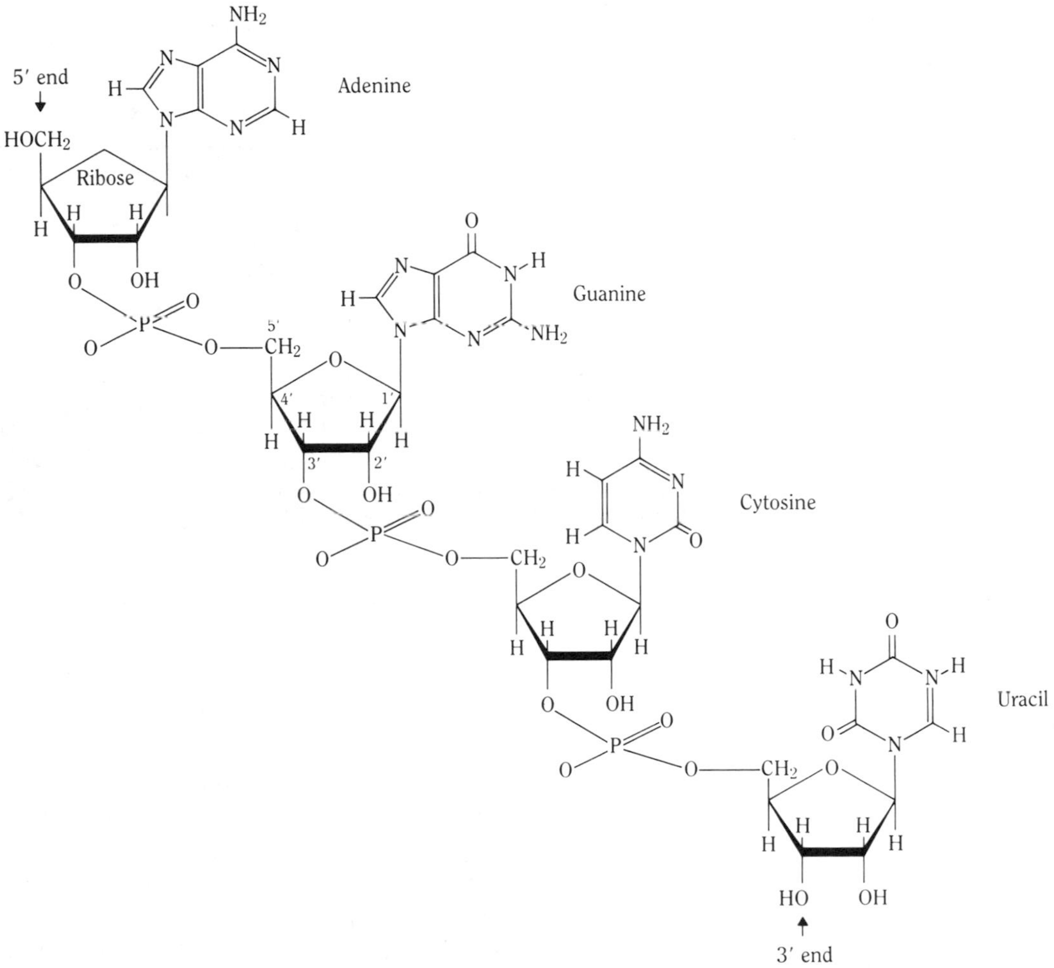

FIGURE 8-1 The chemical structure of RNA. The four different bases in RNA are adenine, guanine, cytosine, and uracil (replaces thymine found in DNA). The sugar in an RNA nucleotide is ribose (replaces deoxyribose found in DNA). Most RNAs are single strands instead of the double-stranded helices characteristic of DNAs.

genes in DNA and are used subsequently in the translation process to produce proteins, many of which are enzymes. The other two types of RNA are **ribosomal RNA** (rRNA) and **transfer RNA** (tRNA); although they are not translated into proteins, they do perform other vital functions in the translation process.

The different types of RNA in cells can be distinguished by both size and function (Table 8-2). The sizes of the three bacterial ribosomal RNAs are known with considerable precision. The largest of these, called 23S (the S value refers to its sedimentation rate during centrifugation), contains about 2,900 bases. The 16S rRNA is smaller, with about 1,500 bases. The smallest is 5S rRNA, which has about 120 bases. These three rRNAs are transcribed from genes in DNA and become part of the protein-

TABLE 8-2   The approximate sizes and functions of the three classes of RNA molecules found in cells.

Type of RNA	Approximate size (number of bases)	Function
Transfer (tRNA)	80	Attaches to and transfers each amino acid to its correct position along an mRNA molecule where it is joined to a growing polypeptide chain
Ribosomal (rRNA)	Three different kinds: 120, 1,500, and 2,900	Provides the structural components for the construction of ribosomes
Messenger (mRNA)	Varies between 100 and 5,000	Contains the information from a gene in a sequence of codons that determine the sequence of amino acids in protein

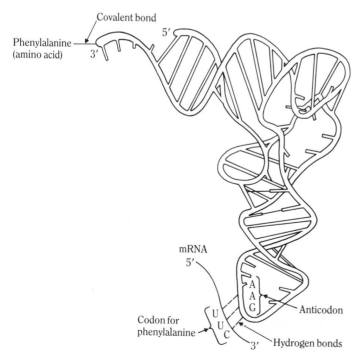

synthesizing particles called **ribosomes** (to be discussed shortly). None of these three rRNAs contains genetic information that is translated into proteins, but all are required for protein synthesis and play both structural and functional roles in ribosomes. Whereas a bacterium contains many thousands of ribosomes, a human cell may contain millions of them, each slightly larger than a bacterial ribosome.

The smallest RNA molecules are transfer RNAs; they contain only about 80 bases but have a complex three-dimensional shape that enables them to perform a crucial function in the process of translation (Figure 8-2). Transfer

FIGURE 8-2   Structure of a tRNA molecule. Each tRNA has a three-base anticodon that hydrogen bonds to its complementary codon in mRNA. It is specific for a particular amino acid, which is attached to one end of the molecule. Transfer RNAs are used to "read" the genetic code.

RNA molecules are often called *adapter* molecules because they convert the information (sequence of bases) in the mRNA molecules into the information (sequence of amino acids) in proteins.

Each transfer RNA molecule can chemically unite with one of the twenty different amino acids and with a particular sequence of three bases (called a **codon**) in a messenger RNA molecule. A transfer RNA delivers the amino acid that it carries to a site on a ribosome where it joins other amino acids to form a protein. After releasing the amino acid at the ribosome, the tRNA is released and is free to attach to another amino acid of the same kind—a cycle that is endlessly repeated throughout the life of the cell. In all organisms, from bacteria to human beings, only from thirty to forty different kinds of tRNA molecules are used to "read" the various codons in different mRNAs. The particular sequence of codons in each mRNA constitutes the "message" in messenger RNA and directs the sequence of amino acids that are linked together during protein synthesis (the process of translation).

Messenger RNA molecules are the most diverse of the three classes of RNA found in cells. They are transcribed from the thousands of different genes in DNA and are subsequently translated into the thousands of different proteins that carry out the cells' functions. Each mRNA transcribed from a gene is "read" by each ribosome that attaches to it and moves along its length; thus each mRNA directs the synthesis of hundreds or thousands of identical proteins depending on the number of ribosomes that become attached. Different proteins are synthesized from different mRNAs, which is why so many different kinds of mRNAs are present in a cell.

All three classes of RNA molecules—rRNA, tRNA, and mRNA—are synthesized by the process of transcription. All are required to manufacture proteins in the process of translation. However, only messenger RNA carries the genetic information for making proteins that is encoded in the sequence of bases in DNA. Transfer RNAs and ribosomal RNAs are essential in processing the information carried in mRNA and converting that information into proteins.

Mechanisms of Transcription

DNA molecules can be thought of as enormously long tapes of information. The process of transcription is the first step in retrieving the information from specific regions (genes) of the DNA tape. The problems of transcription are similar to those that might be encountered when you first insert a cassette into an audiotape machine. If you want to listen to a particular song on the tape, either you have to know precisely where the song is located or you have to search for it by running the tape backward and forward. To transcribe a gene or a group of genes from a DNA tape, the enzyme that synthesizes RNA molecules, called **RNA polymerase**, must be able to recognize where to begin transcription of the DNA and where to terminate it; that is, transcriptional "start" and "stop" signals are needed.

Two basic mechanisms of transcription allow the information carried in any group of genes to be selected and transcribed from any location in DNA. At the beginning of a gene or group of genes in DNA is a site, called a **promoter**, that serves as a start signal. Such a site consists of a short sequence of bases (about thirty-five in bacteria) that is recognized by RNA polymerase, allowing it to initiate transcription of an RNA molecule there. At the end of the gene or genes is another sequence of bases, the **terminator site**, that signals the RNA polymerase to stop.

Because the two strands of a DNA molecule contain different, but complementary, base sequences, the genetic information in each strand is different. The promoter region determines not only where transcription is to begin, but also the strand ("sense" strand) that is to be transcribed. RNA polymerase attaches to the promoter on the sense strand and uses the bases in that strand as the template for synthesis of the complementary strand of RNA (Figure 8-3). In different regions of the DNA, the promoter site (and genes) may be on either one strand or the other of the DNA molecule.

Every cell has many thousands of RNA polymerase molecules; so many genes can be transcribed from its DNA at the same time. Like other enzymes, RNA polymerase molecules can be recycled because, although they catalyze the chemical reactions used to synthesize RNA molecules, they are not themselves used up or destroyed in those reactions. As we will see later, only those genes whose functions are needed are transcribed at a given time—that is, gene expression is regulated in all cells (discussed in Chapter 10).

Not all kinds of RNA molecules are synthesized in the same quantities; their transcription from DNA depends in part on the kind of promoter sequence providing access to the gene. For example, in *E. coli* several clusters of ribosomal RNA genes are scattered throughout its DNA molecule, and the rate at which these rRNA genes are transcribed is generally much greater than the rates observed for other genes whose products are used in lesser amounts. Furthermore, the rate of rRNA transcription varies with the environment in which the bacteria grow.

Transcription in Animal Cells

The transcription process in animal cells is not nearly as simple as it is in bacteria. This is partly

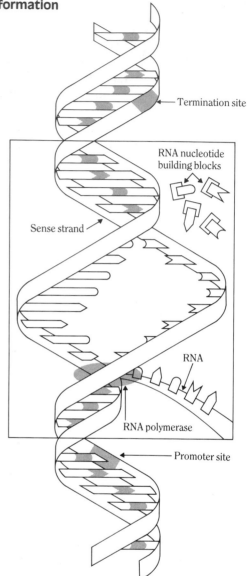

FIGURE 8-3 The process of transcription. RNA polymerase attaches to a promoter site at the beginning of a gene and transcribes an RNA molecule. The RNA nucleotides that are used to construct the RNA molecule are synthesized by other enzymes in the cytoplasm of cells. The correct nucleotide is paired up with its complementary base in the sense strand of the DNA and joined to the RNA by RNA polymerase. Each RNA polymerase begins transcription at a promoter site, copies the sense strand of the DNA, and finishes transcribing at the terminator. When transcription is completed, the RNA molecule is released from the DNA and the RNA polymerase is free to start transcribing another gene.

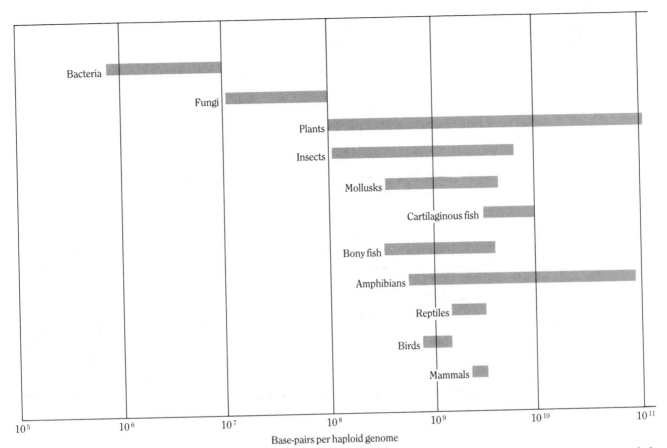

FIGURE 8-4 The number of base-pairs per haploid genome in different organisms. Bacteria have the smallest number. Reptiles, birds, and mammals vary over a narrow range and have similar numbers. Some plants have more than a hundred times as much DNA in cells as do human beings. The amount of DNA is not an indication of the amount of genetic information that it carries. For example, as much as 90 percent of human DNA is said to be "junk" DNA because it does not code for proteins or have any other function that has been identified.

because of the large size and complex structure of animal (and plant) chromosomes (Figure 8-4). In addition, the primary RNA transcribed in an animal cell's nucleus must be modified to become the mRNA that is transported to the cytoplasm, where proteins are synthesized.

In contrast with bacteria, which use a single RNA polymerase for transcription, a human cell has three different RNA polymerases for transcribing the three different kinds of RNA— mRNA, tRNA, and rRNA. Each kind of RNA polymerase recognizes different "start" signals in the chromosomes. Then, as the mRNAs are transcribed from different genes, the first base of each mRNA is "capped" with an unusual guanine nucleotide that is chemically different from the guanine nucleotides used to synthesize the mRNAs. When synthesis of each mRNA is finished, a tail ranging in length from 100 to 200 adenine nucleotides is added to the last base.

However, the most surprising and puzzling feature of the mRNAs synthesized in animal

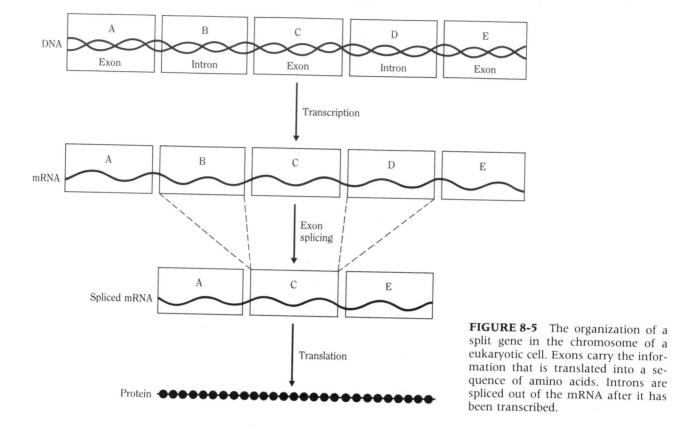

FIGURE 8-5 The organization of a split gene in the chromosome of a eukaryotic cell. Exons carry the information that is translated into a sequence of amino acids. Introns are spliced out of the mRNA after it has been transcribed.

cells is that much of the sequence of bases in the mRNAs is not needed and is not used when proteins are synthesized from them. The observation that most genes in animal cells contain intervening sequences of bases led to the discovery of *split* genes. This organization of genetic information is quite different from what had been learned by studying bacteria.

Split Genes and RNA Splicing

In prokaryotic cells, the information in a gene is contained in a continuous sequence of bases in DNA that is transcribed into a continuous sequence of bases in mRNA. Until 1977, there was no reason to think that the situation is any different in eukaryotic cells. In that year, however, researchers made the surprising discovery that in eukaryotic cells (such as human cells) the information in many genes is contained in discontinuous segments of DNA that are separated from one another by sequences of bases whose information is not expressed. This organization of genetic information in eukaryotic cells is referred to as **split genes**.

An easy way to grasp the concept of split genes is to imagine a human gene that codes for a hypothetical protein called "information." If the gene (word) is organized so that "information" is written as inxxxxforxxmaxxxxtion, the gene (word) is split. The segments of DNA containing information that codes for a sequence of amino acids are called **exons** (expressed regions); the pieces of noncoding DNA separating the information-containing segments are called **introns** (Figure 8-5).

When an mRNA has been transcribed from a split gene, the extra bases must be removed before the mRNA is read; otherwise, proteins with extra and incorrect amino acids will be synthesized. For example, the gene that codes for the β-polypeptide of human hemoglobin consists of three exons and two introns. The β-hemoglobin mRNA is transcribed with all of the information—exons and introns—that is in the gene. But, before the mRNA is translated, **RNA splicing enzymes** remove the bases corresponding to the introns from the mRNA and rejoin the remaining RNA fragments correctly. The reconstructed mRNA can now be translated into the correct amino acid sequence. Of the eukaryotic genes that have been analyzed so far, most contain both introns and exons in the DNA.

Why are genes split in animal cells and why are genes continuous in bacteria? How many different RNA splicing enzymes are needed to ensure accurate splicing of all mRNA? Have introns and exons played important roles in the evolution of eukaryotic organisms? Are split genes or continuous genes the more-primitive form of gene organization? These are but a few of the as yet unanswered questions that have been raised by the discovery of split genes. A few preliminary experiments suggest that each exon might have originally directed the synthesis of short chains of amino acids that functioned as simple enzymes. In the course of millions of years of evolution, chance recombination of segments of DNA may have joined various exons and introns together. The techniques of recombinant DNA (Chapter 12) are powerful tools for studying the organization and regulation of eukaryotic genes, because segments of DNA from any organism, including human beings, can be inserted into bacteria where it can be conveniently studied.

Translation

Translation is the process by which the information in mRNA molecules is decoded and converted into the information carried by the different sequences of amino acids that make up different proteins. Translation is basically quite similar in all cells, but once again the mechanism has been worked out in greatest detail in bacteria; so that is the one presented here.

Translation is much more complex than transcription because in this process genetic information must be *decoded*; that is, information encoded in one molecular form (the sequence of bases) must be converted into another molecular form (the sequence of amino acids) that assigns to proteins their particular functions. Translation, then, is the cellular process that makes the genetic information accessible to the cell in the form of enzymes, which can carry out a variety of chemical reactions.

The translation process is analogous to that of converting information contained in the dots and dashes of the Morse Code into the information contained in the sequence of letters and words in sentences. In the Morse Code, the conversion is from a code of two elements (dot and dash) into the code of twenty-six letters in the alphabet. In the process of translation, the conversion is from a genetic code of bases into a protein alphabet of twenty different amino acids. (The genetic code is discussed in Chapter 9.)

Translation includes several components: mRNAs, ribosomes, tRNAs, activating enzymes, amino acids, and many other enzymes. The construction of even a single protein in a cell requires a coordinated effort by all of these

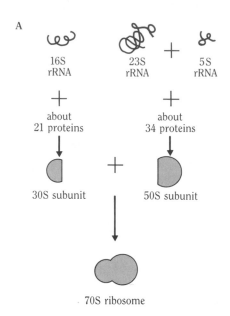

FIGURE 8-6   Ribosomes.
A. Each subunit of a ribosome consists of at least one RNA molecule and many different proteins. The 30S and 50S subunits join to make the 70S ribosome, which is active in protein synthesis. The sedimentation (S) values are not exactly additive when molecules of different sizes are centrifuged.
B. Model of a ribosome (side and front views), showing the 30S subunit in black and the 50S subunit in white. This model was constructed from many electron micrograph pictures of ribosomes extracted from cells. (Photographs courtesy of Miloslaw Boublik, Roche Institute of Molecular Biology.)

component parts of the translation system. We will first examine the properties and functions of each of the components; then we will look at how they function together.

Ribosomes

Ribosomes are the components in cells that are designed exclusively to carry out protein synthesis. They are complex particles consisting of three different classes of ribosomal RNAs—23S, 16S, and 5S—together with about fifty-five different *ribosomal proteins* (Figure 8-6A). These particular proteins are found only in ribosomes and together perform special functions in the process of translation. (The rRNAs of eukaryotic organisms have various other S values, depending on their size.)

Bacterial ribosomes are constructed from two subunits that are easily separated into their component parts, called the 30S and 50S subunits (Figure 8-6B). To begin translation of an mRNA molecule, the two ribosomal subunits must first recognize and bind to a particular sequence of three bases located near the 5' end of the mRNA called the **translation initiation site**. The mRNA chain is somehow threaded through each ribosome that attaches to the initiation site, and then the ribosomes move along the mRNA much like beads strung on a string. As each ribosome moves down the mRNA, other ribosomes can attach to the newly exposed initiation site. Thus, each mRNA is capable of making as many polypeptides as the number of ribosomes that attach to it. An mRNA continues to synthesize proteins until it is eventually destroyed. Any ribosome can attach to any mRNA molecule; therefore, the kind of protein that is synthesized is determined solely by the particular mRNA molecule, whose information is, in turn, determined by the sequence of bases in the gene from which it has been transcribed.

Transfer RNA

Ribosomes cannot, by themselves, accomplish the decoding of an mRNA molecule. Rather, transfer RNA molecules (adapter molecules)

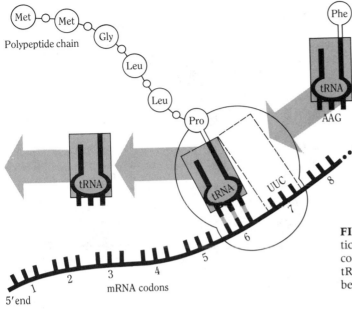

FIGURE 8-7   The function of tRNA. The Phe-tRNA anticodon (AAG) hydrogen bonds to the phenylalanine codon (UUC) in mRNA. Attached to the other end of the tRNA molecule is the amino acid phenylalanine, ready to be joined to the growing protein chain.

convert the information from one coding system into the other. As shown in Figure 8-2, one end of a tRNA molecule is covalently bonded to a particular amino acid—in this illustration, phenylalanine (Phe). At the other end of the Phe-tRNA molecule are three bases called the **anticodon**, which can form hydrogen bonds with the corresponding three bases in the mRNA (the codon). In this way, each mRNA codon specifies a particular amino acid and the tRNA molecule delivers that amino acid to the correct position.

Each of the thirty to forty tRNAs has a unique three-dimensional structure that is recognized by one of twenty different **activating enzymes** in the cell (biochemists call these enzymes aminoacyl-tRNA synthetases). The activating enzymes attach the different amino acids to their corresponding tRNAs. For example, the phenylalanine activating enzyme recognizes both the Phe-tRNA and the amino acid phenylalanine and joins the two together by a covalent bond. This group is now ready to attach to an mRNA-ribosome complex (Figure 8-7). The

phenylalanine codon in an mRNA is recognized by the Phe-tRNA anticodon and the codon and anticodon are joined by hydrogen bonds. At the same time, a ribosome helps to position the phenylalanine so that it can be joined to the end of the polypeptide that is being synthesized by that particular ribosome.

Figure 8-8 summarizes the overall series of steps in the process of translation. In the cell cytoplasm, amino acids are attached to tRNAs by activating enzymes. Each amino-acid–carrying tRNA moves about in the cell until a three-base anticodon in the tRNA pairs with its complementary codon in an mRNA. Ribosomes position adjacent tRNAs along the mRNA so that enzymes can catalyze the formation of a bond between the amino acids. As each ribosome moves along an mRNA, a protein is synthesized.

Proteins are synthesized at a remarkably fast pace in cells. In *E. coli*, each ribosome can chemically link amino acids together at a rate of about 15 per second. If we make the reasonable assumptions that an average protein contains

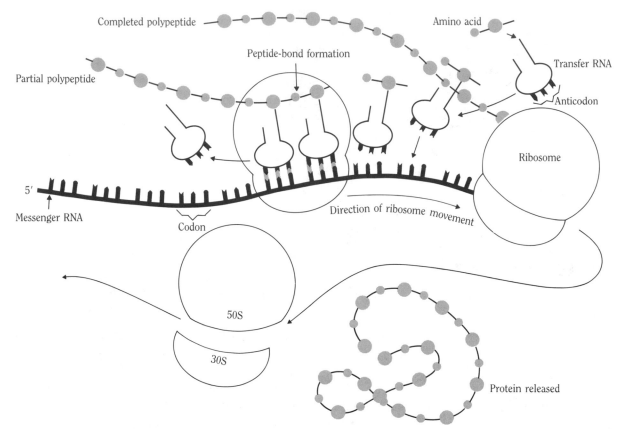

FIGURE 8-8 The process of translation. In the cytoplasm of cells, amino acids are attached to their appropriate tRNAs by the various activating enzymes. The tRNAs recognize specific codons in mRNAs and line up amino acids in the proper sequence. Ribosomes move along the mRNA forming peptide bonds between amino acids until a complete polypeptide is synthesized. The protein is released, and the ribosomes separate into their two subunits and reattach to another mRNA.

about 225 amino acids and that there are about 15,000 ribosomes in each bacterium, we can calculate that about 1,000 protein molecules are synthesized every second, or 60,000 proteins per minute per bacterium. Because proteins are needed to perform all enzymatic cellular functions, the process of translation, even at this high speed, must proceed with very few mistakes if the cell is to survive and function normally.

Human Disease and Defective Proteins

The observations of Archibald Garrod led him to suggest in 1909 that human metabolic diseases such as albinism were due to a lack of a particular enzyme (Box 3-2). We now know that thousands of human diseases, both mild and serious ones, are due to defective proteins synthesized from abnormal genes that are inherited from one or both parents (more on human

FIGURE 8-9   A hemoglobin protein molecule. Each hemoglobin consists of two α-chains and two β-chains. The black rectangles represent the heme group, the part of the molecule where oxygen is bound.

hereditary diseases in Chapter 14). Studies of altered human proteins, especially hemoglobin, have taught us a great deal about human diseases, as well as about mechanisms of protein synthesis in cells.

As mentioned in Chapter 5, the protein *hemoglobin* is found in red blood cells (erythrocytes) of all animals. Its function is to bind oxygen as blood circulates through the lungs and to exchange the oxygen for carbon dioxide as blood flows to other organs of the body. Each hemoglobin molecule consists of four polypeptides, two alpha (α)-globin and two beta (β)-globin chains, that combine with yet another molecule called heme to form active hemoglobin (Figure 8-9). The α-globin chain has 141 amino acids and the β-globin chain 146; a change of one amino acid in either chain, in principle, can result in a nonfunctional protein.

Adult human hemoglobin is the expression of two different genes: one directs the synthesis of the α-globin chains and the other directs the synthesis of the β-chains. Other globin chains are synthesized in the course of human embryonic and fetal development. Their synthesis is directed by genes other than α- and β-globin genes that are switched off after birth. There are other globin genes in adults that are never expressed (called pseudogenes) or that may be switched on in certain people. The number of different globin genes and their regulation during development explains certain aspects of hemoglobin diseases.

By the mid-1960s, the complete amino acid sequence for the β-chain had been determined by biochemical techniques. The first seven amino acids of the β-chain of normal human hemoglobin are shown in Figure 8-10, together with the corresponding amino acid sequences of hemoglobins isolated from the blood of anemic patients—that is, people whose blood is deficient in functional red blood cells. In both abnormal hemoglobin proteins, there is a single amino acid substitution at position 6 of the β-chain. People with hemoglobin S have valine inserted at position 6 instead of glutamic acid; they suffer from sickle-cell anemia. People with hemoglobin C have a lysine substitution at position 6, and they suffer from severe anemia. These single amino acid substitutions in the β-chain of hemoglobin reduce the hemoglobin's ability to bind oxygen and thus cause serious anemia. Several hundred variations of human hemoglobin caused by mutations have been detected.

The chemical differences between hemoglobin S, hemoglobin C, and normal hemoglobin were first demonstrated by Nobel laureate chemist Linus Pauling in 1949. Pauling used an analytical technique called *electrophoresis*, which separates similar protein molecules by exploiting the fact that different proteins migrate at different rates on a moist piece of paper or in a gellike material when an electrical current is passed through it. The more electrically charged the proteins are (charges form on amino acids in the proteins when they are placed in solutions of salts), the more rapidly the proteins move in the electrical field created by the current. Figure 8-11 shows a characteristic electrophoretic separation of hemoglobin molecules from people having normal or abnormal hemoglobins. Those who are homozygous (from the Greek *homos*, one and the same) for normal or mutant hemoglobin genes produce only one kind of hemoglobin (AA, normal; SS, CC, mutant in Figure 8-11). Those who are heterozygous (from the Greek *heteros*, different) produce both normal and mutant forms of hemoglobin (SA) or only mutant forms (SC).

Thalassemias are another serious form of anemia in human beings caused by the abnormal synthesis of either the α- or the β-globin chains. In some cases of thalassemia, the amount of one or the other globin chain is below normal; in other cases, one of the globin chains may be missing completely. If the defect is in the α-chain, the fetus usually dies before birth because that chain is essential during fetal development. However, defects in the β-chain do not affect a child until several months after birth because that is when the β-globin gene is

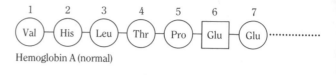

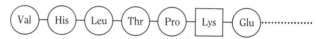

FIGURE 8-10   The normal function of human hemoglobins is lost by substituting one amino acid for another in the β-chain. In hemoglobin A, the predominant normal adult hemoglobin, glutamic acid resides at position 6 of the β-chain. In hemoglobin S, valine substitutes for glutamic acid at position 6, causing sickle-cell anemia if the abnormal gene is inherited from both parents or sickle-cell trait if inherited from one parent. In hemoglobin C, the substitution of lysine for glutamic acid at position 6 causes another form of severe anemia.

switched on and β-chains begin to be synthesized (Figure 8-12).

The kinds of mutations that have been observed in different thalassemias include changes in the promoter region of the β-globin gene (transcription is abnormal) and those that prevent the correct splicing of the exons constituting the mRNA after the introns are removed

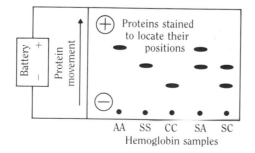

FIGURE 8-11   The use of electrophoresis to separate and identify different hemoglobins. Five hemoglobin samples differ from one another by single amino acid substitutions that change the electrical charge on the hemoglobin molecules, causing them to move at different velocities in the electrical field. People who are homozygous for hemoglobin genes produce only one kind of hemoglobin—AA, SS, or CC in the samples shown. Normal and mutant alleles can be distinguished by the different electrophoretic patterns of the hemoglobin proteins, as shown for samples SA and SC.

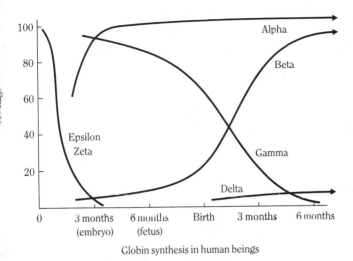

Globin synthesis in human beings

FIGURE 8-12 Synthesis of human hemoglobins at different stages of development. Embryonic hemoglobin consists of epsilon- and zeta-chains; adult hemoglobin consists of alpha- and beta-chains. Synthesis of gamma-chains is shut off shortly after birth; delta-chain synthesis is negligible in all but a few persons. The various hemoglobin chains are encoded by a family of different hemoglobin genes.

(translation is abnormal), as well as amino acid substitutions. You might be wondering why mutations that produce such serious, even lethal, diseases have persisted in human populations. The answer to this question came from studying various human populations in which many people have sickle-cell anemia or some form of thalassemia. It turns out that sickle cell anemia is found primarily in populations in certain regions of Africa (and in American Blacks) and that thalassemia is very common in Greek and Italian (Mediterranean) populations. Members of these populations survive better in malarial environments if they carry one of the defective globin genes even though homozygous people die prematurely of anemia (see Box 8-2). These studies show quite clearly that the advantage or disadvantage of having a particular gene or set of genes cannot be determined unless the

environment is specified and the gene-environment interactions are understood.

Proteins Made-to-Order

In the future, it may be possible to design and manufacture desired proteins outside of cells. Proteins might be engineered that would degrade pesticides or hazardous chemicals or that would function in detergents even in hot water containing bleach. Enzymes might be designed that would produce all sorts of industrial chemicals and drugs more cheaply than by present methods. The prospects of such protein technology have prompted a humorous revision of the central dogma described at the beginning of this chapter.

DNA → RNA → PROTEIN → MONEY

Although optimism is great, there are still major obstacles to be overcome before made-to-order proteins become a reality. We know that the amino acid sequence of a protein determines its shape and its specific function. We have also seen how a change in just one amino acid in hemoglobin can abolish its function. What protein engineers are trying to understand is how a particular amino acid change might *improve* a protein's function or how it might work under a new and useful set of conditions. Will a particular amino acid substitution make the protein ten times as active? Will the change make it more stable in an acid solution? Will it improve the fermentation process in a vat of crushed grapes? Protein scientists would like to be able to answer such questions from theoretical models of protein structure and chemistry. However, answers are still a long way off.

The ultimate goal is to develop mathematical and computer models that can predict how the shape of a protein will be affected by changing

The Connection Between Sickle-Cell Anemia and Malaria

BOX 8-2

Natural selection usually works to eliminate from a population those genes that cause serious disease, because individuals carrying these mutant genes often do not survive long enough to reproduce. There are exceptions, however. Certain complex gene-environment interactions sometimes permit the maintenance of otherwise lethal genes in human populations. Sickle-cell anemia and β-thalassemia (Cooley's anemia) are good examples (see the accompanying electron micrographs). These serious human diseases are caused by mutations that reduce the oxygen-carrying capacity of hemoglobin molecules in blood cells, yet they are maintained in certain human populations because they are caused by mutations that also enhance the survival and reproduction of certain people in those populations.

Sickle-cell anemia and β-thalassemia are caused by different mutations arising in the gene that codes for the β-chain of human hemoglobin. This gene is carried on human chromosome 11; if both chromosomes carry mutant alleles, the person is homozygous for the mutation and will suffer from severe anemia. Without medical help, the homozygous person is likely to die quite young. However, if a person is heterozygous for that allele—that is, if the cells have one normal and one mutant allele—then half the hemoglobin molecules will function normally, and only half will be defective. Because of the amount of normal hemoglobin, heterozygous persons rarely suffer from anemia. In some regions of the world, especially in those parts of Africa where sickle-cell anemia is prevalent and along the southern coast of the Mediterranean Sea where β-thalassemia is common, malaria is responsible for as many as half of all childhood deaths. Interestingly enough, however, children with one mutant hemoglobin allele tend to be among the survivors—they are less affected by malaria than are children with normal hemoglobin. What does malaria have to do with hemoglobin?

Malaria is caused by a protozoan that reproduces in human

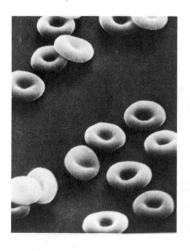 

Normal blood cells (left) compared with sickled cells (right) from patients with sickle-cell anemia. The characteristic sickle shape is produced when the supply of oxygen in the blood is low. People with this blood disease are unable to transport sufficient oxygen to other body tissues because of the altered hemoglobin's inability to function normally. (Courtesy of George Brewer, University of Michigan Medical School, and Marion I. Barnhart, Wayne State University Medical School.)

red blood cells, causing the cells to be destroyed and resulting in anemia and even death. Recent research has shown that the malaria protozoa cannot grow nearly as well in red blood cells having mutant hemoglobin molecules—such as those found in people with sickle-cell anemia—as they can in red blood cells whose hemoglobin is normal. Apparently, the altered shape of the red blood cell reduces the amount of potassium in the cell, thereby preventing reproduction of the parasite. In people who are heterozygous, then, the effects of malaria infections are reduced without their experiencing a change in their red-blood-cell functions serious enough to cause symptoms of anemia.

Thus, a deleterious mutation is maintained in the DNA of certain human populations because it allows people to survive in an environment that otherwise might kill them. In areas of the world (such as the United States) where malaria has been eradicated, there is no selective advantage in possessing sickle-cell genes. But neither is there a strong selection against them, and so these mutant genes persist generation after generation.

The advantage of having defective hemoglobins in certain environments shows why it is difficult to predict with confidence what the consequences of a particular genetic change will be. The interaction of genes and the environment is very complex; today's lethal mutation may be tomorrow's salvation.

READINGS: Milton J. Friedman and William Trager, "The Biochemistry of Resistance to Malaria," *Scientific American*. March, 1981. Louis W. Sullivan, "The Risks of Sickle-Cell Trait," *New England Journal of Medicine*. September 24, 1987.

---

just one amino acid in a particular position. And, even more important, how will the function of the protein be affected by the amino acid change? So far, the models have been only marginally successful because of the large number of different shapes that a protein can assume with each amino acid change. The structure of proteins is studied by x-ray crystallography (which was used to solve the structure of DNA) and by a newer technique called nuclear magnetic resonance (NMR; also used to detect tumors and other diseases in people). For each amino acid that is changed in a protein, a new structure must be determined—a task that is very tedious and expensive.

If simple rules governing protein structures and functions can be developed, the technology already exists for synthesizing proteins by machine and for changing specific amino acids at any position. We just need to be able to predict how a particular sequence of amino acids will function in a particular environment. Because of the huge market potential, made-to-order proteins continue to be a very active area of research.

Protein Patterns and Health Risks

Except for identical twins, every person carries a unique set of genes, making him or her genetically and biochemically distinct from every other person. The proteins produced in the cells of each organ in the body are unique to that particular organ, and many proteins vary from person to person. Because proteins are the products of genes, analyzing the kinds of proteins found in particular cells enables researchers to determine the specific alleles that different people carry in their chromosomes.

Biochemical techniques have been developed that allow the separation of more than 2,000

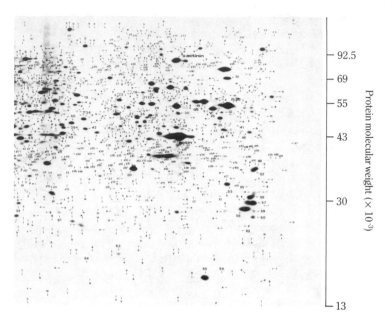

Protein molecular weight ($\times 10^{-3}$)

— 92.5

— 69

— 55

— 43

— 30

— 13

FIGURE 8-13 Biochemical analysis of human proteins. Two-dimensional electrophoresis can separate thousands of different proteins extracted from a few hundred human cells. Each protein is assigned a number, and its position and quantity is entered into a computer. This newly developed technique allows protein comparisons to be made between cells, individuals, or species. In this example, human cancer cells from a line called HeLa were grown *in vitro* and the proteins were analyzed. (Courtesy of J.E. Celis, Aarhus University, Aarhus, Denmark.)

different human proteins from a sample containing just a few hundred cells. The separation of all of these proteins by techniques of electrophoresis is so accurate and reproducible that the position and quantity of each protein can be recorded in a computer by automated protein scanning devices (Figure 8-13). Using this modern electrophoretic technique for the biochemical analysis of proteins (and indirectly of genes), researchers can compare any protein of an individual with the normal average value of that protein in a healthy human population. Should the value be much too low or much too high, abnormal gene expression and abnormal cellular functions can be detected.

These advances may eventually make it possible to establish a computerized file of each person's particular proteins. If this file were recorded at birth, a comparison of each newborn infant's protein profile with some standard profile might provide an early indication of abnormalities that would not otherwise be detectable.

Recording a baby's protein profile may eventually become as routine as recording its sex and weight at birth. So far, the cellular functions of only a few of the protein spots revealed by electrophoresis have been identified; most spots represent proteins (and genes) whose cellular functions are unknown. Applications of this new technique of protein identification are still years in the future; nevertheless, the use of protein profiles for screening for health risks seems promising.

Summary

DNA, RNA, and proteins are the informational macromolecules of cells. The expression of genes in DNA occurs in two steps: transcription, in which the genetic information contained in sequences of bases in DNA (a gene) is converted into a complementary sequence of bases in RNA; and translation, in which the genetic information contained in the sequence of bases in messenger RNA (the largest and most-diverse class of RNAs) is converted into a sequence of amino acids in a polypeptide. Information flows from DNA to RNA to protein in all organisms except for certain RNA viruses in which it flows from RNA to DNA.

Like DNA, RNA consists of nucleotides, each of which is made up of a sugar (ribose), phosphate, and a base (adenine, cytosine, guanine, and uracil). The three functionally different types of RNA molecules are messenger RNA, ribosomal RNA, and transfer RNA. Messenger RNA is translated into proteins, whereas ribosomal RNA and transfer RNA are not; they are incorporated into ribosomes where they participate in translation. Ribosomal RNAs have both structural and functional roles in ribosomes, the structures in which proteins are synthesized. Transfer RNAs convert the genetic information (sequences of bases) in messenger RNA into the functional information (sequence of amino acids) in proteins.

RNA polymerase is the enzyme that catalyzes the synthesis of RNA from DNA. It initiates transcription at the promoter site in DNA (the beginning of a gene) and stops it at the terminator site. Animal cells have three different kinds of RNA polymerase, one for each of the three types of RNA.

Many eukaryotic genes contain expressed segments of DNA that are separated by unexpressed bases. Such genes are called split genes; the expressed segments are exons and those unexpressed are introns. Thus, the primary RNA molecule transcribed in the nucleus of a eukaryotic cell must be modified by the removal of introns and the splicing together of exons to become the messenger RNA transported to the cytoplasm for protein synthesis. Other modifications to the messenger RNA also are made in the nucleus.

In translation, genetic information is decoded: the sequence of bases in RNA is translated into the sequence of amino acids in a protein. Ribosomal subunits begin translation by binding to a sequence of three bases (the translation initiation site) in messenger RNA. The ribosome then moves along the mRNA molecule, which enables another ribosome to attach to the vacated initiation site. Transfer RNAs bring amino acids to the ribosome-mRNA complex. An amino acid covalently binds to one end of a molecule of tRNA, a reaction catalyzed by an activating enzyme. Three bases (an anticodon) at the other end of the tRNA molecule are then hydrogen-bonded to three corresponding bases (a codon) in mRNA. The sequential addition of amino acids to the ribosome-mRNA complex eventually leads to the formation of a protein.

Key Words

activating enzymes The different enzymes responsible for attaching tRNAs to their corresponding amino acids.

adapter molecule A transfer RNA molecule.

anticodon Three adjacent bases in transfer RNA that pair with three complementary bases in a codon.

central dogma The rules that govern the exchange of information between DNA, RNA, and protein molecules.

codon Three adjacent bases in mRNA that specify each of the twenty amino acids.

electrophoresis A technique used for the separation of nearly identical proteins based on differences in their electrical charge.

exons The discontinuous segments of DNA in eukaryotic genes that carry information that is translated into the sequence of amino acids in proteins.

introns The untranslated segments of DNA in eukaryotic genes that separate exons.

messenger RNA (mRNA) An RNA molecule whose sequence of bases is translated into a specific sequence of amino acids (a polypeptide).

promoter (*p* site)   A sequence of base pairs in DNA to which RNA polymerase enzymes attach to initiate transcription of structural genes.

ribosomal RNA (rRNA)   The RNA molecule that is the main structural component of ribosomes.

ribosome   Structure in the cytoplasm of cells on which protein synthesis occurs.

RNA polymerase   The enzyme used to synthesize rRNAs, tRNAs, and mRNAs from genes in DNA.

RNA splicing enzyme   An enzyme that removes introns from eukaryotic mRNA and that splices the exons together in mRNA before translation.

split genes   Eukaryotic genes in which genetic information is encoded in the DNA in discontinuous segments.

terminator site   A sequence of bases in DNA where transcription of RNA molecules is terminated.

transcription   The process of synthesizing RNA molecules from specific segments of DNA molecules.

transfer RNA (tRNA)   A small RNA molecule that helps line up amino acids in the proper sequence by serving as a link between an mRNA codon and the amino acid that it codes for.

translation   The process of converting the information in the sequence of bases in a messenger RNA molecule into a sequence of amino acids in a polypeptide.

translation initiation site   Sequence of bases in mRNA where polypeptide synthesis begins.

Additional Reading

Chambon, P. "Split Genes." *Scientific American*, May, 1980.

Darnell, J.E., Jr. "RNA." *Scientific American*, October, 1985.

Doolittle, R.F. "Proteins." *Scientific American*, October, 1985.

Judson, H.F. *The Eighth Day of Creation: Makers of the Revolution in Biology*, Simon and Schuster, 1979.

Lake, J.A. "The Ribosome." *Scientific American*, August, 1981.

Moffat, A.S. "Protein Engineering." *Mosaic*, Summer, 1987.

Pederson, T. "Messenger RNA Biosynthesis and Nuclear Structure." *American Scientist*, January-February, 1981.

Vigue, C.L. "Murphey's Law and the Human Beta-Globin Gene." *American Biology Teacher*, February, 1987.

Study Questions

1   What are the functions of the three different classes of RNA in cells?

2   How does transcription of a gene begin in bacteria?

3   How does synthesis of a protein begin in bacteria?

4   What is an exon? About how many exons are present in bacterial DNA?

5   How many different proteins can be synthesized on one ribosome at any one time?

6   How does messenger RNA in human cells differ from that in bacteria?

7   What molecule is defective in people affected by sickle-cell anemia? How does the altered molecule differ from normal?

Essay Topics

1   Discuss why the stability of mRNAs in prokaryotes differs from that of mRNAs in eukaryotes.

2   Compare the structural arrangement of genes in prokaryotes with that of genes in eukaryotes.

3   Discuss the different kinds of mutations that can result in defective hemoglobin molecules in human cells.

9

The Genetic Code and Mutation

How Genetic Information Changes

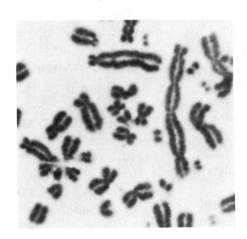

If a man will begin with certainties, he shall end in doubts; but if he will be content to begin with doubts he shall end in certainties.

FRANCIS BACON, *philosopher*

By THE EARLY 1960s, great strides had been made in understanding the chemistry of DNA and how information in genes is expressed. Following the announcement of the structure of DNA in 1953 by Watson and Crick, scientists from many fields increased their efforts to figure out precisely how genetic information is stored in DNA and how that information is transferred to other molecules in cells. By 1961, the existence of messenger RNAs had been confirmed. Until then, these information-containing RNA molecules had been difficult to isolate and analyze in bacteria because of their biological instability. Along with discovery of mRNA molecules came the first concrete idea of how hereditary information is encoded in genes in DNA and how genetic information is expressed as proteins.

By 1961, Francis Crick and Sidney Brenner in England had proved that the genetic code is a *triplet code*—in other words, that a sequence of three bases in an mRNA molecule codes for each amino acid in a protein. However, the most-fundamental question of molecular genetics still remained unanswered: Precisely how do genes direct the synthesis of proteins? More specifically, what *are* the codons—the three-base combinations that code for each different amino acid?

It began to appear, after the initial discovery of the triplet nature of the genetic code, that the deciphering of the code might prove to be an exasperatingly slow task. Efforts to "crack" the genetic code were hampered by a lack of suitable experimental techniques. But, finally, toward the end of 1961, a couple of unexpected experimental breakthroughs were made, and the exciting scientific race to discover the base sequence of each of the codons began. By the mid-1960s, the complete genetic code had been cracked. The "dictionary of life" had been discovered.

It may be difficult for most people not engaged in scientific research to envisage serious, methodical scientists racing against one another to make discoveries and to announce them first.* But cracking the genetic code would mean fame and fortune for the successful scientists—academic promotions, trips to international scientific meetings, and, for the most fortunate among them, a Nobel Prize. As codon after codon became known, Francis Crick arranged them into a meaningful order that eventually led to the standard genetic-code dictionary—the code that is used by every living organism. This chapter explains how the genetic-code dictionary is read and what conclusions regarding evolution can be drawn from the information carried by the code itself. It also describes how mutations change genetic information.

What Is the Genetic Code?

The information in the genetic code (Figure 9-1) is similar to that found in a translation dictionary (say, English–German): it allows one molecular language—the sequences of bases in RNA—to be translated into another molecular language—the sequence of amino acids in proteins. The genetic code is a triplet code consisting of four different bases arranged in groups of three. Because proteins are synthesized from

---

*For an engrossing account of such a race, read James Watson's controversial little book *The Double Helix*, which recounts Watson's and Crick's race to be the first to discover the molecular structure of DNA.

mRNAs, the genetic code in Figure 9-1 refers to messenger RNA bases—uracil (U), cytosine (C), adenine (A), and guanine (G). However, it should be remembered that DNA is the permanent storehouse of all genetic information and that the bases in mRNAs are synthesized from the complementary bases in DNA (Figure 9-2). The facts and important inferences drawn from the complete genetic code (Figure 9-1) are as follows:

1 The genetic code has a well-defined structure; that is, the codons are not randomly assigned to the various amino acids. For those amino acids that are specified by more than one codon, the first two bases are usually the same; base differences occur only in the third position of the codon for most amino acids (the exceptions are leucine, serine, and arginine).

2 In the third position of each codon, the bases U and C carry the same meaning; for example, UUU and UUC code for phenylalanine, and AGU and AGC code for serine.

3 The genetic code is **degenerate**. This means that most of the amino acids are coded for by

	Second base in codon				
	U	C	A	G	
U	UUU } Phe UUC UUA } Leu UUG	UCU ⌐ UCC UCA ⌐ Ser UCG ⌐	UAU } Tyr UAC UAA } Stop UAG	UGU } Cys UGC UGA Stop UGG Trp	U C A G
C	CUU ⌐ CUC CUA ⌐ Leu CUG	CCU ⌐ CCC CCA ⌐ Pro CCG	CAU } His CAC CAA } Gln CAG	CGU ⌐ CGC CGA ⌐ Arg CGG	U C A G
A	AUU ⌐ AUC ⌐ Ile AUA ⌐ AUG (start)	ACU ⌐ ACC ACA ⌐ Thr ACG	AAU } Asn AAC AAA } Lys AAG	AGU } Ser AGC AGA } Arg AGG	U C A G
G	GUU ⌐ GUC GUA ⌐ Val GUG	GCU ⌐ GCC GCA ⌐ Ala GCG	GAU } Asp GAC GAA } Glu GAG	GGU ⌐ GGC GGA ⌐ Gly GGG	U C A G

First base in codon (left), Third base in codon (right)

FIGURE 9-1 The genetic code. The code shows the correspondence between each three-base codon in mRNA and the amino acid for which it codes. The first base in each codon is closest to the 5′ end of the mRNA molecule and the third base in closest to the 3′ end. Uracil (U) in the third position often translates the same as cytosine (C); adenine (A) in the third position often translates the same as guanidine (G). The codon most commonly used to start synthesis of a polypeptide is AUG. Three codons (UAA, UGA, and UAG) are used to terminate polypeptide synthesis; these "stop" codons do not code for any amino acid. Figure 5-9 gives the complete names and structures for the amino acids.

several different codons. For example, phenylalanine has two codons, UUU and UUC, and leucine has six codons, UUA, UUG, CUU, CUC, CUA, and CUG. An important consequence of code degeneracy is that the base composition—that is, the relative amounts of the four bases—in the DNA of different or-

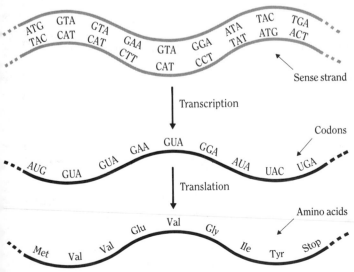

FIGURE 9-2 The flow of information from DNA to mRNA to protein. The bases in one of the two DNA strands (the sense strand) are transcribed into complementary bases in an mRNA molecule. The mRNA codons are translated into the sequence of amino acids in a protein.

The Genetic Code Is Universal, BUT . . .

BOX 9-1

When presented with a general rule by a teacher, some students feel challenged to find an experiment or fact that will contradict the generality of the principle that is being taught. This questioning attitude is healthy and shows that students are trying to solve things for themselves. Some teachers enjoy being questioned by students; others bristle at what seems to be a threat to their expertise or classroom authority. The maxim "Exceptions prove the rule" is often used to stifle further questions or debate. But, in fact, many scientific breakthroughs derive from paying attention to exceptions to the rule. Exceptions and unexplained observations can be sources of new insights; they can stimulate new hypotheses and eventually result in new and more-accurate theories.

The use of the genetic code in the DNA of mitochondria is one exception proving the rule that the "universal" genetic code is not quite universal. Mitochondria have several biochemical properties that, in some respects, more closely resemble those of primitive bacteria than those of animal cells. Some scientists suggest that, millions of years ago, eukaryotic cells "swallowed" some prokaryotic cells and that eventually the prokaryote and the eukaryote established a symbiotic relation. Ultimately, the "ingested" cells may have evolved into mitochondria performing specialized functions.

Mitochondria take a few liberties in reading the universal genetic code. In human mitochondria, the stop codon UGA is read as tryptophan, AGA and AGG mean "stop" instead of coding for arginine, and the codon AUA is read as methionine instead of isoleucine. The few tRNAs that are transcribed from genes in the mitochondrial DNA are much smaller than normal, and they have lost some of their codon specificity. In fact, some of the mitochondrial tRNAs are able to recognize four different codons instead of only one. Another unusual feature of mitochondrial DNA is that genes are contained in both strands, which are simultaneously transcribed into RNAs. This means that a change in just one base pair causes mutations in two genes. This is a most-unusual situation, and it is difficult to understand how such interdependent genes could evolve independently.

Natural selection may have changed the genetic code in mitochondrial DNA for reasons that have yet to be discovered. It is curious that one of the smallest and least complex of eukaryotic cells, yeast, has the largest known mitochondrial DNA (78,000 base pairs), whereas human cells have the smallest—a paradox that has yet to be explained. Comparing mitochondrial DNAs from different organisms and determining the extent to which the genetic code is interpreted differently by different mitochondria provide another powerful means of analyzing the evolutionary mechanisms that have resulted in biological diversity.

READING: Robert Reid, "Genes Break the Rules in Mitochondria," *New Scientist*. December 11, 1980.

ganisms can be quite variable, yet the amino acid composition of their proteins may be virtually identical.

4 The genetic code is **universal**. This means that in every living organism each codon specifies the same amino acid. For example, the codon ACG specifies threonine in cells of bacteria, wheat, whales, and human beings. Thus, mRNA molecules from any organism can, in principle, be accurately translated by

the protein-synthesizing machinery of cells from any other organism. For example, mRNAs carrying genetic information for human hemoglobin have been inserted into frog cells and into bacteria, and in each instance human hemoglobin polypeptides have been synthesized. To date, the only exceptions to the universality of the genetic code have been found in mammalian mitochondria (see Box 9-1).

5 The genetic code contains codons that are used to signal the beginning and end of genes, just as capital letters and punctuation marks are used to begin and end sentences. Four codons serve as start and stop signals for the translation process. The AUG codon, which codes for methionine, starts the synthesis of most proteins in bacteria and probably does the same in eukaryotic cells. In principle, every protein should have methionine as the first amino acid. However, it turns out that the initiating methionine is usually removed from newly synthesized proteins by enzymes. Three stop codons, UAA, UAG, and UGA, are reserved exclusively for termination of translation—that is, for stopping protein synthesis; none of these three codons specifies any amino acid. Normally, no tRNA molecules recognize the stop codons, and therefore no amino acid can be inserted into the growing polypeptide chain when these codons appear in the mRNA.

Why the Code Must Be a Triplet Code

Now that we have reviewed the important features of the genetic code as it was eventually deciphered, we will go back in time to examine some of the ideas and experiments that led to the complete cracking of the code. Early theoretical arguments showed that the genetic code could not possibly consist of just one or even two bases. Because only four different bases are used (A, C, G, and T in DNA and A, C, G, and U in RNA), taken singly these four letters could code for only four amino acids. And, if the four letters were used two at a time, still only sixteen different amino acids could be specified (4^2). Because there are twenty different amino acids, the four bases must be used in groups of at least three—and, in fact, four bases taken three at a time can code for as many as sixty-four amino acids (4^3). This realization reduced the problem to determining which amino acid is coded for by each triplet. The extra coding information must also be accounted for because, in principle, only twenty triplet codons are needed although sixty-four are available.

It is one thing to hypothesize that the genetic code is a triplet code consisting of sixty-four codons, but it is quite another matter to prove it experimentally. In 1961, Francis Crick, Sidney Brenner, and their coworkers devised a series of experiments with which they were able to demonstrate that bases in mRNA are indeed read in groups of three. They reasoned that, if bases are read three at a time when the code is deciphered, then adding or removing one or two bases will cause the reading of the bases to be shifted so that the triplet codons no longer make sense.

To understand what happens when a reading frame is shifted, notice what happens to a sentence if a letter is removed and the same grouping of letters in the words is maintained. For example, do ouu nderstandn oww hati mean? The addition or deletion of one or two bases in mRNA can be accomplished by the insertion or removal of bases in DNA, causing **frameshift mutations** (Figure 9-3).

Crick, Brenner, and their coworkers were able to produce a large number of different frameshift mutations by exposing the DNA of T4 viruses to a particular chemical that causes the

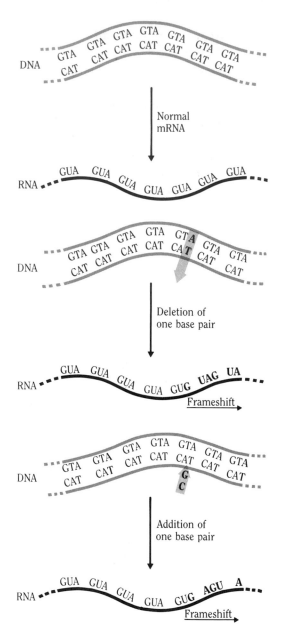

FIGURE 9-3 Frameshift mutations are due to the deletion or addition of one (or two) base pairs in DNA. As a result, the reading of codons following the frameshift mutation is out of phase and the wrong amino acids are inserted into the protein synthesized from the mRNA.

deletion or addition of base pairs. Although it is impossible to observe such a small change in DNA directly, these mutations can be detected indirectly by observing whether or not the mutant T-viruses are able to grow in bacteria. By employing some ingenious genetic manipulations, the researchers were able to determine the exact number of base additions or deletions in each of the mutant T-viruses. The key discovery in the experiments was that the addition or deletion of any three base pairs in the DNA would still permit the T-viruses to grow. As a result of the base changes, a small region of the mRNA transcribed from the gene having the three mutations was usually misread, but enough of the mRNA was translated correctly and produced some functional protein so that the T-viruses were able to grow. These results, complicated as they may seem, were the first experimental confirmation of the triplet code. For genes to be expressed correctly, precise groups of three bases must be read sequentially from one end of the mRNA to the other just as words in a sentence are read from beginning to end.

Deciphering the Code

Even though the fact that the genetic code is read in groups of three bases had been proved experimentally, the correct three bases (codon) for each amino acid still had to be assigned. An important technical breakthrough made by Marshall Nirenberg and Heinrich Matthei in the early 1960s finally led to the cracking of the genetic code—the assignment of codons to amino acids. Nirenberg and Matthei show that an artificial, chemically constructed mRNA molecule consisting only of uracil bases (polyuridine) could be successfully translated into a polypeptide in an *in vitro* protein-synthesizing system. In such a system, the cellular components necessary for synthesizing proteins—ribosomes, tRNAs, activating enzymes, amino acids, and other essential enzymes—are first isolated from cells. Then the individual components are combined in a test tube. If messenger RNAs are added, they are translated by the same chemical reactions that normally take place inside the cell.

Nirenberg and Matthei isolated the cellular components of the translation process and then carefully removed all remaining cellular mRNA molecules from the preparation in order to test the capacity of the artificial mRNA for directing the synthesis of a protein. When the polyuridine mRNA was added to the *in vitro* protein-synthesizing system, a polypeptide chain was produced. It proved to be polyphenylalanine, a polypeptide made up solely of the amino acid phenylalanine. The researchers thus concluded that UUU codons were being read as phenylalanine.

Artificial mRNAs composed of other bases were constructed and used to establish two other codons: AAA was found to code for lysine and CCC for proline. However, synthesis of artificial mRNAs was a difficult task for chemists at that time and, in general, gave unambiguous codon assignments only when all of the bases in the artificial mRNAs were identical.

As a result of another experimental breakthrough made by Nirenberg in 1964, within a year or so the complete genetic code was cracked—one of the more-remarkable accomplishments of molecular genetics. Nirenberg discoverd that specific codons—just three bases that could be constructed chemically—would function as mini-mRNAs in the *in vitro* protein-synthesizing system. In a series of experiments, Nirenberg and scientists in other laboratories were able to show that each three-base codon would attach to one, and only one, tRNA carrying a particular radioactively labeled amino

acid. Because the codon—tRNA—radioactive-amino-acid complex became attached to a ribosome, its presence was easily detected by pouring the contents of each experimental test tube through a filter that trapped the complex of molecules and the radioactivity. (The principle is the same as using a coffee filter to trap the coffee grounds while letting the coffee-flavored water run through.) Because each different chemically constructed three-base codon would attach to only one particular radioactive amino acid in each tube, each codon could be assigned unambiguously to each amino acid. Thus, the code was rapidly cracked and led to the codon—amino-acid assignments shown in Figure 9-1.

Mutations

Mutations in cells are analogous to automobile accidents. It is impossible to predict who will be in the 15 million or so automobile accidents that occur each year in the United States and whether they will be hurt, killed, or remain unscathed. However, it *is* possible to identify persons who are at risk of being in an accident. For example, many studies show that drivers who consistently speed or who drink substantial amounts of alcohol are more likely to be in accidents. Speeding and drinking do not directly cause accidents, but they do increase the chance of an accident. Similarly, although the overall frequency of mutations can be estimated, it is impossible to predict which DNA molecule will change or what the effects of the mutations will be. As with auto accidents, certain factors increase the chance that a mutation will occur. These factors, called **mutagens**, include such environmental agents as x-rays, radioactive materials, ultraviolet light, and chemicals that interact with molecules of DNA and increase the frequency with which bases are changed. The

TABLE 9-1 Rates of spontaneous mutations of particular genes in various organisms.

Organism and traits	Mutation rate (mutations per genome per generation)
E. coli (bacterium)	
Resistance to streptomycin	4×10^{-10}
Resistance to infection by T-viruses	1×10^{-9}
Inability to metabolize lactose	2×10^{-7}
Neurospora crassa (bread mold)	
Ability to synthesize adenine	4×10^{-8}
Zea mays (corn)	
Shrunken seeds	1×10^{-6}
Purple seeds	1×10^{-5}
Drosophila melanogaster (fruit fly)	
White eye	4×10^{-5}
Yellow body	1×10^{-4}
Homo sapiens (human beings)	
Hemophilia type A (bleeding disease)	3×10^{-5}
Achondroplasia (dwarfism)	4×10^{-5}
Retinoblastoma (tumor of eye)	1×10^{-5}
Huntington's chorea (nerve degeneration)	1×10^{-6}

greater the exposure to mutagenic agents, the more likely it is that one or more mutations will occur.

Even with minimal exposure to such highly mutagenic agents as x-rays or such weak mutagenic agents as ultraviolet radiation in sunlight, there is still some probability for a gene in any cell to mutate. Spontaneous mutations seem to be unavoidable and arise from occa-sional errors in base-pairing in the course of DNA replication or from unavoidable exposure to naturally occurring radiation or to naturally occurring chemicals in the environment and in food that can damage DNA. The frequencies of spontaneous mutations vary enormously from gene to gene and from organism to organism (Table 9-1). When the frequency of mutations rises significantly above these spontaneous mutation rates, the additional mutations are assumed to be caused by some agent—radioactivity, x-rays, chemical mutagens, and so forth. However, the kinds of mutations are the same whether the mutations are spontaneous (unattributable to specific causes) or induced (caused by an environmental agent).

To evaluate the effects of mutations in causing disease, in producing genetic defects, and in creating new varieties of organisms, it is first necessary to understand the various kinds of mutations that can occur in DNA.

Point Mutations

Mutations are classified according to the kinds of changes that occur in DNA molecules. The simplest mutation, called a **point mutation**, consists of a change in a single base pair. Mutations may or may not affect the functions of molecules in cells, depending on whether or not the activity of a protein is altered.

Point mutations can be neutral with respect to amino acid changes in proteins or they can cause changes that are called missense or nonsense mutations (Figure 9-4). A **neutral mutation** changes a base pair in DNA and a base in one codon of mRNA. However, this codon change does not affect the phenotype of the organism or the function of the protein. In the neutral mutation illustrated in Figure 9-4, the base-pair change in DNA merely results in threonine being substituted for serine in the protein.

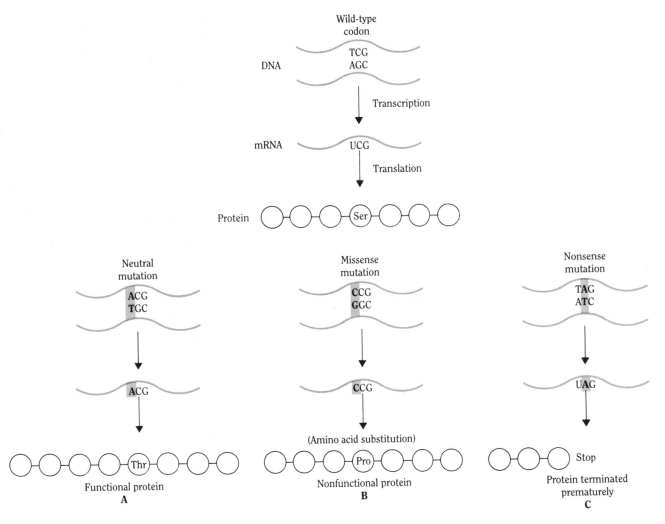

FIGURE 9-4 The effect of a neutral, missense, or nonsense mutation on the amino acid sequence in a protein: (A) the change in the base pair (G–C to C–G) does not change the amino acid; (B) the change in the base pair (T–A to C–G) changes the amino acid from serine to proline; (C) the change in the base pair (C–G to A–T) produces a stop codon and protein synthesis terminates prematurely at the stop codon.

The overall effect is no change in the function of the protein even though bases in DNA and RNA have been altered.

Point mutations causing codon changes in mRNA that result in amino acid substitutions in proteins are called **missense mutations**. Quite often the function of the protein will be changed by the amino acid substitution (recall the example of sickle-cell anemia caused by hemoglobin S). Other point mutations may change a

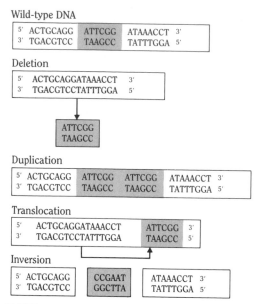

FIGURE 9-5   Kinds of large, or gross, mutations. In a deletion, several (even thousands) of base pairs are lost. In a duplication several (even thousands) of base pairs are added. If a segment of DNA moves (translocates) from one chromosomal location to another, the mutation is called a translocation. If a segment of DNA is cut out and reinserted in an inverted orientation, the mutation is called an inversion.

codon from one specifying an amino acid to any one of the three terminating, or stop, codons. These kinds of point mutations are called **nonsense mutations**. The appearance of a nonsense codon anywhere in an mRNA molecule causes premature termination of protein synthesis, and so an essential cellular function is usually lost.

Mutations in More Than One Base Pair

Many mutations cause more than just a change in a single base pair. (As already discussed, an addition or deletion of just one base pair can cause the misreading of an mRNA molecule and the loss of protein function.) Mutations may change many bases in DNA or they may cause the addition, removal, or relocation of parts of

chromosomes or even whole chromosomes. The four categories of these more-extensive kinds of mutations are deletions, duplications, translocations, and inversions (Figure 9-5).

Deletions result not only from the loss of at least one base pair in DNA, but also more generally from the loss of hundreds or thousands of bases. If many bases are lost, there is virtually no chance that all of them will be replaced correctly; therefore, deletion mutations are not correctable. Point mutations, on the other hand, may be corrected; that is, the original base pair may be restored in the DNA in the course of replication.

Duplications result when a segment of DNA is added that is identical with one that already exists. Geneticists believe that the duplication of genes has played a particularly important role in the evolution of genetic information. By duplicating genes and having extra copies present in a chromosome, the organism would, so to speak, have "spares" of the genes. Thus, if a gene on one copy were to mutate (sometimes to produce a useful new protein, sometimes not), the unchanged gene would remain intact and continue to function in its essential role. Under certain environmental conditions, the duplicated genes might enhance the survival of the organism and, in this way, be preserved and passed on to progeny.

When a segment of DNA is moved from its normal location in the chromosome to a new site, the mutation is called a **translocation**. Segments of DNA can translocate to other sites in the same chromosome or, in cells with many chromosomes, from one chromosome to another.

If a segment of DNA is removed, rotated end for end so as to maintain the correct chemical polarity, and then reinserted into the DNA, the mutation is called an **inversion**. In such a case, bases have not been added to or deleted from the DNA, but the genetic information has been changed because the sequence of bases is different. Inversions have evolutionary significance because they help keep particularly useful groups of genes intact when they are passed on to progeny. In insects, inversions may provide a mechanism by which an organism can preserve essential groups of genes for many generations. In human beings, however, it is not clear that inversions confer any advantage.

Individual base pairs, the arrangements of large segments of DNA, and even the number of chromosomes in cells may be changed by mutations. To help you understand the various kinds of mutations that arise in DNA, Table 9-2 provides analogies in the form of sentences. You can think of each sentence as a molecule of DNA, each word as a gene, and letters as base pairs. Keep in mind that in a cell the change caused by a mutation first arises in DNA, is transcribed into mRNA, then appears as an amino acid change in a protein, and finally results in a change in phenotype. The organism may look different, function differently, or exhibit different behavior as a consequence of a mutation (see Box 9-2).

Mutations from Radiation

Rates of spontaneous mutations are low under normal conditions, but certain environmental agents can interact with DNA and dramatically increase mutation rates. For example, **ionizing radiation**, such as gamma rays, x-rays, and ultraviolet rays, and particles emitted from radioactive materials damage DNA and thus increase the frequency of mutations.

Ultraviolet rays have enough energy to penetrate the skin and damage skin cells. The ultraviolet radiation in sunlight causes tanning — and, in some people who are out in the sun a great deal, skin cancer. X-rays are considerably

TABLE 9-2   Analogies for the various kinds of mutations that occur in DNA. Assume that each word is a gene and the whole sentence is a chromosome.

DNA	Kind of mutation
This is an accurate statement	Wild-type
This is a*m* accurate statement	Point
This is *not* an accurate statement	Addition
This is an *an* accurate statement	Gene duplication
This is an *in* accurate statement	Gene duplication followed by a point mutation
This is an accurate statement *This is an accurate statement*	Chromosome duplication
This *statement* is an accurate	Translocation
This is an *etarucca* statement	Inversion
Thii sa na ccurates tatement	Frameshift

more energetic than ultraviolet radiation and can pass completely through the body. That is why an x-ray sensitive film placed behind the body forms an image when the body is exposed to x-rays. Denser structures such as bones absorb more of the x-rays than the softer body parts do; so an image of the skeleton appears on the film. However, in passing through the body tissues, the x-rays also damage DNA, as well as other molecules and cells. If the x-rays damage the DNA of somatic (body) cells, mutations may occur in those cells that may eventually transform them into cancer cells (see Chapter 11). If the x-rays damage the DNA in egg or sperm cells, they may cause heritable mutations — that is, mutations that can be passed from generation to generation.

In 1927, Hermann J. Müller discovered that x-rays cause mutations in *Drosophila* (fruit flies). By irradiating the sperm of male flies, Müller was able to increase the mutation rates for

Favism
When It Is Dangerous to Eat Beans

BOX 9-2

Most mutations that arise in organisms nowadays are likely to be harmful or at best to have little or no effect on survival. The widely accepted explanation for this is that natural selection has kept all the beneficial mutations that arose in the past. The plant or animal surviving today is the one that is "fittest" in its present environment. But a mutation that is adaptive in one environment may be quite maladapted in another. A case in point is a group of mutant alleles that persist in human populations because they enhance people's survival in one kind of environment yet may cause a lethal disease under a different set of environmental conditions.

These mutant alleles are of a gene on the human X chromosome that codes for the enzyme glucose-6-phosphate dehydrogenase (G6PD), which participates in the metabolism of glucose. More than eighty different mutant alleles of this gene have been detected in human populations. About one person in forty carries one of the mutant G6PD alleles, and cells of these persons show some deficiency in G6PD enzymatic activity. For most people,

the altered enzymes function well enough that no disease is detectable, and they are unaware that they carry the mutation.

However, for certain people the generally innocuous presence of a mutant G6PD gene can abruptly prove fatal. If such people eat fava beans, a popular food in Mediterranean countries, they may suffer from *hemolytic anemia*—destruction of red blood cells. If a sufficiently large fraction of their red blood cells is destroyed, the victims die. In people susceptible to this disease, called *favism*, substances from the digested fava beans alter the chemistry of their red blood cells. If the G6PD enzyme level is sufficiently low, the chemical changes may be sufficient to destroy the red blood cells. If affected persons never eat fava beans, they are usually without symptoms because their red blood cells maintain some G6PD activity.

Why is such a potentially harmful mutation maintained in human populations? Like

the altered hemoglobins discussed in Box 8-2, the G6PD mutations make people more resistant to malaria infections. As might be expected, most of the mutant G6PD genes—and most people suffering from favism—are found in malaria-infested countries, which include those bordering on the Mediterranean Sea. It has been possible to demonstrate that the malaria-causing protozoa have difficulty growing in red blood cells taken from people with favism or from people with other G6PD mutations. Thus, the resistance to malaria conferred by G6PD mutations more than compensates for the lowering of cellular G6PD activity in terms of survival.

Favism demonstrates once again why it is impossible to predict the consequences of any mutation without a detailed understanding of the environment in which the mutation is expressed. Do you think that persons with mutant alleles of G6PD should be classified as genetically defective?

READING: Milton J. Friedman and William Trager, "The Biochemistry of Resistance to Malaria," *Scientific American*. March, 1981.

various *Drosophila* genes more than a hundredfold. The mutant genes were passed on to succeeding generations of flies, proving for the first time that x-rays do indeed cause hereditary mutations. Since Müller's original experiments, it has been shown that x-rays and other forms of ionizing radiation increase the frequency of mutations in all living organisms, including human beings.

Subatomic particles (electrons, protons, neutrons) are emitted from radioactive atoms as their nuclei decay into other elements. These subatomic particles also have sufficient energy to penetrate cells, damage DNA molecules, and cause mutations. The physical damage produced in DNA by such radioactivity and by ionizing radiation can be observed by examining irradiated chromosomes under the microscope. Figure 9-6 shows the kind of physical chromosome damage that is caused by low doses of x-rays, for example. Because the chemistry of DNA molecules is the same in all organisms, it is logical to expect that the genetic damage caused by ionizing radiation would be the same in all cells. This expectation is borne out by data showing that the number of x-ray–induced mutations in various species is proportional to the amount of DNA in their cells (Figure 9-7).

(a)

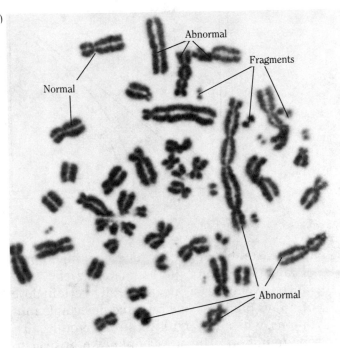

FIGURE 9-6 Human chromosomes damaged by x-rays. In this photograph, metaphase chromosomes show abnormalities caused by x-rays. (Courtesy of Judy Bodycote and Sheldon Wolff, University of California, San Francisco.)

Although the evidence that x-rays and other forms of ionizing radiation cause mutations is undisputed, nobody knows just how harmful very low levels of radiation are to living organisms. At extremely low doses of radiation, mutation rates are virtually impossible to measure accurately. Thus, the dispute centers on the question of whether it is fair to extrapolate from

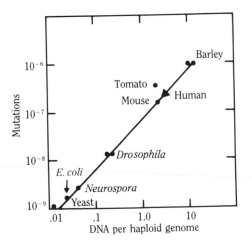

FIGURE 9-7 Relation between the frequency of mutations produced by ionizing radiation and the amount of DNA in various organisms. All of the points except for human have been determined experimentally. The human value is estimated from the DNA content of human cells. [Adapted from S. Abrahamson, M.A. Bender, A.D. Conger, and S. Wolff, "Uniformity of Radiation-Induced Mutation Rates among Different Species," *Nature* 245 (1973).]

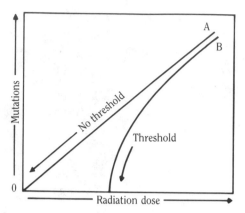

FIGURE 9-8 Two theoretical models showing mutations as a function of radiation dose. Curve A shows a linear relation down to zero dose. This model implies no threshold; mutations occur even at very low doses. Curve B shows a model with a threshold below which mutations do not occur. Note that at high doses the models converge and predict the same number of mutations per dose.

the measurable mutagenic effects of high doses of radiation in order to estimate the effects of much lower doses. For example, if a certain dose of x-rays produces one hundred detectable mutations, will a dose 1/100 as strong produce just one mutation? The as yet unknown answer to this question has important consequences for anyone who ever undergoes an x-ray—and for radiologists, x-ray technicians, nuclear-reactor technicians, and others who work with and are exposed to low-level ionizing radiation.

The problem of low-level radiation doses is exemplified in two different theoretical interpretations of how mutation rates vary as a function of dose, as shown in Figure 9-8. In both theories the number of mutations is proportional to the radiation dose. However, the mutagenic effects predicted for low doses of radiation are quite different. Curve A extrapolates to zero, which means that regardless of how low the radiation dose is, some mutations will still be produced. Curve B, on the other hand, has a *threshold* below which no additional mutations arise. Because accurate experimental data on the number of mutations produced by very low doses of radiation are difficult to obtain, especially for human beings, it is not possible to determine which relation—the one indicated by curve A or

the one indicated by curve B—is the one that occurs in nature. If, in fact, there is no threshold, then any exposure to ionizing radiation is potentially harmful.

Periodically, a committee of experts of the U.S. National Academy of Sciences surveys all the data pertaining to radiation and publishes reports on the biological effects of ionizing radiation (BEIR Reports). At this writing, the BEIR guideline for individual exposure to ionizing radiation recommends that such exposure not exceed 5 rems over a thirty-year period. (A rem is a standard measure of radiation and stands for roentgen equivalent in man.) Natural sources of radiation such as cosmic rays and radioactivity contribute about 2.5 rems to the average person over a thirty-year span, and dental and medical x-rays are sources of an additional 2 rems. Thus, 5 rems is generally regarded as the amount of unavoidable exposure to ionizing radiation.

This guideline does not say or mean that this amount of radiation is safe; it is simply what the committee feels constitutes an acceptable and unavoidable risk to the public based on current data. It is worth noting that in the past thirty years the maximum recommended dose of radiation considered acceptable has been steadily reduced. Although the effects of small doses of radiation are still being debated, the trend in recent years has been to reduce human exposure to radiation and to assume that all ionizing radiation—whatever the dose—is potentially harmful. As with other aspects of life, the risks and benefits of radiation exposure must continually be weighed by each of us, depending on the circumstances. A person who might readily agree to an x-ray of a broken bone might object to a routine chest x-ray, especially if one had been taken just a short while ago.

For many people, the word mutation conjures up images of physical deformities or late-show movie monsters. But a mutation is simply any change that arises in the genetic information of cells. The precise phenotypic consequences of any mutation are impossible to predict because the expression of any mutation depends on the organism's internal and external environment. Mutations are chance events and arise spontaneously in all DNA molecules. Mutations are neither good nor bad; they are a natural consequence of normal cellular processes. There is no basis for the belief, still held by some people, that human genetic defects are some sort of punishment or retribution. Mutations are not caused by immorality or unethical conduct; rather, they are a natural consequence of physical and biological processes.

All hereditary changes are ultimately caused by mutations. A mutation may cause a lethal defect in an organism or it may suppress an otherwise lethal change. Mutations are sources of new species and the causes of the extinction of others. Extinction of a species is not a sign of failure of that species, any more than a person's death signifies that his or her life has been a failure. Deaths of individual members of a species and the extinction of an entire species are part of the normal and necessary processes of biology and evolution. Nobody can predict with any assurance what organisms will inhabit the earth a million years from now. It is quite conceivable—even likely, judging from the past history of life forms on earth—that the human species may not be the most-complex or even the most-intelligent form of life that will ultimately inhabit the earth. However, it is quite safe to predict that whatever forms of life arise in the distant future, they will be composed of cells that will utilize this same genetic code to store genetic information and that the same cellular mechanisms of transcription and translation will be used to express it.

Summary

Genes express their information as proteins through a triplet code: a sequence of three bases (a codon) codes for an amino acid. The genetic code employs only four bases: adenine, cytosine, guanine, and thymine (in DNA) or uracil (in RNA). It is degenerate (most of the twenty or more amino acids are encoded by several different codons) and universal (each codon specifies the same amino acid in all organisms). Codons also signal the beginning of a gene and its termination in the DNA molecule.

Mutations are chance events and arise spontaneously in DNA. They are classified according to the kinds of molecular changes that occur in the DNA molecule. A point mutation is the simplest because it consists of a change in only one base pair. Neutral, missense, and nonsense mutations are point mutations. More-extensive mutations, which result from changes in more than a single base pair or from the addition, removal, or relocation of parts of chromosomes or even entire chromosomes, are classified as deletions, duplications, translocations, and inversions.

Although the overall frequency of mutations in a population can be estimated, it is impossible to predict which base or bases in DNA will be changed or what the effects of the mutations will be. Environmental factors called mutagens (chemicals, radiation, and viruses) increase the chances for mutations. Reducing one's exposure to mutagens can benefit one's health.

Key Words

degenerate code   Refers to the fact that each amino acid (except methionine and tryptophan) is specified by more than one codon.

deletion   Loss of one or more base pairs from DNA in a chromosome.

duplication   Addition of one or more base pairs to DNA in a chromosome.

frameshift mutation   A mutation that results from the insertion or deletion of one or two base pairs in DNA.

inversion   A chromosome segment that has been rotated by 180 degrees such that the order of genes is inverted with respect to the rest of the chromosome.

ionizing radiation   High-energy radiation such as gamma rays, x-rays, ultraviolet light, and particles emitted by radioactive materials.

missense mutation   A single base-pair change that causes the substitution of one amino acid for another in the protein.

mutagen   Any environmental agent that increases the frequency of mutations.

neutral mutation   A single base-pair change that does not change the amino acid in a protein and, hence, does not change the protein's function or the organism's phenotype.

nonsense mutation   A single base-pair change that generates a stop codon and terminates the polypeptide chain.

point mutation   A change in only one base pair in DNA.

universal code   Refers to the fact that each codon specifies the identical amino acid in all organisms from bacteria to human beings.

Additional Reading

Beebe, G.W. "Ionizing Radiation and Health." *American Scientist*, January-February, 1982.

Broad, S. "Human Harm to Human DNA." *The CoEvolution Quarterly*, Spring, 1979.

Crick, F.H.C. "The Genetic Code." *Scientific American*, October, 1962, and October, 1966.

Kohn, K.W. "DNA Damage in Mammalian Cells." *BioScience*, September, 1981.

Miller, O.L. "The Visualization of Genes in Action." *Scientific American*, March, 1973.

Rich, A. "Bits of Life." *The Sciences*, October, 1980.

Sagan, D., and L. Margulis. "Bacterial Bedfellows." *Natural History*, March, 1987.

Study Questions

1   What is meant when we say that the genetic code is universal? That it is degenerate?

2   What codons are used as start and stop signals in protein synthesis?

3   How many codons mean leucine (Leu)? How many mean tryptophan (Trp)?

4   What distinguishes a frameshift mutation from either a missense or a nonsense mutation?

5   What are three different classes of mutagenic agents found in the environment?

6   Give one example of a neutral mutation.

7   Can a point mutation affect more than one gene?

Essay Topics

1   Discuss some ways in which ionizing radiation can affect health.

2   Explain why the genetic code consists of sixty-four codons.

3   Discuss how neutral mutations might have contributed to the evolution of new species.

10

Gene Regulation

Switching Genes On and Off

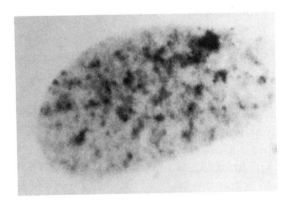

Are the stimulation of transcription, the silencing of transcription, and the initiation of DNA replication all implemented in eukaryotes by proteins which spend their time, like characters from the film *Tampopo*, searching for the perfect noodle?

N.J. SHORT, *biologist*

EVERY MOMENT, many thousands of chemical reactions are taking place in every cell of your body. The kinds of chemical reactions and the rate at which they occur must be regulated with exquisite precision if your tissues and organs are to function properly. The functions performed by each organ in the body are different, and it is unlikely that precisely the same reactions take place at the same moment even in identical cells in the same organ. How are all of the chemical reactions in all of the different kinds of cells regulated?

The particular biochemical activities of cells are determined by the particular genes that are being expressed in those cells. Most genes in a cell code for the enzymes that carry out the cell's chemical reactions. The enzyme-coding genes, in turn, are often regulated by other proteins or by regulatory sites in the DNA itself. During the development of an organism, its cells must differentiate; that is, particular genes must be switched on or off according to the functions that the cells must perform in a particular stage of development. Cellular **differentiation** determines the sort of plant or animal that will be formed and is the process by which cells undergo a change (usually irreversible) from a relatively unspecialized state to more-specialized functions as an organism develops.

The cells of a multicellular organism vary greatly in size, shape, and function (Figure 10-1). A human brain cell differs from a muscle cell in many respects, yet both cells contain precisely the same number of genes and the same number of chromosomes. What causes cells to be so different from one another is the way in which their genes are regulated and expressed—which genes are switched on and which ones are switched off.

Even after cells have differentiated and become integrated into a particular tissue, where they perform specific functions, they must still be able to adapt to changing conditions. For example, the tissue may become damaged or infected, the water or nutrient supply to the cells may change, or hormone or nervous-system signals may require changes in gene expression. As with all dynamic processes, conditions both inside and outside the cell must be monitored, and the expression of genes must be continually adjusted in response to stimuli.

To further appreciate the complexity of gene expression in animals, think of your body as a

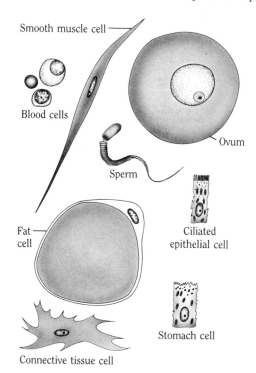

FIGURE 10-1 Schematic diagram of various human cells. Cells vary in shape, size, and function in different organs and tissues. Different genes are expressed in the various cell types.

biological orchestra made up of billions of instruments (cells). The kind of music that the orchestra plays will depend on which instruments are being played at any given moment, as well as on the particular notes, the intensity or loudness of the sounds, and the tone of the sounds. Clearly, in such a biological orchestra, the possibilities for different kinds of sounds are limitless. To even begin to understand how such an orchestra functions, it is first necessary to understand in detail how each individual instrument plays and responds to the instructions of the conductor (the brain). Because this is such an enormous task, the simpler the instrument one chooses to study, the more likely it is that one may begin to understand how the biological orchestra is regulated.

Bacteria have been chosen as model systems for studying gene regulation because they are the simplest of all organisms. But, even in these simple, single-celled organisms, the ways in which the expression of a single gene is regulated may be extremely complex. However, knowledge of some of the details of gene regulation in bacteria makes it easier to formulate ideas about gene regulation in the complex cells in the human body.

Levels of Gene Regulation

Enzymes catalyze the different chemical reactions that take place in cells. As explained in Chapter 8, genetic information flows from DNA to RNA to protein. So, in principle, at least three levels exist for regulating gene expression and enzyme activities in cells; actually, at least five mechanisms have been identified (Figure 10-2). Preventing the transcription of genes is the most-economical method of gene regulation because it saves the most cellular energy, given that neither RNA nor proteins are made. This is

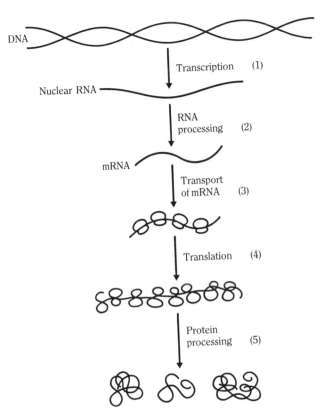

FIGURE 10-2 Five levels of regulation at which the expression of a gene can be affected: (1) regulation of transcription by sequences of bases in DNA such as promoters, operators, and enhancers; (2) the transcribed mRNA is processed by the addition and the removal of sequences; (3) the processed mRNA is transported or degraded; (4) the transported mRNA may or may not be translated in the cytoplasm; (5) polypeptides translated from the mRNA may be further processed to become active proteins.

called regulation of transcription, through which synthesis of RNA is prevented. If a gene is not transcribed into a molecule of RNA, the information carried by that gene remains unexpressed in the cell. If, however, a molecule of RNA is transcribed, it still must be processed before becoming a functional messenger (mRNA) capable of synthesizing a protein. In human cells, transcribed RNAs are processed by the removal of introns (noncoding sequences of bases) and by the splicing together of exons (the bases that are translated). Also, a special nucleotide caps one end of the RNA and a string of adenine nucleotides attaches to the other end. Each modification gives the cell an opportunity to select those RNAs that will be used and those that will be broken down and recycled. One human genetic disease (hemophilia B) is caused by a mutation that changes a single base pair at an intron-exon splice site.

Next, the finished mRNAs must be transported from the nucleus of the cell to the appropriate sites in the cytoplasm where the proteins will be synthesized. Several nuclear proteins complex with the mRNAs and facilitate their processing and transport to the cytoplasm. So transport is an important way that cells can regulate gene expression even if genes have been transcribed and RNA has been processed. Once the mRNAs have reached their destinations in the cell, other mechanisms determine whether they will be translated or degraded. Finally, even if the mRNAs are translated, the proteins themselves may be regulated in various ways. To become active, they may have to be processed by the removal of certain amino acids. Or small molecules may attach to functional proteins and inactivate them. Examples of these two kinds of protein regulation are described next.

Insulin Synthesis

Human **insulin** is a polypeptide hormone whose synthesis and activity can be regulated in numerous ways. First, insulin is synthesized only in certain cells of the pancreas even though insulin genes are present in every cell of the body. So mechanisms exist that keep the insulin genes switched off in all cells except for particular pancreatic cells. After having been synthesized, insulin is secreted from the pancreas,

TABLE 10-1   Effects of insulin after secretion from the pancreas.

Within seconds:

Insulin binds to receptor in cells.
The insulin-receptor protein is activated and becomes chemically changed.

Within minutes:

The pattern of gene transcription is changed.
Enzyme activities are changed.
The transport of sugar and minerals into the cell is changed.

Within hours:

Synthesis of proteins, RNA, and lipids in various organs is changed.
Cell growth is changed.

enters the blood, and circulates throughout the body performing numerous different functions (Table 10-1).

An important function of insulin is to regulate the level of glucose in the blood; loss of this regulation causes the disease **diabetes**, which is in fact many different diseases. Those that occur in childhood are called insulin-dependent diabetes (Type I). Diabetes that occurs later in life can often be controlled by diet or drugs other than insulin and is known as insulin-independent diabetes (Type II). The causes of diabetes are numerous and include viral infection of the pancreas, obesity, too much dietary sugar, and hereditary defects (see Box 10-1).

Once the insulin genes have been activated in the cells of the pancreas, the RNA is transcribed, processed, transported, and translated as shown in Figure 10-2. However, the polypeptide that is synthesized, called preproinsulin, must be cleaved by enzymes first to proinsulin and then to insulin molecules (Figure 10-3). A sequence of nineteen amino acids (the signal sequence) in preproinsulin does not appear in the insulin molecule; this sequence

Diabetes
A Disease Caused by Lifestyle or Genes?

BOX 10-1

Diabetes mellitus, called diabetes for short, is one of the best examples of a disease that can be termed either a lifestyle disease or a genetic disease, depending on your particular viewpoint. An estimated 10 million Americans have some form of diabetes and about half of them are unaware of their condition because symptoms are absent. The primary symptom of diabetes is failure in the regulation of blood glucose levels. Abnormal levels of glucose in the blood may or may not be due to abnormalities in production of the hormone insulin, which regulates the rate of sugar metabolism.

Type I (insulin-dependent) diabetes accounts for about 10 percent of cases and is observed most commonly in children. Daily insulin injections are required to control the symptoms. Type II (insulin-independent) diabetes is primarily a disease of adults and older persons; often overweight is a contributing factor. About 90 percent of diabetics have some form of Type II diabetes and do not necessarily require insulin to control their blood sugar levels. Quite often diet and weight loss are sufficient to control the diabetic condition.

Diabetes tends to "run in families" and quite often is described as a genetic disease by physicians in discussing diabetes with their patients. However, the genetic basis of diabetes, particularly for the vast majority of Type II diabetics, is far from an established fact. No defective genes have been identified for diabetes, as they have for thousands of other Mendelian genetic disorders. Human twin studies, which are the only other means of establishing the heritability (Chapter 20) of diabetes, give ambiguous results. Among monozygotic twin pairs, if one twin has Type I diabetes, the other twin has a 20-

to-50-percent chance of acquiring diabetes also. This would suggest that some "genetic susceptibility" exists, but it also indicates that environmental factors are equally important in the development of this form of diabetes. For Type II diabetes, the influence of heredity is also unclear. Among monozygotic twin pairs, if one twin becomes diabetic, the other twin is likely to acquire diabetes also, even though it may be many years after the first twin comes down with the disease. At first glance, this strong association between identical twins would seem to support a genetic basis for the disease. However, other studies show that environmental factors are very important because the frequency of Type II

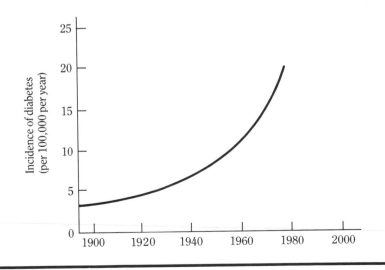

diabetes can vary severalfold in the same population from one generation to the next, presumably because of lifestyle changes. Diabetes also varies dramatically from country to country: France has one of the lowest rates, whereas Finland has a rate ten times as great.

The incidence of diabetes among whites in the United States has increased dramatically in this century (see the accompanying graph). In this brief period, the frequency of "diabetic genes" cannot have changed significantly. Thus, the only warranted conclusion is that neither form of diabe-

tes is due to purely genetic causes. This fact is important because it means that, to a large degree, diabetes can be controlled by lifestyle, which includes the diet.

READING: J.I. Rotter and D.L. Rimoin, "The Genetics of Diabetes," *Hospital Practice*. May 15, 1987.

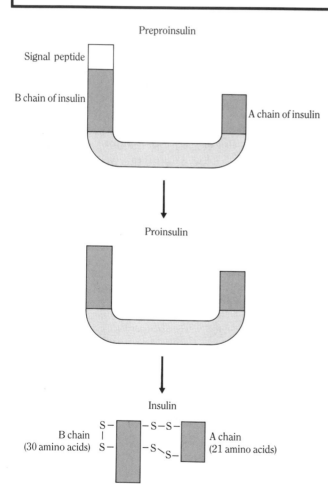

FIGURE 10-3 Processing of preproinsulin is necessary to produce insulin. The signal peptide is cleaved from preproinsulin after the entire polypeptide has been synthesized in the pancreas. Proinsulin is subsequently cleaved (shaded area) and insulin is released.

functions in the subsequent secretion of insulin from the pancreas. Also, thirty amino acids in proinsulin are not part of active insulin; they function to correctly align the A and B chains of the insulin molecule.

Mutation in even a single base pair in DNA can block any one of the critical steps in the synthesis of insulin. For example, a mutation might prevent the conversion of proinsulin into insulin. Or a mutation might cause the signal sequence to become nonfunctional. Or a mutation can cause the insertion of an incorrect amino acid into either the A or the B chains.

There are still other ways that the activity of insulin can be blocked. Insulin exerts its many effects by binding to insulin-receptor proteins on the surfaces of cells in other organs (such as the liver). Consequently, if these receptor proteins are defective owing to mutations in the insulin-receptor gene, then even normal insulin molecules cannot bind to the receptors and insulin function is lost. These few examples reveal the enormous complexity of chemical communication between cells, proteins, and hormones in different organs of the human body. Any one of several mutations in insulin-receptor genes or biochemical defects in insulin itself can produce different diseases, all of which exhibit symptoms resembling those of diabetes.

Feedback Inhibition

Another way of regulating a protein's activity has been discovered in bacteria. In *E. coli*, for example, many enzymes, such as those that catalyze reactions leading to the synthesis of amino acids, are regulated by a mechanism called **feedback inhibition**. If an amino acid such as tryptophan is added to the growth medium, it is taken up by the cell and blocks the activity of the first enzyme in the pathway used to synthesize tryptophan, as shown in Figure 10-4. Other amino acids behave in a similar manner by inhibiting the first enzymes in the pathways leading to their synthesis. Feedback inhibition keeps bacteria from producing substances that are already present in excess in their environment. As a result, they can spend more energy on synthesizing those substances that they do need. (Can you see how such a regulation might have resulted from a chance mutation and natural selection of bacteria that had acquired feedback inhibition of certain enzymes?)

Feedback inhibition is quite similar to feedback mechanisms used in manufacturing plants or in electronics. In an automobile assembly line, if more automobiles roll off the line than can be handled, information is "fed back" so that the assembly process is slowed down. By the same token, if the assembly is too slow, that part of the process that is causing the delay can be located and speeded up. Circuits used in radio and television receivers also use positive or negative feedback to adjust the amplification of the receiver so that the output is constant regardless of the strength of the input signal.

Having examined some of the ways that the activities of proteins and polypeptide hormones are regulated in human and bacterial cells, we now turn our attention to the primary and most-economical level for regulating the expres-

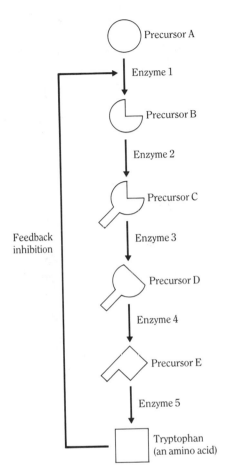

FIGURE 10-4   Feedback inhibition of amino acid biosynthetic pathways in bacteria. The final product of the pathway (in this example, the amino acid tryptophan) can inhibit the activity of the first enzyme in the pathway (Enzyme 1). Thus, when tryptophan is present in the cell in excess, its further synthesis is prevented.

sion of genetic information. As mentioned earlier, this is the regulation of transcription—the switching on or off of genes in DNA. Once again, the best-understood examples have been worked out in bacteria. How human genes are switched on or off is still pretty much a mystery; however, that which is known about regulatory mechanisms for human genes will be mentioned later in this chapter.

Genes Regulate Bacterial Growth

Escherichia coli normally grows in the human colon but does not contribute to digestion, which is finished in the small intestine. Outside of the human colon, this bacterium can grow in a wide range of conditions; it does so by expressing only the particular genes that are needed for optimal growth in each habitat.

In the early part of this century, a simple experiment was performed that showed how *E. coli* bacteria are able to grow in a nutrient medium consisting only of essential minerals and two different sugars, a mixture of glucose and lactose. Not until almost fifty years later, however, did researchers figure out the mechanisms of gene regulation that produce this particular pattern of growth. If bacteria are grown in a

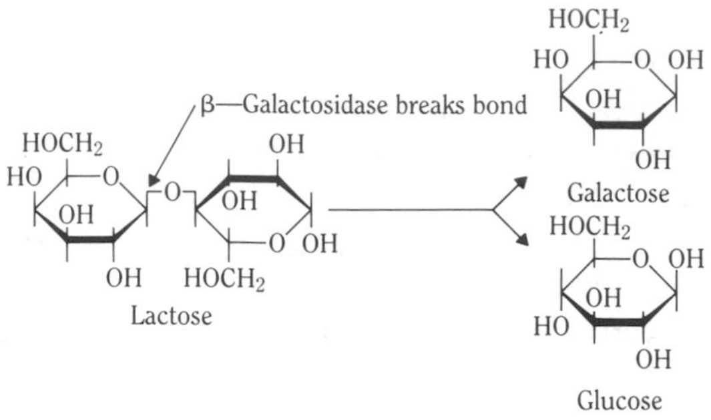

FIGURE 10-6   The chemical reaction that is catalyzed by the enzyme β-galactosidase. Lactose is broken down into the two simple sugars glucose and galactose, which is then converted into glucose by other enzymes.

solution containing a small amount of glucose and a large amount of lactose, they grow at first by utilizing only the glucose; the lactose in the medium is not touched until all the glucose is used up (Figure 10-5). When the glucose is depleted, the cells abruptly stop growing for a brief period. They then resume growing by metabolizing the lactose, which up to this point has remained unused.

An enzyme called **beta-galactosidase** must be synthesized by the bacteria to break down the lactose into the two simpler sugars glucose and galactose (Figure 10-6). (Other enzymes are needed to convert the galactose into glucose, but these enzymes will not be considered here.) Most plant and animal cells use glucose as sources of energy and of the carbon atoms needed to synthesize other molecules. Ultimately, all sugars must be converted into glucose before their atoms can be used.

What is not explained by the growth pattern shown in Figure 10-5 is why β-galactosidase is synthesized by bacteria only after the glucose has been used up. Given that much more lactose than glucose is available, why would not at least some of the lactose be used along with the glucose? Pursuing the answer to this question

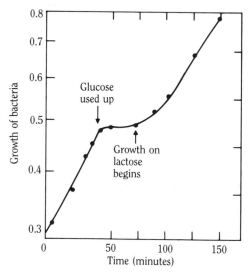

FIGURE 10-5   Growth pattern of *E. coli* in a medium containing both glucose and lactose. Bacteria grow until they use up all the glucose. At that point, other genes are switched on and the enzyme β-galactosidase is synthesized. Bacteria need this enzyme to utilize the sugar lactose. When enough β-galactosidase has accumulated in the bacteria, they can metabolize the lactose and begin to grow again.

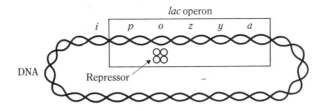

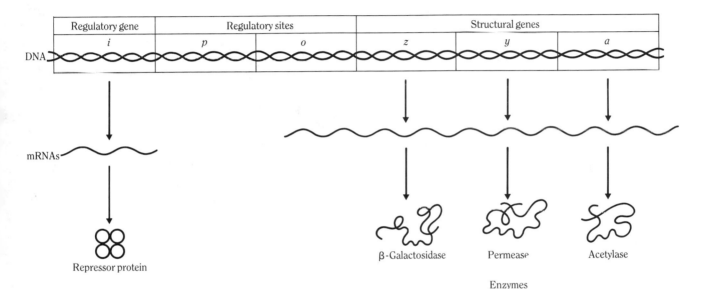

FIGURE 10-7 The lactose (*lac*) operon. To utilize lactose, bacteria have three genes that code for the synthesis of enzymes and three genetic regulatory elements in the DNA. The genetic elements of the *lac* operon are the *z* gene (structural gene coding for the synthesis of β-galactosidase), the *y* gene (structural gene coding for the synthesis of permease), the *a* gene (structural gene coding for the synthesis of acetylase), the *i* gene (regulatory gene coding for the synthesis of repressor proteins), the *o* site (operator site in the DNA to which repressor proteins attach), and the *p* site (promoter site in the DNA to which RNA polymerase attaches).

eventually led scientists to discover how the gene coding for the synthesis of the enzyme β-galactosidase is regulated in bacteria.

The Lactose Operon: Negative Regulation

In 1961, Jacques Monod and François Jacob at the Pasteur Institute in Paris proposed an unusual model to explain how the expression of genes is regulated in bacteria. These researchers had isolated numerous bacterial mutants in which the normal synthesis of β-galactosidase was changed. In particular, they isolated mutants in which the synthesis of the enzyme could not be switched on under any condition, as well as mutants in which the enzyme was continually synthesized and could not be switched off. The most-novel feature of the Monod-Jacob model for the regulation of β-galactosidase synthesis was the idea that a repressor protein is synthesized that normally

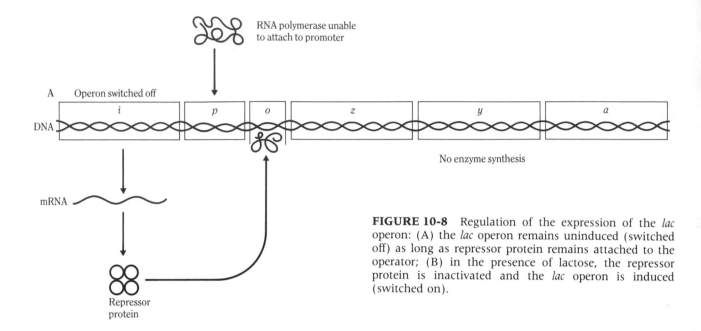

A   Operon switched off

DNA

mRNA

Repressor
protein

RNA polymerase unable
to attach to promoter

No enzyme synthesis

FIGURE 10-8   Regulation of the expression of the *lac* operon: (A) the *lac* operon remains uninduced (switched off) as long as repressor protein remains attached to the operator; (B) in the presence of lactose, the repressor protein is inactivated and the *lac* operon is induced (switched on).

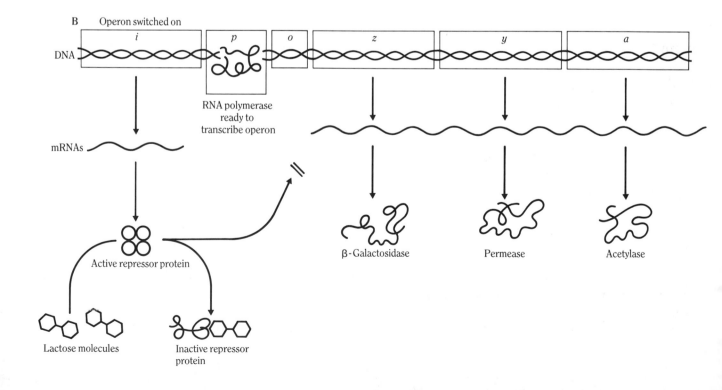

B   Operon switched on

DNA

mRNAs

RNA polymerase
ready to
transcribe operon

β-Galactosidase   Permease   Acetylase

Active repressor protein

Lactose molecules

Inactive repressor
protein

keeps the genes of the lactose (*lac*) operon switched off. An **operon** is a segment of DNA consisting of two or more adjacent genes whose expression is jointly regulated by sites in the DNA situated close to the **structural genes**—that is, genes that direct the synthesis of polypeptides (Figure 10-7).

The *lac* operon consists of three structural genes that code for enzymes required for the utilization of lactose and three regulatory genes that control expression of the structural genes. The *z* gene codes for β-galactosidase, the enzyme that splits lactose into glucose and galactose. The *y* gene codes for the enzyme *permease*, which is able to trap the lactose molecules and thereby build up the concentration of lactose inside the cell. The *a* gene codes for the enzyme *acetylase*, whose function is not essential for lactose utilization.

As originally proposed by Monod and Jacob, the *lac* operon (so named because all of the genetic elements participate in the metabolism of lactose) contains three genetic elements that regulate the expression of the three structural genes of the operon. One of these regulatory elements—the *i* gene—codes for the synthesis of **repressor proteins** that normally keep the *lac* operon structural genes switched off. The other two regulatory elements are called *sites* in the DNA because these genetic elements are not transcribed into RNA. The **operator**, or *o* site, consists of a stretch of about thirty base pairs in the DNA that are recognized by the repressor protein. Normally, *lac* repressor proteins are attached to the *o* site in the DNA, which keeps the *lac* operon switched off and prevents transcription of the *z*, *y*, and *a* structural genes (negative regulation). As mentioned in Chapter 8, the **promoter**, or *p* site, consists of a slightly longer sequence of base pairs in the DNA that is recognized by the enzyme RNA polymerase, which catalyzes the transcription of RNA molecules.

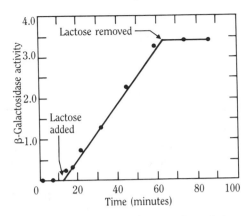

FIGURE 10-9 Lactose-induced expression of the *z* gene, which codes for β-galactosidase. Until lactose is added to cells, no β-galactosidase is synthesized. When the inducer lactose is added, β-galactosidase is synthesized as long as lactose is present. When lactose is removed, the repressor proteins become active and β-galactosidase synthesis stops.

When the *o* site is occupied by a repressor protein, RNA polymerase is physically blocked from attaching to the *p* site and thus transcription cannot occur.

Lactose molecules act like switches that turn on the *lac* operon. Lactose is referred to as an **inducer** because it induces the expression of genes that had been switched off. There are other inducers in addition to lactose; these small molecules are able to switch on the expression of genes by a variety of mechanisms. Lactose functions as an inducer by attaching to repressor proteins, thereby preventing them from attaching to the *lac* operator (Figure 10-8A). When the lactose molecules attach to the repressor, the physical shape of the repressor proteins is changed so that they cannot attach to the *o* site (Figure 10-8B).

Thus, the *lac* operon exists in either of two conditions in bacteria: switched off or switched on. Both conditions can be demonstrated experimentally by measuring the synthesis of β-galactosidase in a culture of bacteria in either the presence or absence of lactose (Figure 10-9).

Human *lac* Mutants

BOX 10-2

Mother's milk is an essential food for newborn infants. Milk is the principal food in an infant's diet for the first few months, even years, of its life. About 7 or 8 percent of human milk is the sugar lactose. Human babies are born with an active enzyme (lactase) in the small intestine that is essential for the digestion and metabolism of lactose, which it splits into the simpler sugars glucose and galactose. After the child reaches about four years of age, the gene that directs the synthesis of lactase is switched off, and a new gene, the adult lactase gene, is switched on. More correctly, it is switched on in *some* adults—and therein lies an interesting story about human nutrition.

Many human adults, probably a majority of the world's population, are intolerant of the sugar lactose. If these persons drink milk, they get diarrhea to varying degrees, and in the most severe cases they can dehydrate to the point of death. Only white American and European populations tend to have active adult lactase genes; most other human populations, including Africans, Chinese, and Thais, for example, are lactose-intolerant (see graph).

The ability to digest lactose is an inherited trait, although the particular human genes coding for the various lactose enzymes have not been identified. Most of the genetic and anthropological evidence indicates that millions of years ago all human adults were lactose-intolerant. Milk was probably not generally included in the adult diet until about 10,000 years ago, when goats, cattle, sheep, and reindeer were first domesticated. If a mutation occurred in human beings that permitted adults to drink large quantities of milk, those people with the mutant allele probably had an advantage over lactose-intolerant adults because they could adapt to a wider variety of environments and were therefore more likely to survive until reproductive age and leave more progeny. At least this seems to explain why some populations are almost entirely lactose-intolerant, whereas others are almost entirely lactose-tolerant.

Generally, lactose intolerance is not an all-or-nothing condition; adults who are lactose-intolerant can usually drink some milk without ill effects. And drinking even a little milk daily may serve to increase the kinds of bacteria in the gut that help break down lactose. Milk may not be the best food for everyone, but lactose-intolerant persons usually can eat yogurt and other fermented milk products because the bacteria used in the fermenting process have digested the lactose from the milk.

READING: Norman Kretchmer, "Lactose and Lactase," *Scientific American*. October, 1972.

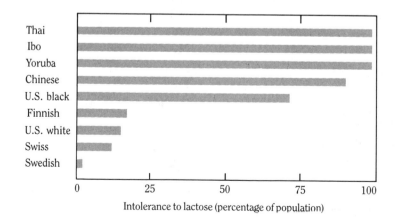

Intolerance to lactose (percentage of population)

If no lactose is present in the medium, β-galactosidase is not synthesized. When lactose is added, β-galactosidase synthesis begins and continues as long as lactose is present. If lactose is removed, β-galactosidase synthesis stops immediately.

The Monod-Jacob model for the regulation of the *lac* operon has been shown to be correct by many experiments and has provided a useful framework for formulating ideas about gene regulation in more-complex organisms, including human beings. Lactose is present in the milk of all mammals and must be broken down by an enzyme similar to β-galactosidase that is found in the stomachs of many (but not all) persons. The survival of human infants, of course, depends on their ability to metabolize the lactose present in mother's milk (See Box 10-2).

Lactose-operon Mutations

Because β-galactosidase is the easiest of the three enzymes to measure, changes in its activity are used to detect mutations that affect expression of the *z* gene and synthesis of β-galactosidase. Table 10-2 summarizes the effects of some of the regulatory mutations that affect synthesis of β-galactosidase in either the presence or absence of lactose. In fact, the changes in β-galactosidase activity of various bacterial regulatory mutants led Monod and Jacob to formulate their original model for regulation of the *lac* operon.

Just as mutations in the *z* gene can destroy the activity of β-galactosidase, mutations in any one of the *lac* operon regulatory sites (*i*, *p*, and *o*) can affect the expression of the *z*, *y*, and *a* structural genes. Mutations can affect the functioning of the *i* gene (which codes for the repressor protein) in a number of different ways. Some mutations destroy the repressor's shape and ability to attach to the operator site (*i*⁻

TABLE 10-2 Regulatory mutations affecting the synthesis of β-galactosidase in *E. coli*.

Bacterial genome	Amounts of β-galactosidase (arbitrary units)	
	Excess lactose	No lactose
$i^+p^+o^+z^+$ (wild-type)	1,000	1*
$i^+p^+o^+z^-$	0	0
$i^-p^+o^+z^+$	1,000	1,000
$i^sp^+o^+z^+$	10	1
$i^+p^+o^-z^+$	1,000	1,000
$i^+p^-o^+z^+$	1	1

*Even with no lactose present, wild-type bacteria have a low level of β-galactosidase because occasionally the operon escapes from repression and an mRNA molecule is transcribed. This level of enzyme activity is called the *basal level*. With lactose present, the enzyme activity increases 1,000 times. Intermediate levels of enzyme activity are also possible as a result of mutant genes. And mutations in the *z* gene itself can prevent any active β-galactosidase from being synthesized. The symbol *i*ˢ refers to a "superrepressor" that binds more tightly to the operater (*o*) than does normal repressor.

mutations), resulting in full expression of β-galactosidase activity. Other *i*-gene mutations change the repressor's sensitivity to lactose (*i*ˢ mutations); because lactose is unable to bind to the repressor in these bacteria, the repressors remain more or less permanently attached to the operator site. In *i*ˢ mutant bacteria, lactose is unable to induce the operon, and only a small amount ("escape synthesis") of β-galactosidase can be synthesized.

Mutations in the operator (*o* site) usually prevent the repressor from attaching to the DNA there. If the repressor is completely unable to attach, full β-galactosidase activity is observed; other operator mutations result in partial enzyme activity because of less-efficient attachment of the repressor to the operator. Mutations in the promoter (*p* site) affect attachment of the RNA polymerase and the rate of transcription of

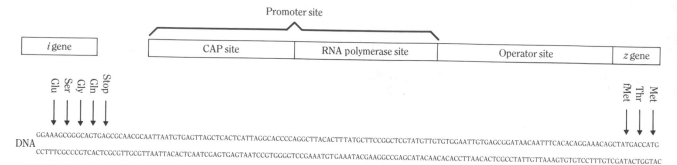

FIGURE 10-10 The base-pair sequence in the *lac* operon. The base pairs corresponding to each site are aligned below the promoter and operator sites. The last four codons of the *i* gene are shown, along with the four amino acids at the end of the repressor protein. The promoter site is recognized by two proteins: RNA polymerase and the catabolite activator protein, or CAP (with cyclic adenosine mono- phosphate attached to it). RNA polymerase cannot attach effectively to the promoter unless the CAP protein is also bound to the promoter. The repressor attaches to the operator. Keep in mind that DNA is not a straight rod but has a complex structure that is recognized by these regulatory proteins. fMet, Thr, and Met are the first three amino acids in β-galactosidase, the product of the *z* gene.

the *lac* operon. Promoter mutations may abolish synthesis of β-galactosidase, lower synthesis partially, or even raise the level of synthesis, depending on whether the attachment of the RNA polymerase to the promoter is decreased or increased.

The complete base sequence of the regulatory region of the *lac* operon DNA and the locations of the promoter and operator sites are shown in Figure 10-10. However, what was still not explained by the model is why lactose will not induce the *lac* operon and why synthesis of β-galactosidase does not occur as long as cells are supplied with glucose (refer to Figure 10-5). The final piece of the puzzle was solved with the discovery of another protein that is required to fully switch on the *z*, *y*, and *a* genes.

The Lactose Operon: Positive Regulation

The interaction of the repressor with the operator provides an all-or-nothing switch for the *lac* operon. When the repressor is attached to the operator, transcription is prevented. As mentioned earlier, this is called negative regulation.

If the repressor is removed from the DNA, the *lac* operon has the capacity for expression, but it still must be positively switched on by another protein called the **catabolite activator protein** (CAP). This is called positive regulation. The CAP protein has the potential for attaching to a section of the promoter, thereby allowing the RNA polymerase to begin transcription.

However, the attachment of CAP to the promoter depends on the presence of another small molecule (about the size of lactose) called **cyclic adenosine monophosphate (cAMP)**, whose chemical structure is shown in Figure 10-11. When cAMP and CAP combine, the complex attaches to the left-hand end of the promoter site and permits the RNA polymerase to begin transcription. If the CAP–cAMP complex is not attached to the promoter site, attachment of the RNA polymerase to the promoter is infrequent, and the operon cannot be induced efficiently even if lactose has inactivated the repressor.

High levels of glucose prevent synthesis of cAMP in bacteria; this now explains why lactose cannot act as an inducer when glucose is also present. The repressor proteins function negatively to regulate the on-off switch for expression of β-galactosidase; the CAP–cAMP complex functions positively to switch on the *lac* operon when glucose, a preferred energy source, is no longer available. This combination of negative and positive gene control is a complicated business. And, as you might guess, regulation of genes in human cells is even more complex than in bacteria. Although we know that human hemoglobin genes and lactase genes are switched on and off at different stages of development, we do not know as yet how these genetic switches are accomplished or what molecules are responsible.

One might regard the study of promoters, operators, and repressors in bacteria as something that would interest only a few molecular geneticists. Of what use is the *lac* operon except to bacteria trying to grow in a milk environment? Well, it turns out that the regulatory elements isolated from bacteria can be used in many ways in the genetic engineering of new products and even of new organisms (discussed in Chapter 12). For example, bacterial regulatory elements play a crucial role in what is now the routine manufacture of human insulin in bacteria.

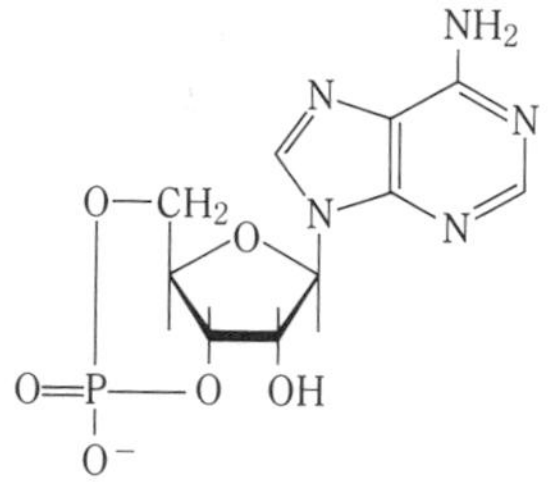

Gene Regulation in Animal Cells

Developing animal embryos produce differentiated cells that perform specialized functions. The different cellular activities are determined by the regulation of gene expression in the various kinds of differentiated cells (Figure 10-12). For example, in muscle cells, the genes that are expressed and the proteins that are synthesized must be different from the genes expressed and the proteins synthesized in skin cells or eye cells. Do these different types of cells contain different genes and amounts of DNA or

FIGURE 10-11 The chemical structure of an important small regulatory molecule, cyclic adenosine monophosphate (cAMP). When cAMP attaches to the CAP protein in bacterial DNA, the *lac* operon is switched on. Cyclic AMP is also found in human cells, where it regulates the expression of human genes.

is the specialization of organs merely due to differences in gene expression of cells with identical chromosomes? This important question was answered in 1964 by John B. Gurdon's experiments with the African toad *Xenopus laevis*.

Gurdon was able to demonstrate **totipotency**—the ability of a cell to proceed through all stages of development and ultimately produce a normal adult animal. He showed that a certain type of cell of an adult animal, in this case the African toad, contains all the genetic information needed for reconstructing the entire animal if the DNA is introduced into an unfertilized egg. Gurdon removed the nucleus from an intestinal cell of a tadpole and injected it into an unfertilized egg whose own nucleus had been deliberately destroyed by a beam of ultraviolet light. The egg with its transplanted nucleus grew through all stages of development until an adult toad was formed (Figure 10-13). Of all the eggs so injected, between about 1 and 2 percent developed into mature toads.

The remarkable thing about this experiment is not that most of the eggs failed to develop properly but rather that every egg that did, did so by means of genetic information in a somatic, not a sex, cell. From this result, Gurdon inferred that every somatic cell contains all the genetic

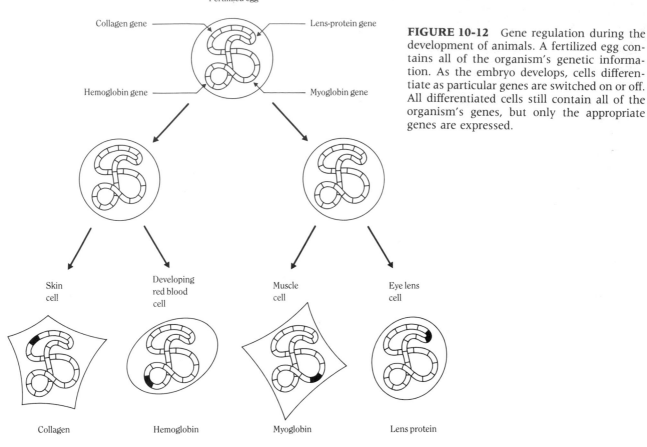

Fertilized egg

Collagen gene

Lens-protein gene

Hemoglobin gene

Myoglobin gene

Skin cell

Developing red blood cell

Muscle cell

Eye lens cell

Collagen

Hemoglobin

Myoglobin

Lens protein

FIGURE 10-12 Gene regulation during the development of animals. A fertilized egg contains all of the organism's genetic information. As the embryo develops, cells differentiate as particular genes are switched on or off. All differentiated cells still contain all of the organism's genes, but only the appropriate genes are expressed.

Unfertilized egg (two-nucleoli strain)

Tadpole (one-nucleolus strain)

UV radiation destroys nucleus

Epithelial cells from tadpole intestine

A nucleus is removed

Nucleus is injected into egg

No development

Blastula

In some cases development proceeds as far as an abnormal embryo

Blastula

Tadpole

Occasionally a mature adult toad develops

FIGURE 10-13 Totipotency in the African toad *Xenopus laevis*. This was the first animal to be cloned, proving that the genetic information in the nucleus of every somatic cell is complete and potentially capable of directing development of a complete animal. All the adult toad's cells have one nucleolus, proving that its genes came from the transplanted nucleus and not from the original nucleus present in the egg.

information required for development to proceed normally to the adult stage.

To show that all of the genes expressed in the egg did indeed come from the transplanted nucleus and not from the egg's own nucleus owing to failure of the ultraviolet light to destroy the egg nucleus, Gurdon used a genetically determined cellular marker. All intestinal cells from which nuclei were extracted came from a strain of toads having only one nucleolus (a substructure of the nucleus), whereas all of the eggs were taken from a strain of toads that had two nucleoli in all of their cells. If the cells in the resultant toads had only one nucleolus, it would prove that the transplanted nucleus had supplied all the genetic information, including that which determines the number of nucleoli. And, indeed, all of the toads that developed from the transplanted eggs did have only one nucleolus in their cells. Gurdon's experiments not only showed that all of a toad's genetic information was carried in a functional form in the animal's somatic cells, but also demonstrated that the cloning of certain animals is possible. (Human identical—that is, monozygous—twins are naturally occurring clones because they also have identical genes.)

Cloning frogs and toads is much easier than cloning mammals. A toad develops outside the mother's body, whereas a mammal develops only inside the female's uterus. In 1978, a book by David M. Rorvik, titled *In His Image: The Cloning of a Man*, was published that purported to be a true account of how a scientist, in return for a handsome payment by a millionaire, produced a son that was a clone of the millionaire by a technique similar to the one used to clone toads. A British geneticist whose name appeared in the book read the story, realized that it was a fabrication, and sued the author. In 1981, the U.S. District Court in Philadelphia ruled that the book was "a fraud and a hoax"

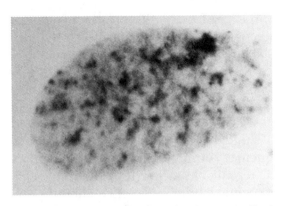

FIGURE 10-14 Photograph of a human cell showing a Barr body (the dark area at the upper right). An inactive X chromosome is observable as a Barr body in female cells. (Courtesy of A.J.R. de Jonge and D. Bootsma, Erasmus University, Rotterdam.)

and ordered the author and publisher to pay damages.

X-Chromosome Inactivation

In the late 1940s, two Canadian scientists, Murray L. Barr and Ewart G. Bertram, were staining the chromosomes in the nuclei of animal cells with dyes so that they could study them under the microscope. They repeatedly noticed a particularly dark staining spot in the nuclei of cells from female animals that did not appear in the nuclei of male cells. Such dark-stained nuclear spots, which are characteristic of female cells, became known as **Barr bodies** (Figure 10-14).

Some years later, it was noticed that somatic cells taken from men with Klinefelter syndrome (XXY) also have Barr bodies, whereas somatic cells from normal men do not. This fact, and the detection of several Barr bodies in abnormal cells with multiple X chromosomes, prompted a British geneticist, Mary Lyon, to suggest that each Barr body is, in reality, a condensed, inactive X chromosome. She reasoned that, because women have two X chromosomes and men have only one X chromosome, one of the X chromosomes in every female cell must be inactivated to compensate for gene-dosage effects. If both X chromosomes were active in female cells, she argued, there would be twice as much gene expression for the hundreds of genes on the X chromosome in women as there is in men. Because it was known that too little or too much of even a single gene product could have serious consequences during animal development, it seemed reasonable to her that X-chromosome inactivation is nature's way of equilibrating the expression of genes on the X chromosome.

The **Lyon hypothesis**, as these ideas came to be called, proposes that only one X chromosome is allowed to be active in any cell; additional X chromosomes must be inactivated so that none of the genes will be expressed. The Lyon hypothesis also predicts that inactivation is a random event in cells and that early in development either one of the two X chromosomes in each cell can be inactivated. Once the choice is made, however, all progeny cells produced by cell division will have the same X chromosome inactivated. The inactivated X-chromosome affects only gene expression; its movement in mitosis and cell division is normal. The Lyon hypothesis has been shown to be essentially correct, and each Barr body does represent an inactivated X chromosome.

X-Chromosome Mosaicism

One of the testable predictions of the Lyon hypothesis is that, if the inactivation of X chromosomes is random in cells, females should be **mosaic** (show different allele activities in different cells) for traits governed by genes on the X chromosome. For example, a woman who is heterozygous for genes on the X chromosomes that code for the synthesis of the enzyme glucose-6-phosphate dehydrogenase (G6PD) could have two forms of the enzyme in every

TABLE 10-3 The number of Barr bodies observed in human cells as a function of the number of X chromosomes in the cells.

Sex-chromosome constitution	Number of Barr bodies
XX	1
XY	0
XO	0
XXY	1
XXX	2
XYY	0
XXXY	2
XXXXY	3

cell, because she has two alleles for G6PD. However, if one X chromosome is randomly inactivated in each cell as her tissues develop, her cells should express either one allele or the other, but both forms of the enzyme should not be present in a single cell.

When individual cells from a woman heterozygous for G6PD are analyzed, one enzyme activity or the other is indeed found, and both forms of the enzyme are never found in the same cell. This experimental evidence strongly supports the idea that most, if not all, genes are inactivated on one of the two X chromosomes in female cells, although it is not known precisely when in female development the inactivation occurs. In animals that have an abnormal number of X chromosomes, the number of Barr bodies in the cell nuclei increases in proportion to the number of X chromosomes, indicating that only a single X chromosome remains active in any given cell (Table 10-3). (At this point you might be wondering why all women do not have Turner syndrome if only one X chromosome is active in a normal woman's cells. The explanation is that X-chromosome inactivation occurs after sex determination (discussed in Chapter

13) has been initiated. Exactly how or when the X chromosomes are randomly inactivated in cells after female sex development has been initiated is not understood.)

A practical application of the correspondence between Barr bodies and X-chromosome inactivation is its use in determining the sex of athletes in international competitions. Women atheletes are required to have a Barr body in their cells; male athletes should have none. If a female athlete's cells do not show a Barr body, she is not allowed to compete as a woman. Olympic athletes are now routinely examined for Barr bodies through the use of a test known as a *buccal smear*, in which cells scraped from the inner lining of the athlete's cheek are stained and examined. This test is deemed necessary because on various occasions sports officials have had good reason to believe that some athletes in women's events were actually males. Indeed, several Olympic athletes competing as women have been disqualified because their cells failed to show any Barr bodies, demonstrating that they had male genotypes.

Histones, Hormones, and Enhancers

Although relatively little is known about the specific molecules that regulate gene expression in animal cells, various classes of molecules are known to be associated with the DNA in chromosomes. These molecules affect the transcription of DNA, which suggests that they regulate gene expression. **Histones** are a class of proteins that are attached to chromosomes in large quantities. Histones are thought to be important in determining chromosomal structure, but relatively little is known about their actual functions in cells. About all that can be said with certainty is that five distinct kinds of histones can be isolated from animal cells and that they help determine the structure of chromosomes.

Somewhat more is known about **hormones**, a large class of organic molecules produced in one tissue or organ of a plant or animal and carried to another part of the organism, where they regulate gene expression and other physiological processes (see Table 10-4).

Earlier, the regulatory effects of insulin, a polypeptide hormone, were discussed at some length. Other hormones regulate gene expression in animal cells by interacting either with cellular proteins or with DNA. They are usually specific for certain tissues and are often found in cell nuclei. Hormones interact with chromosomal proteins and the resulting hormone-protein complex can activate a particular gene or group of genes in DNA.

Superficially, the interaction of a hormone with a protein seems analogous to the interaction of lactose with a repressor protein or of cAMP with the CAP protein. However, little is known about the mechanism of hormone-protein interaction, and even less about the regulation of gene expression by these complexes in human cells. All one can say with certainty is that regulatory molecules are found associated with chromosomes and regulate the expression of genes by mechanisms that remained to be discovered.

Another recently discovered regulatory element in DNA also influences the transcription of genes in animal cells. These are called **enhancers** — repeated sequences of from 50 to 200 nucleotides that are located on either side of a promoter (recall that this is the site where transcription of a gene begins). Enhancers exert their effects at a considerable distance — that is, they are located many thousands of bases away on either side of the gene whose regulation they affect (Figure 10-15A). Enhancer sequencers are so named because they increase the activity of a gene by as much as one thousand times.

The fact that an enhancer sequence can be on

TABLE 10-4   Some of the hormones that regulate gene expression in human cells, organs, and tissues.

Hormone	Site of synthesis	Site of action	Major physiological role
Thyrotropin (TSH)	Pituitary	Thyroid gland	Stimulates synthesis and secretion of thyroid hormone
Adrenocorticotropin (ACTH)	Pituitary	Adrenal medulla	Stimulates synthesis of adrenal steroids
Growth hormone (somatotropin)	Pituitary	Many tissues	Promotes protein synthesis and skeletal growth
Luteinizing hormone (LH)	Pituitary	Gonads	Stimulates synthesis of progesterone in ovaries and testosterone in testes; causes ovulation
Follicle-stimulating hormone (FSH)	Pituitary	Gonads	Stimulates growth of ovarian follicles and of Sertoli cells of testes
Prolactin	Pituitary	Breast	Stimulates synthesis of milk proteins and growth of breast
Insulin	Pancreas	Many tissues	Promotes transport of glucose and amino acids into certain cells; synthesis of fatty acids in adipose cells and in liver; glycolysis of glucose; protein synthesis
Glucagon	Pancreas	Liver	Stimulates glycogenolysis and gluconeogenesis in liver
Antidiuretic hormone (vasopressin)	Hypothalamus	Kidney	Prevents loss of water and NaCl and controls blood pressure
Oxytocin	Hypothalamus	Milk glands	Stimulates secretion of milk and uterine contractions
Parathyroid hormone	Parathyroid gland	Bone, kidney	Acts to increase Ca^{2+} in blood
Calcitonin	Thyroid C cells	Bone, kidney	Acts to decrease Ca^{2+} in blood
Cortisol	Adrenal glands	Liver and peripheral tissue	Stimulates glycogenolysis and synthesis of certain liver proteins; promotes protein breakdown in peripheral tissues
Aldosterone	Adrenal glands	Kidney	Aids in NaCl retention

TABLE 10-4 *continued*

Hormone	Site of synthesis	Site of action	Major physiological role
Estradiol	Ovary	Uterus, breast	Stimulates development of secondary female sex characteristics
Testosterone	Testis	Spermatogonia	Promotes sperm synthesis; stimulates development of secondary male sex characteristics
Progesterone	Ovary, placenta	Uterus, breast	Helps preserve pregnancy
Vitamin D_3	Skin	Bone, kidney	Stimulates Ca^{2+} transport by small intestine; acts on bone to increase CA^{2+} in blood
Epinephrine	Adrenal glands	Liver, heart, adipose tissue	Stimulates glycogenolysis and breakdown of fats; increases cardiac output
Norepinephrine	Adrenal glands	Heart, adipose tissue	Increases blood pressure; acts as neurotransmitter
Thyroid hormones	Thyroid gland	Many tissues	Increases respiration; required for nervous-system growth in fetus and young child; stimulates synthesis of certain enzymes

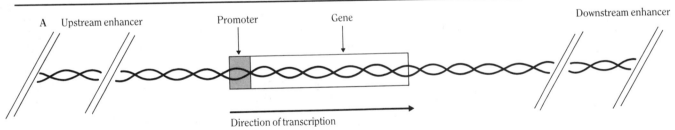

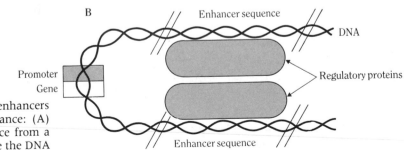

FIGURE 10-15 A hypothetical model of how enhancers might affect the expression of genes at a distance: (A) enhancer sequences are located at some distance from a gene on either side of it; (B) the enhancers cause the DNA to fold in such a way that regulatory proteins can bind and activate expression of the gene.

either side of a gene and yet affect its transcription is puzzling. In the *lac* operon, the binding of the cAMP-CAP complex right next to the promoter stimulates binding of the RNA polymerase, thereby increasing transcription. But how can we explain the action-at-a-distance of enhancers, particularly when they are located at any position relative to the gene that they regulate? One possible model is shown in Figure 10-15B. Regulatory proteins bind to enhancer sequences and thus cause the chromosome to loop. The DNA in that looped region is then more exposed to the RNA polymerases that bind to the promoter for the exposed gene. Thus, enhancers change the structure of the DNA in localized regions and enhance transcription of newly exposed genes.

The opposite of an enhancer is a **silencer**, which is a sequence of bases that acts to diminish or silence the activity of a gene. If a silencer sequence were to become associated with a regulatory protein, it could result in turning off the expression of a gene or group of genes. Experiments designed to understand how hormones, enhancers, silencers, and regulatory proteins interact to regulate gene transcription in human cells are just beginning. The ability to clone human genes and regulatory sequences of DNA in bacteria has greatly facilitated this research (discussed in Chapter 12).

Regulation of the expression of genes is complex enough in bacteria as we have seen, but it is even more so in human cells. Understanding how gene expression is regulated is the key to understanding how organisms develop and how cells differentiate and become specialized. Gene regulation is also the key to understanding the causes of human diseases, particularly those caused by such abnormal cell growth as cancer. How mutations cause changes in gene regulation and lead to the development of cancer is the topic of the next chapter.

Summary

The cells of a multicellular organism are differentiated—that is, they vary not only in size and shape, but also in function. Yet all the cells in that organism contain the same genetic information. Therefore, cellular differentiation is governed by the regulation of genes and their expression, as is the adaptation of differentiated cells to changing conditions: genes can be switched on and off.

Genes are regulated in at least five ways: (1) in transcription by promoters, operators, and enhancers; (2) in the processing of mRNA by the addition or removal of bases; (3) in the transport of processed mRNA into the cytoplasm; (4) in the translation of mRNA in the cytoplasm; and (5) in the processing of proteins (for example, human insulin).

The regulation of transcription is the primary means of regulating the expression of genes. Little is known about the mechanisms for switching human genes on and off, but bacterial research has yielded a great deal of information about gene regulation. In the *lac* operon of bacteria, repressor proteins (encoded by the *i* gene of the operon) exercise negative control by attaching to the operator site and blocking the attachment of RNA polymerase to the promoter site, thus preventing transcription. The presence of an inducer (in this case, lactose) can prevent the repressor from attaching to the operator, giving the operon the capacity for expression. However, switching it on requires positive regulation: a complex of catabolite activator protein and cyclic adenosine monophosphate attaches to the promoter and enables RNA polymerase to begin transcription.

Various classes of molecules are known to affect transcription in plant or animal cells or both. Histones are proteins that help determine chromosomal structure, but how they function to regulate gene expression is not known. Hormones (also proteins) regulate gene expression by interacting either with cellular proteins or with DNA. An enhancer is a sequence of nucleotides in DNA that increases the activity of a gene, though it may be thousands of bases away from either end of the gene. Conversely, a silencer is a sequence of nucleotides in DNA that decreases a gene's activity.

Key Words

Barr body The condensed X chromosome observed in the nucleus of somatic cells of women and other female mammals.

beta-galactosidase An enzyme that breaks apart lactose into the sugars glucose and galactose.

catabolite activator protein (CAP) The positive controlling element for glucose-sensitive operons in bacteria.

clone A group of genetically identical cells or organisms. Monozygous human twins are clones, as are bacteria in a single colony growing on a surface.

cyclic adenosine monophosphate (cAMP) An important small regulatory molecule in prokaryotic and eukaryotic cells.

diabetes A disease caused by abnormal insulin production and blood glucose levels.

enhancer A sequence of bases in the chromosomes of eukaryotic cells that increases the expression of a gene at some distant location from it.

feedback inhibition A mechanism by which the final product of a biosynthetic pathway (say, an amino acid) inhibits the activity of the first enzyme in that pathway.

hormones A large class of different molecules in plants and animals that regulate gene expression and other physiological processes.

inducer Any small molecule (for example, lactose) that is able to switch on the expression of one or more genes.

insulin A protein hormone produced in the pancreas that regulates the level of glucose in the blood.

Lyon hypothesis Only one X chromosome is expressed in each female cell; the other X chromosome is inactivated.

mosaicism A condition in which cell lines in an individual organism have different genetic expressions or chromosome constitutions; women are mosaic for heterozygous loci on the X chromosome.

operator (o site) A sequence of base pairs in DNA to which repressor proteins attach, preventing transcription of structural genes of the operon.

operon A segment of DNA consisting of two or more adjacent genes capable of synthesizing polypeptides, along with the regulatory sites that govern their expression. The genes in an operon are transcribed together into continuous mRNA molecules.

repressor A protein that "turns off" or prevents a gene or a group of genes in DNA from being expressed.

silencer A sequence of bases in the chromosomes of eukaryotic cells that decreases the expression of a gene at some distant location from it.

structural gene One that codes for the synthesis of an enzyme or structural protein.

totipotency The ability of a cell to proceed through all the stages of development, producing a normal adult organism. The nucleus from a single cell, in principle, contains all of the information for reconstructing the complete organism.

Additional Reading

Check, W. "The Regulation of Gene Expression." *Mosaic*, November-December, 1982.

Eisenstein, B.I. "Pathogenic Mechanisms of *Legionella pneumophilia* and *Escherichia coli*." *American Society of Microbiology (ASM) News*, 53, 621 (1987).

Gurdon, J.B. "Transplanted Nuclei and Cell Differentiation." *Scientific American*, December, 1968.

Inouye, M., and O. Pines. "Antisense RNA Regulation in Prokaryotes." *Trends in Genetics*, November, 1986.

Piper, P.W. "How Cells Respond and Adapt to Heat Stress Through Alteration in Gene Expression." *Scientific Progress*, 71, 531 (1987).

Ptashne, M., A.D. Johnson, and C.O. Pabo. "A Genetic Switch in a Bacterial Virus." *Scientific American*, November, 1982.

Rotter, J.I., and D.L. Rimoin. "The Genetics of Diabetes." *Hospital Practice*, May 15, 1987.

Study Questions

1 Why is it not safe for some people to drink milk?

2 List all of the different molecules that are used in *E. coli* to regulate the expression of the *lac* operon.

3 What are three different mechanisms used to regulate gene activity in eukaryotic cells?

4 What molecule is used to regulate the amount of sugar in the blood?

5 Can an entire adult animal be grown from just one somatic cell of another animal of the same species?

6 Are all genes active in all of the chromosomes of an organism?

7 What are the regulatory functions of inducer molecules, enhancers, and hormones?

Essay Topics

1 Compare the regulation of gene activity in bacteria and in animal cells.

2 Explain why it is necessary that an X chromosome be inactivated in female cells but not in male cells.

3 Look up additional information on hormones and discuss how a particular human hormone regulates the activity of cells and tissues.

11

Cancer
Somatic Mutations and Oncogenes

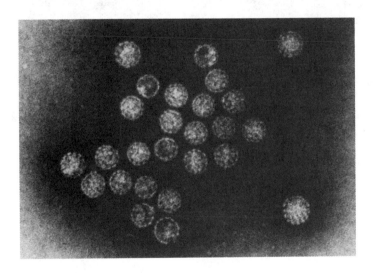

By now it is abundantly clear that the incidence of the common human cancers is determined by various controllable, external factors. This is surely the single most important fact to come out of all cancer research; for it means that cancer is a preventable disease.

JOHN CAIRNS, *biologist*

FTER HEART DISEASE, the second most-frequent cause of death in the United States is cancer. One of every three Americans will develop some form of cancer in his or her lifetime, and one in five Americans will die from it. Many people worry about getting cancer, especially if some close family member has died from it. Although treatments for various cancers have improved, most cancers are still incurable.

Certain people may have a genetic predisposition to cancer, and family members do share a significant number of their genes. But family members also share a common environment, and many environmental factors—ionizing radiation, exposure to mutagenic chemicals and viruses, nutritional excesses or deficiencies— contribute to increased risks of cancer formation.

The World Health Organization and the National Cancer Institute have estimated that as much as 90 percent of all forms of cancer is attributable to specific environmental factors (Figure 11-1). Because exposure to these environmental factors can, in principle, be controlled, most cancers could be prevented. Thus, the maxim "prevention is the best cure" is particularly appropriate for cancer. The mistaken but commonly held idea that cancer is a hered-

FIGURE 11-1 Approximate incidence of cancer in the United States attributed to specific environmental factors. Such factors account for approximately 80 percent of all causes. Reducing or eliminating exposure to environmental carcinogens would dramatically reduce the prevalence of cancer in the United States.

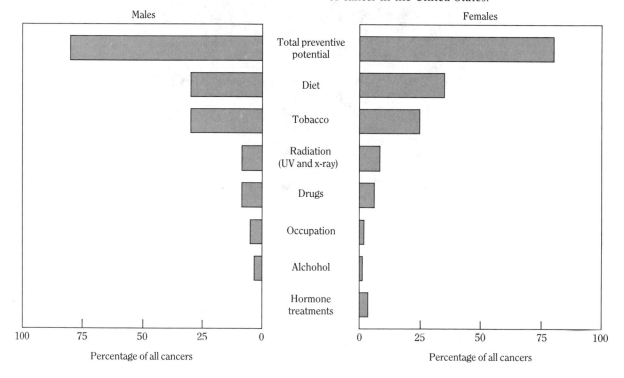

Lymphoma
About 5% of all cancers, the most common being Hodgkin disease
Cancer similar to leukemia; abnormal production of white blood cells
by the spleen and lymph system

Leukemia
About 4% of all cancers
Cancer of the organs and tissues (lymph glands, bone marrow,
and so forth) that form blood cells, causing an overproduction
of immature white blood cells

Carcinoma
From 80% to 90% of all cancers
Cancer originating from epithelial tissues, such as skin,
membranes around glands, nerves, breasts, linings of respiratory,
urinary, and gastrointestinal tracts

Sarcoma
About 2% of all cancers
Cancer originating from connective tissues, bone, muscles, fat,
and blood vessels

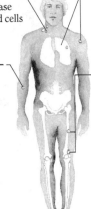

FIGURE 11-2   Classification and frequencies of the major kinds of cancer.

itary disease is due to a lack of understanding of how environmental agents cause mutations and initiate cancers. The mutations that lead to the formation of cancer cells arise in somatic cells and are not inherited. This chapter discusses how somatic-cell mutations are thought to trigger the initiation of cancerous diseases.

What Is Cancer?

Cancer can be defined as the unregulated growth and reproduction of cells in higher animals and in some plants. Often the unregulated growth is due to incomplete differentiation of particular cells. The term cancer is used as a general description for more than a hundred kinds of human diseases that are caused by the accumulation of abnormal cells into masses (**tumors**) in various organs of the body. Cancerous diseases can be subdivided into **leukemias** and **lymphomas**, cancer of white blood cells; **sarcomas**, cancer of bone and muscle cells; and **carcinomas**, cancer of skin and membrane cells (Figure 11-2). **Benign tumors**

are noncancerous growths; they can usually be removed surgically. **Malignant tumors**, on the other hand, may spread throughout the body, causing numerous other tumors to appear in vital organs and thus threatening the patient's life.

Cancer cells, unlike normal cells, have been changed by one or more somatic mutations in a way that causes them to lose the capacity to respond to the chemical signals that normally regulate cell growth and reproduction. The abnormal growth of cancer cells can be demonstrated *in vitro* by comparing their growth with the growth of normal cells in a liquid medium containing all of the necessary nutrients. Unlike bacteria, which grow freely in a liquid medium, most animal cells grow only if they attach to the surface of the container. For example, if normal human skin cells are placed in a plastic dish or flask and are covered with liquid nutrient medium, they will attach to the surface and grow until a layer just one cell in thickness uniformly covers the bottom of the dish. If cancer cells are allowed to grow under the same conditions, however, they will rapidly cover the surface and

TABLE 11-1   The hypothesis of cancer formation formulated by Theodor Boveri* in 1914 still appears to be correct.

1   Cancer cells arise from normal cells.
2   Tumors (masses of cancer cells) arise from a single cancer cell—that is, all of the cells in a tumor are genetically identical (are of clonal origin).
3   The cancer cells are chromosomally abnormal.
4   The effect of the chromosomal (genetic) abnormality is to cause cells to grow in an unlimited manner.
5   The factors that contribute to the chance of a cancer developing are (a) aging, (b) exposure to x-rays or chemicals, and (c) genetic predisposition.

*Boveri's 1914 book in German; *The Origin of Malignant Tumors* was published in English in 1929 by Williams and Wilkins, Baltimore, Maryland.

continue to grow and reproduce, eventually piling up on top of one another forming a thick, uneven layer of cells. Thus, a distinguishing characteristic of all cancer cells is their failure to correctly regulate cell growth.

There is now abundant experimental evidence to support the idea that most cancers begin with one or several genetic changes in a single somatic cell and that these mutations result from exposure to environmental agents. Mutations that may lead to formation of a cancer cell can be caused by:

1   Exposure to high-energy radiation.
2   Exposure to mutagenic chemicals.
3   Exposure to RNA or DNA tumor viruses.

One of the most-insightful scientists in the history of cancer research was Theodor Boveri, a German biologist whose experimental work was done in the early part of this century. In 1914, Boveri formulated a hypothesis explaining the origin and causes of cancer that has been confirmed in its basic tenets by more than a half century of research (Table 11-1). Despite intensive research efforts and the "war on cancer" that has been waged in the United States since

the 1960s, the overwhelming majority of human cancers remain incurable by any existing treatment (surgery, radiation, or chemotherapy). This fact does not reflect negatively on the quality of cancer research or on the great progress that has been made in understanding cancer. Rather it testifies to the enormous complexity of the genetic and cellular events that convert a normal cell into a cancer cell.

It is important to realize that not all mutations in animal cells result in cancer. Only if the mutations alter genes that regulate cell growth and division will the cell begin to grow uncontrollably and possibly form a tumor. And it seems that most cells, if they are able to become cancerous, must not only carry certain mutations but also be exposed to other substances that affect growth (Figure 11-3).

The smallest tumor detectable already contains many millions of cells; large tumors contain hundreds of millions of cancerous cells. Cells from malignant tumors are often dislodged from the original mass of cells and carried either by the lymphatic system or by the bloodstream to other parts of the body by a process called **metastasis**. As the cancer cells migrate through the body, they penetrate other organs and develop into new tumors. It is the

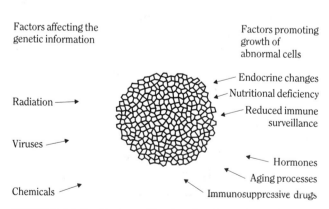

Factors affecting the genetic information

Factors promoting growth of abnormal cells

Radiation ⟶

Viruses ⟶

Chemicals ⟶

Endocrine changes
Nutritional deficiency
Reduced immune surveillance

Hormones
Aging processes
Immunosuppressive drugs

FIGURE 11-3   Factors affecting the formation of a cancer cell and a tumor.

process of metastasis that causes cancer to spread (metastasize) and eventually destroy vital body organs.

Cancer and Heredity

Human cancers, except for a few rare exceptions, are not classified as hereditary diseases because they are not passed on from one generation to the next. Nor are cancers contagious—that is, they are not transmitted from person to person by physical contact or other means. Moreover, cancers, by and large, are species specific; people do not contract cancer by eating plants with tumors, and they do not "catch" cancer from other animals. In the laboratory, cancers can be transferred between animals of the same species and sometimes between animals of different species or genera, but this process requires the inoculation of an animal with hundreds of thousands or millions of tumor cells, a few of which will become established in the test animal and eventually grow into new tumors.

Cells growing in malignant tumors have been genetically altered in one or more ways. The evidence is now overwhelming that somatic mutations of various kinds can initiate the conversion of a normal cell into a cancer cell, although the various mechanisms responsible for the genetic and cellular changes are only partly understood. As mentioned earlier, somatic-cell mutations are not passed on to progeny; so it is not possible that most cancers can be inherited.

The health of thousands of survivors of the atomic bombing of Hiroshima and Nagasaki and their progeny has been followed for more than forty years. Among those who were exposed to and survived large amounts of radiation, the incidence of cancer, particularly leukemia, increased. However, no increase has been observed in the incidence of cancer among their progeny. Long-term studies such as these and other epidemiological evidence support the view that most cancers are *not* inherited (see Box 11-1).

The strongest evidence that cancer is not inherited comes from human twin studies. As is well known, identical monozygotic twins are genetically identical, whereas fraternal (dizygotic) twins can be as genetically different as any other pair of siblings. If cancer were caused by defective genes that are inherited, then one would expect to find a higher concordance for cancer among identical twins than among fraternal twins. **Concordance** refers to the fraction of twin pairs in which both twins exhibit the trait. For example, among pairs of identical twins hair color is always the same (concordance = 100%), whereas among fraternal twins hair color may be the same in only half of the pairs (concordance = 50%).

Many pairs of twins, both identical and fraternal, served in the U.S. military in World War II. The health of about 15,000 twin pairs of both types was followed for many years after the war. No difference in cancer rates were observed when the concordance for cancer between identical and fraternal twin pairs was compared. This can only be interpreted to mean that cancer is not inherited.

It should be noted that there are some rare human diseases that *are* hereditary and that may *predispose* the affected person to develop some form of cancer, even though the cancer itself is not inherited. According to John J. Mulvihill of the National Cancer Institute, there might be as many as two hundred rare hereditary human diseases that predispose their victims toward the development of some form of cancer, but taken all together they probably account for less than 5 percent of all human cancers.

Epidemiological Studies Show That Most Cancers Are Environmental in Origin

BOX 11-1

Epidemiology is that branch of medical science that investigates the frequency and geographical distribution of diseases and then tries to identify factors that cause them. Various kinds of epidemiological studies support the conclusion that most cancers are caused by environmental agents and are not hereditary. Such studies have shown that the frequency with which different cancers occur varies greatly from country to country. For example, cancer of the mouth makes up about 3 percent of all cancers in the United States, but in parts of Asia where people chew tobacco and betel nuts it can account for as much as 35 percent of all cancers diagnosed. Breast cancer varies by a factor of six from nation to nation, Japan having the lowest incidence and Holland the highest.

Comparison of the incidence of various cancers among Japanese citizens, American citizens, and Japanese who have emigrated to the United States shows quite convincingly that most cancers are attributable to environmental factors. The accompanying table shows that death rates from various cancers in Japan are dramatically different from those for Japanese who have moved to the United States. The incidence of cancer among Japanese Americans born and raised in the United States tends toward the frequencies found among U.S. whites, indicating that genetic differences are not significant in the development of these cancers.

Epidemiological studies have been carried out among workers in different industries as well as among people in different countries. In some industries, workers are found who suffer from cancers that virtually never arise in the general population. For example,

mesothelioma is a rare form of lung cancer that occurs only among workers engaged in the mining and manufacturing of asbestos. It is now known that exposure to microscopic asbestos fibers from any asbestos-containing material increases the risk of this kind of lung cancer (and of other respiratory diseases as well.) Vinyl chloride (VC) is an important industrial chemical that is the basic ingredient in a wide range of polyvinylchloride (PVC) plastics used in phonograph records, food containers, electrical insulation, garden hoses, plastic pipe, and many other products. The finished plastic products themselves do not cause cancer, but the starting material does. By the 1970s, it became apparent from numerous epidemiological and laboratory studies that unpolymerized vinyl chloride, which escapes from PVC products, is a carcinogen—it does increase the risk of developing cancer, particularly of *angiosarcoma*, an unusual form of liver cancer.

Two conclusions consistently emerge from epidemiological studies of cancer: (1) most cancers are environmentally caused, not hereditary, diseases; (2) most cancers are preventable.

| Cancer location | *Relative cancer mortality rates* | | |
	Japanese	Offspring of migrants	U.S. whites
Stomach	100	38	17
Colon	100	288	489
Pancreas	100	167	274
Lung	100	166	316
Leukemia	100	146	265

SOURCE: W. Haenszel and M. Kurihara, "Studies of Japanese Migrants. I. Mortality from Cancer and Other Diseases among Japanese in the United States." *Journal of the National Cancer Institute* 40 (1968), 43–68.

READING: John Cairns, "The Treatment of Diseases and the War Against Cancer," *Scientific American*, November, 1985.

Saying that people are genetically predisposed to getting cancer often conveys a false idea about the relation of heredity and cancer. Perhaps the relation can be clarified with the example of fair skin. People with fair skin are more likely to get sunburn or skin cancer. Fair skin is inherited, but sunburn and skin cancer are caused by overexposure to sunlight, which can be avoided if people realize their increased susceptibility.

Two extremely rare inherited human diseases, *xeroderma pigmentosum* and *ataxia telangiectasia*, lend biochemical support to the idea that cancer begins with genetic changes and that these changes result in the synthesis of defective enzymes that may affect cell growth. The skin cells of persons suffering from **xeroderma** contain a mutant gene and consequently their cells lack the enzymes required to repair DNA that has been damaged by ultraviolet (UV) light. People with this hereditary disease who are exposed to even slight amounts of sunlight invariably develop skin tumors because the UV-damaged DNA in their skin cells cannot be repaired.

People suffering from **ataxia** are particularly sensitive to x-rays. In this disease, also due to an inherited mutation, the person's cells lack enzymes that normally repair x-ray–damaged DNA. As a consequence, the person is predisposed to developing leukemia because new blood cells, which are continually produced, may be genetically damaged. Thus, in both of these rare inherited diseases, a genetic change in a somatic cell is *indirectly* responsible for the eventual cancer formation.

The fact that even genetically normal people vary in their capacity to repair DNA damaged by various environmental agents means that their cells may be more or less susceptible to mutations and to the eventual formation of cancer. This fact has given rise to the controversial idea of screening for cancer-resistant workers, particularly in occupations in which people are exposed to radiation or mutagenic chemicals (see Box 11-2).

Agents That Cause Cancer

Cells possess many different mechanisms and enzymes for repairing DNA that has been altered by a variety of mutagenic agents. Exposure of DNA to certain chemicals can produce changes that are quite different from those caused by radiation. Different enzymes are needed to repair the various kinds of chemical alterations in DNA. Viruses too can damage DNA after cells have been infected. As we will see later, DNA damage caused by viruses may be harder or impossible to correct.

Radiation Is a Carcinogen

The significantly higher than normal incidence of leukemia and other forms of cancer among survivors of the Hiroshima and Nagasaki atomic-bomb blasts leaves no doubt that atomic radiation is a potent **carcinogen**—that is, an agent that can cause cancer. More recently in our own country, it has been demonstrated that the number of leukemia deaths among children born in southern Utah (who were exposed to fallout from aboveground atomic-bomb tests in Nevada) was two to three times that among children born in the years before and after the atomic tests (Figure 11-4). In a landmark legal decision, a federal court ruled in 1984 that the U.S. government was negligent in conducting those atomic-bomb tests in Nevada in the 1950s that released radioactive material into the atmosphere. The court ruled that the families who were exposed to the radioactivity and whose members died as a result of the exposure are entitled to compensation.

Screening for "Cancer-Resistant" Workers

BOX 11-2

Certain occupations entail more risk of developing cancer than others. In some instances, precautions can be taken to reduce this risk, such as shielding x-ray technicians or limiting the exposure of nuclear workers to radioactivity. Industry and government efforts have reduced the exposure of workers to proven carcinogens such as ionizing radiation, asbestos, vinyl chloride, and other agents. Yet there are limits to which this exposure can be controlled by industry or regulated by government agencies.

Are there other solutions to the problem of worker exposure to cancer-causing substances? The answer is yes—but some people have questioned whether the other solutions are ethical or fair. Human beings differ from one another biochemically because they carry different alleles for essential genes and therefore synthesize different enzymes. Even the levels of normal enzymes vary from person to person. So it is quite reasonable to expect that enzymes that repair damage to DNA will be present in greater amounts in some persons than in others. Simple biochemical tests could be used by employers to identify applicants who are more resistant to the effects of the carcinogens that they would be exposed to in a particular job. Such genetic screening by industries could also be used to eliminate "carcinogen-sensitive" workers from jobs.

Manufacturers are aware of the monetary benefits of identifying "cancer-resistant" workers. For one thing, such workers might be able to tolerate higher levels of harmful substances. For another, fewer workers might bring lawsuits claiming that their diseases were caused by occupational exposure to chemicals or claiming negligence on the part of their employer in informing them of the dangers.

The questions now are: Is it ethical or fair for workers to be denied certain jobs simply because of the biochemical and genetic constitution that they inherited? And should exposure of some workers to toxic or cancer-causing substances be permitted just because they are more resistant to biological damage than others?

READINGS: Constance Holden, "Looking at Genes in the Workplace," *Science*. July 23, 1982. Mary P. Lavine, "Industrial Screening Programs for Workers," *Environment*. June, 1982.

FIGURE 11-4 A nuclear-bomb test. The mushroom is 10 miles high and 100 miles wide. (Courtesy of the U.S. Department of Energy.)

Some kinds of electromagnetic radiation such as ultraviolet light or x-rays also act as carcinogens. These kinds of radiation (ionizing) have sufficient energy to knock electrons out of their orbits around atoms and produce mutations in the DNA of cells.

How UV light damages DNA and how this damage is repaired in normal cells is shown in Figure 11-5. The immediate damage to DNA caused by UV light is the formation of *thymine dimers*, the covalent linking together of adjacent thymine bases in a strand of DNA (Figure 11-5A). The DNA-replication machinery cannot replicate thymine dimers, and a mutation will arise at the site of the dimer unless it is removed by DNA-repair enzymes before or during replication of the DNA.

The first step in the removal of the thymine dimer is to break the DNA strand at or near the site of UV-induced damage. This is the function of an endonuclease, an enzyme that breaks the sugar-phosphate backbone of DNA at specific locations (Figure 11-5B). A free end of the DNA strand is then recognized by an exonuclease, an enzyme that removes the damaged thymine bases plus a few others, creating a gap in the DNA strand (Figure 11-5C). This gap is filled in by DNA polymerase I, an enzyme that repairs gaps by using the opposite strand of DNA as a template and inserting the correct bases (Figure 11-5D). Finally, the sugar-phosphate backbone of the DNA strand is again sealed by the enzyme DNA ligase (Figure 11-5E). Repair of radiation-damaged DNA is a continual process in bacteria, as well as in human cells. Loss of any one of the repair enzymes by mutation can be lethal to bacteria that are exposed to UV light, demonstrating that the repair of radiation damage is essential to the survival of cells.

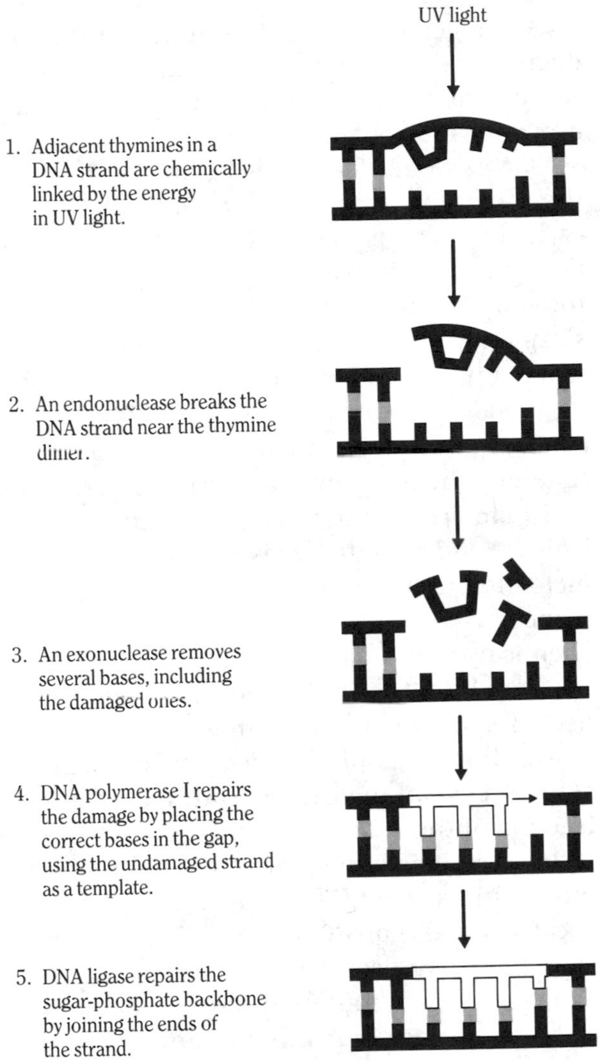

1. Adjacent thymines in a DNA strand are chemically linked by the energy in UV light.

2. An endonuclease breaks the DNA strand near the thymine dimer.

3. An exonuclease removes several bases, including the damaged ones.

4. DNA polymerase I repairs the damage by placing the correct bases in the gap, using the undamaged strand as a template.

5. DNA ligase repairs the sugar-phosphate backbone by joining the ends of the strand.

FIGURE 11-5 Repair of DNA damaged by exposure to ultraviolet (UV) light, which causes the formation of thymine dimers.

Some Chemicals Are Carcinogens

Many chemicals can penetrate cells and interact with the DNA molecules within them, thereby increasing the frequency of mutation. Of the many thousands of chemicals that can cause mutations, some are highly mutagenic, whereas others are only slightly mutagenic. Therefore, it is important to know not only which chemicals are mutagenic and carcinogenic, but also the degree to which they increase mutations—and the extent to which human beings and other animals are exposed to them.

Our modern life-style is intimately linked to the use of manufactured chemicals. Almost everybody uses paints, plastics, pesticides, or pharmaceuticals every day. We depend on styrofoam cups for carrying hot coffee, we sit and sleep on polyurethane foam cushions, we wear nylon, rayon, and polyester clothes, we use plastic combs, cups, and pens. And all of us are exposed to varying amounts of the thousands of different chemicals that are used in agriculture and industry. Furthermore, hundreds of chemicals are added to the foods and beverages in our diet. Although most of the chemicals and substances that we use are perfectly safe, some chemicals are carcinogens.

Since World War II, when the boom in the use of manufactured chemicals really began, many thousands of chemicals have come into common use in the United States. Only a small fraction of these chemicals has been adequately tested for their mutagenic potential (ability to induce mutations). This is a cause for concern because more than 80 percent of the few chemicals that have been tested and that have been shown to cause mutations in microorganisms also cause cancer when fed to animals.

Bacteria have become particularly useful in testing chemicals for their mutagenic potential. Because millions, even billions, of microorganisms can be tested for the occurrence of mutations, accurate mutation rates are easily measured for a wide variety of mutagenic chemicals. Moreover, because DNA is chemically identical in all organisms, any chemical that produces mutations in bacteria is also likely to cause mutations in human cells. For these reasons, microorganisms are usually chosen for the initial screening of potential carcinogens.

A

Histidine-requiring (*his⁻*) mutants of *Salmonella* in glucose-mineral-histidine medium

B   Bacteria spread on Petri plates with only minerals and glucose

Colonies of revertant (*his⁺*) bacteria

FIGURE 11-6   Detection of revertant (*his⁺*) bacteria. A. Histidine-requiring mutants (*his⁻*) of *Salmonella* will grow in a glucose-mineral medium if histidine is added. B. Revertants of the mutant bacteria can be selected by plating large numbers of bacteria on a solid medium that lacks histidine. The histidine-requiring mutants cannot grow there, but bacteria that have undergone a mutation that allows them to synthesize histidine will grow and form a colony. Some of these colonies are true reversions: the wild-type gene has been restored in the DNA. Other revertant colonies (pseudorevertants) result from other kinds of mutations.

The Ames Test   Wild-type *Salmonella* bacteria, like *E. coli*, can grow in a liquid medium containing certain minerals and the sugar glucose. From this medium, the wild-type *Salmonella* cells are able to synthesize all the other molecules that they need for growth—amino acids, vitamins, bases, and so forth. For example, histidine is one of the twenty amino acids that wild-type *Salmonella* synthesizes. A mutation arising in any one of the nine genes that code for the enzymes required for histidine biosynthesis creates a histidine mutant, a bacterium that can no longer synthesize histidine. Histidine-requiring *Salmonella* mutants grow only if histidine is added to the medium. However, if a reverse mutation occurs in the DNA that restores the enzyme activity, these bacterial revertants are again able to grow without added histidine (Figure 11-6). Some of the revertant bacteria are true revertants—that is, the wild-type base sequence has been restored in the DNA. Other revertant bacteria are pseudorevertants and grow without histidine because other mutations compensate for the defect.

In the early 1970s, Bruce N. Ames, professor of biochemistry at the University of California, Berkeley, devised a simple test for chemical mutagens using histidine-requiring mutants of *Salmonella* (Figure 11-7). His hope at the time, which has been borne out to a considerable degree, was that the testing of a mutagenic chemical by bacteria would give an indication of the chemical's carcinogenic potential. In fact, about 85 percent of the chemicals that mutate DNA in *Salmonella* are also able to induce cancers in laboratory test animals, making the Ames test both useful and reliable for its intended purpose. Since the development of the Ames test, this simple bacterial system has been used to screen thousands of potentially dangerous chemicals (see Box 11-3).

In the standard Ames test, four different strains of histidine-requiring *Salmonella* are grown in the presence of varying amounts of the suspected chemical mutagen. Each *Salmonella* strain has a particular kind of mutation (frameshift or missense) in one of the genes in the histidine biosynthetic pathway. None of these

A 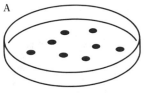 B

Control plate   Chemical test plate

FIGURE 11-7   The Ames test for the mutagenic potency of chemicals.
A. Millions of histidine-requiring mutant bacteria are spread on a medium lacking the amino acid histidine. A few colonies appear because spontaneous mutations produce a few revertant or pseudorevertant bacteria.
B. Approximately the same number of histidine-requiring mutant bacteria are spread on a medium lacking histidine but also containing a suspected chemical mutagen. If more revertant colonies appear on the chemical test plate than on the control plate, the chemical is considered a mutagen. The extent of the increase in revertant colony numbers indicates the chemical's mutagenic potency; some chemicals are millions of times as mutagenic as others.

four mutant strains is able to grow unless the medium is supplemented with histidine. However, in the presence of a mutagenic chemical, both true revertants and pseudorevertants are generated and grow without added histidine. The increase in the number of histidine revertant bacteria after exposure to the chemical indicates the mutagenic potency of the chemical (Table 11-2).

In addition to the Ames test, numerous other biological systems for evaluating mutagens have been devised that use "guinea pigs" ranging from *Drosophila* to yeast to human cells grown *in vitro*. The identification of chemical mutagens is now a relatively easy task; however, the social, political, and economic decisions regarding regulation of the use of chemicals are invariably controversial. For example, when the Environmental Protection Agency (EPA) bans the use of an industrial chemical or pesticide, under existing law it must pay for all the existing stocks of the chemical and for its disposal. Currently, EPA is paying hundreds of millions of dollars to compensate manufactur-

Chemical Mutagens Are Everywhere

BOX 11-3

Since the development of the Ames test and other biological tests for measuring the mutagenic potency of chemicals, it has become increasingly evident that we live in an environment full of mutagens. Millions of pounds of mutagenic chemicals have been widely disseminated in the soil, water, and air in the United States in the past thirty years, and in years to come people will still be exposed to them, like it or not.

Exposure to mutagens cannot be avoided entirely, but there are many substances that *can* be avoided. For example, tris-BP (2,3-dibromopropyl phosphate) is a flame-retardant that was routinely added to all children's sleepwear until its mutagenic activity was discovered. Children absorbed the chemical through their skin while they were asleep, and the chemical was detectable in urine samples collected from them in the morning. About 50 million children were exposed to this chemical before it was finally banned by the Federal Drug Administration (FDA) in 1977.

About 20 million men and women dye their hair in this country. Many hair dyes have been found to be mutagenic through Ames testing. In a study reported in 1975, 150 out of 169 commercial hair-dye preparations were mutagenic to some degree. Although many of the hair dyes are not as strongly mutagenic as tris-BP and certain other chemicals, they are used in large quantities by millions of people for many years. The mutagenic chemicals in hair dyes are absorbed through the skin of the scalp, and once they enter the bloodstream they are carried to all parts of the body.

A final word of caution. Cigarette smoke also causes mutations in bacteria and is a proven carcinogen in human beings (see the accompanying figure). If cigarettes were being introduced for the first time today, they would have to be banned by the FDA to comply with existing laws. Why do you think the sale of cigarettes is still permitted even though the sale of other proven carcinogenic substances is prohibited?

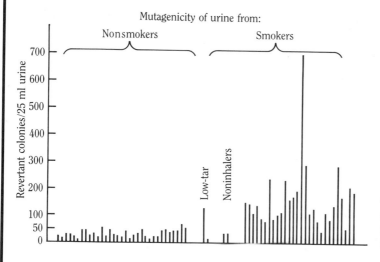

The carcinogenic chemicals in cigarette smoke can be detected in urine. The Ames test shows that smokers' urine contains more mutagens than urine from nonsmokers.

ers for banned chemicals and to store the hazardous substances while safe disposal methods are sought. There is good reason to conclude that EPA has not banned some carcinogenic pesticides simply because it cannot afford to do so.

Exposure to Carcinogenic Chemicals Evidence that certain chemicals cause cancer in human beings is found in the increased incidence of certain types of cancer among workers in particular industries. Ever since an English physician, Sir Percival Pott, first observed an unusually high incidence of scrotal cancer among London chimneysweeps, workers in many industries have been found to have an increased risk of developing certain kinds of cancers because of their occupational exposure to certain substances (Table 11-3). For instance,

TABLE 11-2 Range of sensitivity of mutagen potency that can be detected by the Ames test, which uses bacteria.

Chemical	Revertant ratio*
1,2-Epoxybutane	1 : 1
Benzyl chloride	3 : 1
Methyl methanesulfonate	105 : 1
2-Naphthylamine	1,400 : 1
2-Acetylaminofluorene	18,000 : 1
Aflatoxin B_1	1,200,000 : 1
Furylfuramide	3,500,000 : 1

SOURCE: J. McCann and B.N. Ames in H.H. Hiatt, J.D. Watson, and J.A. Winsten, eds., *Origins of Human Cancer*, Cold Spring Harbor Laboratory, 1977.

*The revertant ratio refers to the number of revertant or pseudo-revertant *Salmonella* colonies able to grow in the absence of the amino acid histidine either with or without the chemical mutagen. For example, the chemical furylfuramide is 3.5 million times as mutagenic as 1,2-epoxybutane.

TABLE 11-3 Occupations and various cancers.

Occupation	Carcinogen identified	Location of cancer
Chimneysweep; manufacturer of coal gas	Polycylic hydrocarbons in soot, tar, oil	Scrotum; skin; bronchus
Chemical worker; rubber worker; manufacterer of coal gas	2-Naphthylamine; 1-naphthylamine	Bladder
Chemical worker	Benzidine; 4-aminobiphenyl	Bladder
Asbestos worker; shipyard and insulation workers	Asbestos	Bronchus; peritoneum
Sheep dip manufacter; gold miner; some vineyard workers and ore smelters	Arsenic	Skin; bronchus
Maker of ion-exchange resins	Bis (chloromethyl) ether	Bronchus
Workers with glues, varnishes, etc.	Benzene	Bone marrow (leukemia)
Poison-gas maker	Mustard gas	Bronchus, larynx, nasal sinuses
PVC manufacturer	Vinyl chloride	Liver
Chromate manufacturer	Chrome ores	Bronchus
Nickel refiner	Nickel ore	Bronchus; nasal sinuses
Isopropylene manufacturer	Isopropyl oil	Nasal sinuses

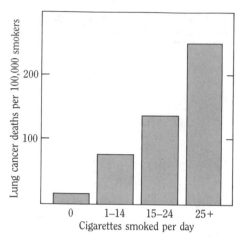

FIGURE 11-8 Number of deaths from lung cancer in relation to number of cigarettes smoked per day.

the incidence of lung cancer is high among workers in asbestos industries. Exposure to vinyl chloride used in the manufacture of (PVC) plastics causes a type of liver cancer among workers that is rarely observed among the general public.

Exposure to carcinogenic chemicals is not restricted to industrial workers. Deaths from lung cancer are almost directy proportional to the number of cigarettes smoked (Figure 11-8). As pointed out by numerous authorities, if people stopped smoking cigarettes completely about one-third of all human cancers, not only lung cancer, would be prevented. In fact, recent statistics indicate that from 20 to 25 percent of all deaths in the United States can be attributed to smoking.

Because many forms of human cancer are caused by environmental agents, they are, in principle, preventable. The solution to the cancer problem in modern industrial societies lies not in more-effective cures (although cures are desirable) but in preventing cancer from developing in the first place. Prevention of disease is not a new or radical idea. Most lethal human infectious diseases — tuberculosis, smallpox, typhoid fever, plague — have been virtually eradicated not be curing people but by instituting health practices that prevent these diseases — vaccination, sanitation, and improved nutrition.

Oncogenic Viruses Are Carcinogens

The genetic information in animal viruses is carried by either RNA or DNA molecules. Either class of virus may cause the development of malignant tumors in animals, in which case the virus is referred to as an **oncogenic** (cancer-causing) **virus**. (Not all viruses are oncogenic.) In 1911, Peyton Rous, working at the Rockefeller Institute (now Rockefeller University) in New York City, showed that tumors could be induced in chickens by injecting them with an RNA virus, later named the *Rous sarcoma virus* (RSV) after its discoverer. Since then, many other oncogenic RNA and DNA viruses have been isolated from animal tumors that again produce tumors when the purified viruses are injected into susceptible animals.

In nature, oncogenic viruses tend to be species specific; that is, viruses that cause cancer in one species generally are harmless in another. For example, feline leukemia virus, which is quite commonly found in household cats, has never been shown to cause leukemia or tumors in other animals, including human beings. Moreover, many cats whose saliva and blood contain great numbers of the feline leukemia virus have no symptoms of leukemia, indicating that oncogenic viruses by themselves do not necessarily produce cancer. In fact, many infected cats live a normal life-span and die of causes other than cancer.

Oncogenic viruses, unlike other animal viruses, usually do not destroy the cells that they infect. Rather, the cells continue to reproduce. When an animal cell is infected by an oncogenic

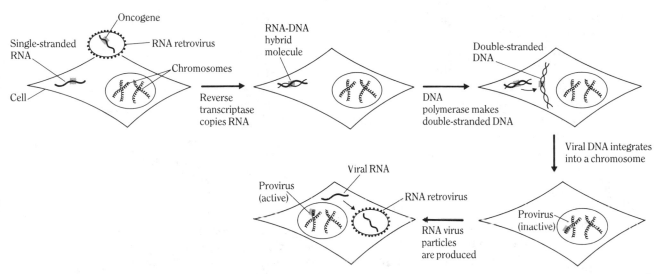

FIGURE 11-9 Infection of an animal cell by an onco-genic RNA retrovirus. The infectious RNA is first copied by the enzyme reverse transcriptase into a complementary strand of DNA and then makes double-stranded viral DNA molecules The viral DNA copies may become integrated into a cell's chromosomes and are passed on to succeeding cell generations as proviruses. New RNA virus particles can be synthesized from proviruses.

DNA virus, the viral DNA is replicated and newly synthesized virus particles are released from the cell, whereupon they may infect other nearby cells and repeat the process. Or, if the invader is an oncogenic RNA retrovirus, the genetic information of the virus is converted into a DNA copy, after which additional RNA viruses can be transcribed from the DNA and infect other cells.

In order for the genetic information of an oncogenic RNA retrovirus to become integrated into a chromosome, the single-stranded viral RNA molecule must first be converted into a DNA molecule (hence the name <u>retro</u>virus: in-formation flows from RNA to DNA instead of the usual flow from DNA to RNA). This conver-sion is carried out by two enzymes. *Reverse tran-scriptase* is responsible for synthesizing a single strand of DNA that is complementary to the infectious strand of RNA, resulting in a hybrid RNA-DNA molecule (Figure 11-9). Next, the

enzyme synthesizes a double-stranded viral DNA. When an animal cell is infected by an oncogenic RNA virus, conversion of viral RNA into viral DNA and the integration of the viral DNA into the cell's chromosome are essential for continued RNA virus reproduction. Once the viral DNA has been integrated into a chromo-some, it is called a **provirus**. The cell may or may not continue to synthesize RNA viruses, although in either case it retains its capacity to do so.

Viral-Oncogene Hypothesis Most biolo-gists believe that viruses must have evolved from cells, possibly in situations in which a few genes became separated from chromosomes and somehow became encapsulated by proteins. One evolutionary consequence of such a process is that it would promote the transfer of genetic information among organisms by accelerating the spread of mutations from their point of origin in the population. In many instances, the acquisition of new genes or the rearrangement of genes caused by the insertion of a provirus would be advantageous; on other occasions, it might change the regulation of expression of genes, leading to uncontrolled cellular growth. If, on the average, the presence of the virus were to increase the survival and reproduction of organisms, the maintenance of such viruses in the population would be ensured.

The **oncogene hypothesis** proposes that millions of years ago viral genes became integrated into the chromosomes of animal reproductive cells and that the viral DNA was passed on, along with normal genes, generation after generation. As long as the viral genes are not transcribed into mRNA or translated into proteins, the provirus DNA causes an organism no harm. However, if ionizing radiation or some other mutagenic agent should damage the DNA and trigger the synthesis of new viruses, cells containing those viruses may begin to grow in an unregulated manner and eventually develop into a tumor.

It is now known that the *src* (pronounced "sark") gene carried by the Rous sarcoma virus is the gene responsible for changing a normal cell into a cancer cell in chickens. The RNA of this oncogenic virus contains only four genes,

one of which is the *src* gene (Figure 11-10). When the proviral DNA is not expressed, the cells function normally. But if the *src* gene becomes turned on, many of the cell's properties change, particularly the capacity to form tumors.

Many other oncogenic RNA viruses have been isolated from tumors in various animals. Each one of these viruses carries a different oncogene that is believed to have been acquired originally from a gene in the cells themselves.

Protooncogenes   Until quite recently, the prevailing view was that the existence of oncogenes in cells was due to infection by viruses and was an abnormal condition. However, in the past few years it has been discovered that both animal and human cells contain genes whose base sequences closely resemble the base sequences in the viral oncogenes. These normal cellular genes are called **protooncogenes**. It is believed that protooncogenes are normally expressed and regulated in cells and therefore cause the organism no harm. However, if protooncogenes become overactive because of some form of chromosomal rearrangement or mutation in the DNA, then a normal cell may become transformed into a cancer cell. Indeed, activation of newly identified cellular protooncogenes has been related to chromosomal translocations observed in a variety of human cancers (Table 11-4).

The relation between cellular oncogenes and their close relatives in viruses is still being worked out. The oncogenes in the viruses are not essential for the growth and reproduction of the small RNA viruses. When the oncogenes are removed from the viruses by techniques of genetic engineering (explained in the next chapter), the viruses can still infect and reproduce in cells, but they do not cause tumors. The evidence so far suggests that viruses pick up oncogenes (and other genes) from the cells' own

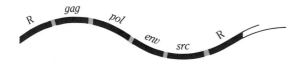

FIGURE 11-10   The single-stranded RNA molecule of the Rous sarcoma virus contains four genes: *gag* codes for a protein in the core of the virus; *pol* codes for reverse transcriptase; *env* codes for a protein that encapsulates the RNA; and *src* codes for a protein that transforms normal cells into cancer cells. The R (terminal repeat) stands for an identical sequence of bases at each end of the RNA; these sequences are essential for replication.

TABLE 11-4 Oncogenes in human cells are identified with some human cancers. An oncogene has a three-letter designation. For each specific cancer indicated below, the oncogene is observed to undergo a translocation (t) from its original chromosomal location (the first number) to another chromosome (the second number). It is thought that the translocation activates the expression of the oncogene, which converts a normal cell into a cancer cell; however, this is still speculative.

Oncogene	Cancer	Percentage of tumors with oncogene expression	Translocation (t)
myc	Burkitt lymphoma	80	t(8:14)
myc	T-cell leukemia	10–20	t(8:14)
bcl-1	Chronic B-cell leukemia	10–30	t(11:14)
bcl-2	Follicular lymphoma	90	t(14:18)
abl	Myelogenous leukemia	90	t(9:22)
abl	Lymphocytic leukemia	10	t(9:22)

chromosomes by recombination. Once a protooncogene has been incorporated into a viral genome, it can change through time by mutation into an active cancer-causing gene.

Changes that increase the reproduction of viruses in cells will be favored by natural selection. It is unfortunate for the host that unregulated cellular growth is what favors increased reproduction and survival of the virus. Cancer may simply be what Darwin described as the continual "struggle of the fittest" in nature.

It is not yet clear that oncogenes really are the cause of any more than a few human cancers. It may be that chromosomal rearrangements are the primary cause for the origin of cancer cells and that altered expression of oncogenes is only a related consequence of the genetic damage. Recent experiments have shown that a chemical in cigarette smoke is converted in cells into a chemical that is able to activate certain oncogenes. However, other evidence argues against cellular oncogenes contributing significantly to human cancer. For example, protooncogenes are often found to be expressed even in normal cells. And cancers that develop in animals in nature rarely contain oncogenic viruses or activated cellular protooncogenes. Only further research can determine what role, if any, the protooncogenes in cells and the oncogenes in viruses play in causing human cancers.

Retroviruses and AIDS

The disease of AIDS (acquired immune deficiency syndrome), which has emerged as a major threat to human health in the 1980s, is caused by a retrovirus. As mentioned earlier in this chapter, retroviruses integrate their genetic information into the chromosomes of cells. Those that infect human beings seem to have a preference for infecting different cells of the immune system. Some retroviruses cause certain white blood cells to proliferate, thereby causing cancers (leukemias or lymphomas). However, a specific kind of retrovirus, the AIDS

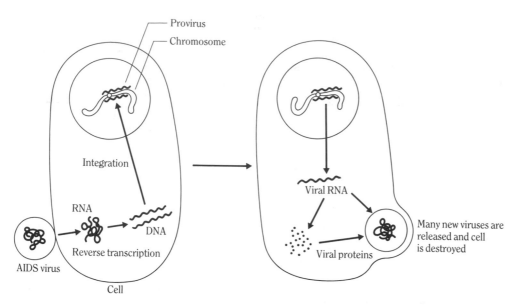

FIGURE 11-11 Life cycle of the AIDS retrovirus. Viral RNA enters the cell and is converted into double-stranded DNA by the enzyme reverse transcriptase. The AIDS virus is integrated into a chromosome where it becomes a provirus.

virus, infects and destroys T-cells of the immune system. The T-cells are largely responsible for protecting us from disease-causing microorganisms; without T-cells people are extremely susceptible to infectious diseases (see Chapter 17). That is why persons infected with the AIDS virus are often victims of opportunistic infections, particularly one caused by a protozoan (*Pneumocystis carinii*) that causes pneumonia. Many also develop a form of cancer called *Kaposi's sarcoma*.

The infectious cycle of the AIDS retrovirus is shown in Figure 11-11. Once the AIDS retrovirus infects a T-cell, the viral RNA is released and eventually converted into viral DNA, which contains the genetic information carried in the single-stranded viral RNA. The DNA is then integrated into one of the cell's chromosomes and replicated along with normal cellular genes. After integration, new virus particles are synthesized

and released from the cell. These AIDS viruses infect other T-cells and the process is repeated.

To date, no drug has been discovered that can cure diseases caused by retroviruses. Drugs such as AZT (azidothymidine) may slow the process of viral proliferation, which is helpful, but the drugs themselves are often toxic, especially with prolonged use. Also, it is not at all certain that an effective vaccine can be developed that will protect uninfected people. For these reasons, the major efforts in trying to stem the AIDS epidemic have been directed toward preventing virus transmission by instituting public education programs.

Moveable Genetic Elements

Retroviruses in animal cells (and viruses in bacteria) are capable of moving their own genes and cellular genes from one organism to another by infection. Chapter 6 described how plasmids in bacteria are transferred from cell to cell and how they foster the exchange of genetic information between microorganisms. Because of their capacity to move genes from one cell to another, both phages and plasmids are referred to as **moveable genetic elements**. Although we tend to think of genes as stable entities in chromosomes, it is clear that many mechanisms exist that allow genes to change locations in chromosomes or to be transferred from cell to cell and organism to organism. However, such movements must be relatively infrequent; otherwise we could not construct reliable genetic maps for bacteria and other organisms.

Transposons

Another remarkable genetic mechanism has been discovered that permits groups of genes to move abruptly from one DNA molecule to another. Initially, it was discovered that a group of genes coding for antibiotic resistance could move from one DNA molecule to another in the same bacterial cell. Because the segment of DNA appeared to "jump" from one DNA molecule to another, these segments of DNA were dubbed "jumping genes." Their formal name is **transposon**, which refers to any group of genes that jumps from one DNA molecule to another without employing the usual mechanisms of recombination. We now know that it is a copy of the transposon that moves—not the original DNA itself.

In bacteria, transposons usually carry genes coding for resistance to various antibiotics, such as penicillin, tetracycline, and chloramphenicol. The transposons responsible for the antibiotic-resistance properties of bacteria were first detected in plasmids (Figure 11-12).

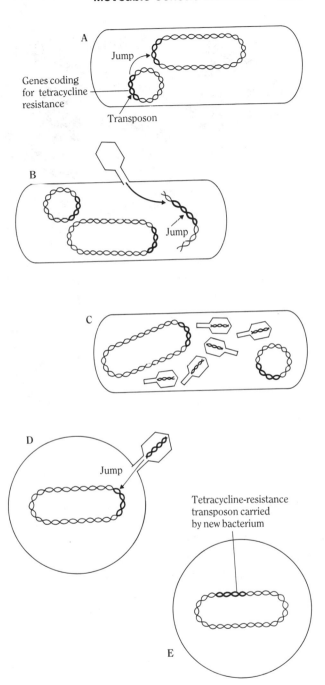

FIGURE 11-12 Transposons are an example of "jumping genes" in bacteria: (A) a group of genes coding for antibiotic resistance jumps from a plasmid into the chromosome; (B) a phage infects the cell and the group of genes (transposon) jumps into the phage DNA; (C) the virus reproduces inside the bacterium, after which the bacterial cell bursts, releasing the phages; (D) a phage carrying the transposon infects another bacterium and the transposon jumps into that cell's chromosome; and (E) the transposon is then carried by the new bacterium.

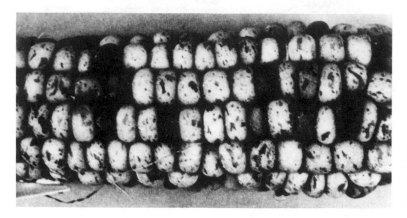

FIGURE 11-13   Variation in the color of corn kernels. Such variation is due to movable genetic elements that alter the expression of genes. (Courtesy of M.G. Neuffer, University of Missouri, and Crop Science Society of America.)

Transposons "jump" from one molecule to another by virtue of special sequences of DNA that are at the end of each transposon. These DNA sequences are called **insertion sequences** (IS) because they determine the jumping properties of each different transposon. As with other cellular mechanisms, enzymes called **transposases** assist in the movement of transposons.

Although transposons were first discovered and characterized in bacteria, other kinds of jumping genes were discovered much earlier. In the 1940s, Barbara McClintock, a plant geneticist, was able to show that the color differences in corn kernels had a genetic basis and resulted from what she called "movable genetic controlling elements" (Figure 11-13). Like many other genetic discoveries, her observations and their importance were generally unappreciated until the discovery of similar genetic elements in bacteria supplied the framework for understanding the movable genetic elements in corn. Many years after her initial discoveries, Dr. McClintock received the Nobel Prize in recognition of her work. It may turn out that movable genetic elements play a more-significant role in the development of cancer cells than was formerly thought.

The importance of viruses, oncogenes, plasmids, transposons, and insertion sequences in increasing genetic diversity in bacteria, plants, and animals is just beginning to be explored and understood. The bits of genetic material that move into and out of chromosomes cause mutations and change the way genetic information is expressed. It may turn out that transposable elements contribute as much to genetic diversity and to the evolution of organisms as do mutations from other sources. Moveable genetic elements from a variety of organisms are essential tools in the present biotechnology industry. This entirely new industry, which emerged in the mid-seventies, rests on our ability to genetically engineer novel DNA molecules, cells, and organisms.

Summary

Somatic-cell mutations trigger the onset of most cancerous diseases. These mutations are caused by environmental agents and, because the affected cells are somatic, the defective genes cannot be inherited. However, some rare heritable human disorders may predispose their victims to the development of cancer.

Cancer is defined as the unregulated growth and reproduction of cells. Human cancers are broadly subdivided into leukemias and lymphomas, sarcomas, and carcinomas.

Mutations that lead to the formation of cancer cells may be caused by exposure to high-energy radiation, mutagenic chemicals, or tumor viruses. Thus, in principle, cancer can be prevented—or the risk reduced—even in people who are predisposed if they are aware of their increased susceptibility.

Cells have mechanisms for repairing DNA that has been damaged by chemical or radioactive carcinogens. The molecular changes in DNA produced by chemical mutagenic agents differ from those caused by radiation; thus the enzymes required for the repair of DNA damaged by different mutagenic agents also differ. If damaged DNA is not repaired before or in the course of DNA replication, the mutation will be carried by subsequent generations of cells. If the genes coding for repair enzymes are themselves altered by mutation, loss of these enzymes can be lethal to the cells and to the organism.

The mutagenic potency of different chemicals varies widely. Potential cancer-inducing chemicals (carcinogens) can be tested in microorganisms by measuring rates of mutation. One such screening device is the Ames test.

Oncogenic viruses are those that cause cancer in animals, probably including people. The genetic information in some viruses is carried in DNA; in others, it is carried in RNA. Some RNA viruses, such as the AIDS virus, are called retroviruses because they integrate their genetic information into cells' chromosomes. Oncogenic viruses generally do not destroy the cells that they infect because they use the cellular machinery to reproduce themselves.

Protooncogenes are normal genes in animal and human cells that are subject to regulation of expression, and, as a result, they cause no harm. If, however, chromosomal rearrangement or mutation causes them to become overactive, they can transform normal cells into cancer cells.

Moveable genetic elements such as proviruses, transposons, and insertion sequences can change location spontaneously in the chromosomes of cells, including bacteria. These moveable genetic elements contribute to genetic diversity. Inadvertently, they may also cause mutations in animal cells that may lead to cancer.

Key Words

ataxia telanglectasia An inherited disorder affecting the ability of human cells to repair damage to DNA caused by x-rays.

benign tumor Unregulated, localized growth of cells that usually does not cause cancer.

cancer Unregulated growth and reproduction of cells.

carcinogen Any agent, such as radiation, a chemical, or a virus, that causes cancer.

carcinoma Cancer of skin and membrane cells.

concordance The degree to which identical (monozygotic) twins or fraternal (dizygotic) twins are alike with respect to a trait.

epidemiology A branch of science that studies the frequency and distribution of diseases in different populations and tries to identify disease causation.

insertion sequences (IS) Small pieces of DNA at the ends of transposons that have identical base sequences. These sequences pair with identical sequences in other DNAs and are responsible for transposon movement.

leukemia Cancer of immature white blood cells.

lymphoma Cancer of white blood cells in the lymphatic system and spleen.

lysogenic bacterium A bacterium that carries a prophage (unexpressed virus) in its DNA.

lysogeny A process by which phage DNA becomes integrated into a bacterial chromosome.

malignant tumor Cancer that can spread through the body and cause death.

metastasis The spread of cancer cells from the original tumor to other sites in the body.

moveable genetic element   Pieces of DNA such as viruses, plasmids, and transposons that can "jump" from one DNA molecule to another and move from one cell to another.

oncogene   A cancer-causing gene.

oncogene hypothesis   The idea that viral genes became integrated into the chromosomes of animal cells millions of years ago. These genes are normally unexpressed but if induced may convert normal cells into tumor cells.

oncogenic virus   A virus capable of causing cancer in animals when the viral genes are expressed.

prophage   An unexpressed phage carried in the DNA of bacteria.

protooncogene   A normal cellular gene that may give rise to an oncogene by mutation.

provirus   An unexpressed viral genome carried in one of the chromosomes of animal cells.

retrovirus   An RNA virus whose RNA is copied (reverse transcription) into DNA in an animal cell; the DNA then becomes integrated into the DNA of a chromosome in the nucleus.

reverse transcriptase   An enzyme present in certain RNA viruses that transcribes the information in the RNA molecule into DNA after the RNA has infected an animal cell.

sarcoma   Cancer of bone and muscle cells.

thymine dimer   Adjacent thymine bases in DNA that are covalently linked by ultraviolet light.

transposase   An enzyme that catalyzes the movement of a transposon.

transposon   A group of genes (often those conferring antibiotic resistance in bacteria) that move as a copied unit from one DNA molecule to another.

tumor   A mass of cells that accumulates at a particular site. If the cells spread to other parts of the body causing disease or death, the tumor is malignant. Benign tumor cells remain at the original site and do not usually cause disease.

xeroderma pigmentosum   An inherited disorder affecting the ability of cells to repair damage to DNA caused by ultraviolet light.

Additional Reading

Bishop, M. "Oncogenes." *Scientific American*, March, 1982.

Cairns, J. *Cancer: Science and Society*. W.H. Freeman, 1980.

Cohen, L. "Diet and Cancer." *Scientific American*, November, 1987.

D'Eustachio, P. "Gene Mapping and Oncogenes." *American Scientist*, January-February, 1984.

Doll, R., and R. Peto. "The Causes of Cancer: Quantitative Estimate of Avoidable Risks of Cancer in the United States Today." *Journal of the National Cancer Institute*, 66 (6), 1195 (1981).

Karpas, A. "Viruses and Leukemia." *American Scientist*, May-June, 1982.

Oppenheimer, S.A. "Advances in Cancer Biology." *American Biology Teacher*, January, 1987.

Patterson, J.T. *Cancer and Modern American Culture*. Harvard University Press, 1987.

Reif, A.E. "The Causes of Cancer." *American Scientist*, July-August, 1981.

Weinberg, R.A. "The Molecular Basis of Cancer." *Scientific American*, November, 1983.

Study Questions

1   What is the main difference between a benign tumor and a malignant one?

2   What is the connection between a mutagen and a carcinogen?

3   What is the role of the enzyme reverse transcriptase in the development of cancer?

4   What carcinogen causes more human cancer deaths than any other?

5   About what fraction of all human cancers are due to hereditary defects?

6   What three classes of environmental agents cause cancer?

7   What environmental agent produces thymine dimers in cells?

Essay Topics

1   Discuss how the Ames test is used to help prevent cancer.

2   Explain the statement: "Most cancers are caused by environmental agents."

3   Discuss whether you think industrial workers should be screened for genes that might predispose them to cancer.

12

Biotechnology
Products from Genetic Engineering

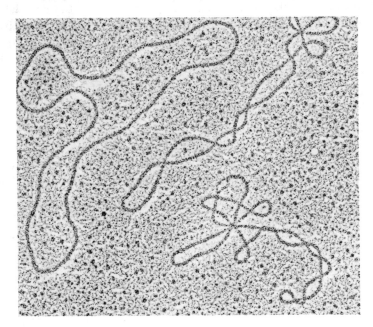

It's a catch-22. If a company behaves in what scientists believe is a socially responsible manner, they can't make a profit.

ROBERT COOK-DEEGAN, *Office of Technology Assessment*

IN BIOLOGY and medicine, recent advances in genetic engineering and biotechnology provide the most-dramatic answer to the question, Of what value is basic biological research to society? Articles in magazines and newspapers daily announce new discoveries emerging from recombinant DNA research. Hundreds of biotechnology companies employ scientists and technicians who are applying new genetic discoveries to produce useful microorganisms, drugs, plants, and chemicals. Valuable human proteins such as insulin, somatostatin, interferon, and growth hormone are already being manufactured and used to treat a variety of human diseases.

Yet few people, even among those who argue the pros and cons of recombinant DNA technology, realize that the techniques for splicing together genes from different organisms have their origins in basic experiments in which phages are grown in bacteria—experiments that were of interest to only a handful of scientists for many years until the potential of their findings suddenly became apparent. Today, recombinant DNA research has fostered an entire biotechnology industry that is less than twenty years old. It has also been a source of insights into gene regulation and gene expression—insights that were unexpected and unpredicted just a few years ago. The pace of current recombinant DNA research and the rate of new discoveries are truly staggering. It is safe to say that our lives will be changed by these discoveries and the products that are developed because of them.

This chapter begins with a description of the original research that led to today's recombinant DNA technology. Next, the techniques that are used to move genes from one kind of organism into another are addressed. This movement and amplification of genes is called **gene cloning**, because billions of identical copies of a particular gene can be produced in bacteria or other kinds of cells. And, finally, the social problems and ethical questions raised by genetic-engineering research and the products and organisms created by the biotechnology industry are considered.

What Is Biotechnology?

We first need to clarify what is meant by biotechnology and other terms that are used in the gene-cloning business. **Biotechnology** refers to the application of scientific knowledge by companies that produce a variety of biological products, such as vaccines, food supplements, diagnostic kits for diseases, monoclonal antibodies, and a wide range of human proteins used in treating and diagnosing human diseases (Table 12-1). Biotechnology includes the production of **transgenic plants and animals**—organisms that have been modified so that they contain genetic information derived from other species (see Box 12-1). Finally, biotechnology also includes reproductive technology in which completely novel animals may be developed by mixing embryonic cells from different species to produce animals such as the goat-sheep chimera (geep) shown in Figure 1-1. Reproductive technology is also used to help infertile couples have children by a variety of procedures (discussed in Chapter 13).

Recombinant DNA technology is a crucial part of biotechnology. It really was the discovery that genes from different DNA molecules could be recombined (spliced together) *in vitro* to create new combinations of genetic information that led to the current revolution in biological research. The terms **recombinant DNA** and **genetic engineering** are often used interchangeably. Both refer to the construction of novel DNA molecules and the insertion of that DNA into cells and organisms. The constructed DNA replicates in cells so that millions of identical copies (clones) of the genes are produced. Recombinant DNA techniques are, in principle, quite simple. However, cloning particular DNA pieces that are scientifically interesting or useful requires both skill and that elusive quality — which even scientists value highly — called luck.

How Genetic Engineering Started

Bacteria that belong to the same species and share many characteristics but that differ from one another in some regard are called **bacterial strains**. For example, some strains of *E. Coli* carry λ-phages, whereas others do not. Special tests can distinguish lysogenic bacteria from cells of the nonlysogenic type.

Lambda phages can infect most strains of *E. coli* (except for those that already carry the λ-prophage in their DNA), and most of the infected bacteria will burst, producing hundreds of new λ-phages. However, a puzzling fact emerges when λ-phages are used to infect two different strains of bacteria, one called *E. coli* K and the other (which carries a plasmid) called *E. coli* R. Lambda phages synthesized in the K strain are labeled λ(K)-phages, whereas those produced in the R strain are labeled λ(R)-phages. After the λ(K)- and λ(R)-phages have been produced, they are again used to infect each bacterial strain separately.

What is observed in these infection experiments is not easily explained. In one case, λ(R)-phages are able to grow in both strains and synthesize new λ-phages. In the other case,

TABLE 12-1 Human proteins currently manufactured by the cloning and expression of human genes in cells grown *in vitro*.

Protein*	Use or treatment
Insulin (Humulin)	Diabetes
HGH (human growth hormone)	Pituitary deficiency; short stature
Alpha, beta, and gamma interferons	Cancer and viral infections
TPA (tissue plasminogen activator)	Heart attacks and stroke
Blood factor VIII	Hemophilia (clotting factor)
Blood factor IX	Hemophilia (clotting factor)
Erythropoetin	Anemia (hormone)
Interleukin-2	Cancer treatment
Atrial natriuretic factor	Blood pressure lowering
Monoclonal antibodies	Diagnosis and treatment of diseases
TNF (tumor necrosis factor)	Cancer treatment

*Some of these proteins have been approved for use; others are still in experimental stages of development. All together, it is estimated that worldwide sales of these protein drugs could range from $15 billion to $20 billion per year in the 1990s.

λ(K)-phages are able to grow *only* in the *E. coli* K strain and are unable to grow in the R strain (Figure 12-1). The initial observation of this phenomenon was made in the early 1950s. By the middle of the 1960s, researchers who had pursued the question had discovered that one particular gene carried in the DNA of the R-plasmid prevents the growth of the λ(K)-phages. Essentially, it became clear that, in bacteria carrying R-plasmids, a gene in the plasmid codes for an enzyme that attacks foreign DNA.

When λ(K)-phage DNA enters bacteria that carry the R plasmid, the λ-DNA is recognized as foreign and the enzyme — now called a **restriction enzyme** — attacks the DNA at specific sites

Should New Life Forms Be Patentable?

BOX 12-1

In June 1980, the U.S. Supreme Court ruled by a slim majority, five votes to four, that "a live, human-made microorganism is patentable." This history-making legal decision involved the case of *Diamond v. Chakrabarty*. Sidney Diamond, the commissioner of patents and trademarks, took the position that patents should not be awarded for living organisms. Anand Chakrabarty, a scientist who worked for the General Electric Company, had constructed a microorganism that was able to use petroleum as its nutrient source. General Electric, which believed that Chakrabarty's organism might have commercial use in cleaning up oil spills, wanted a patent on the microorganism so that it could profit by controlling the microorganism's commercial use.

As a result of the Supreme Court's decision, new microorganisms that are constructed by genetic engineers can now be patented. Although officials at genetic-engineering companies have been pleased with the ruling and view the decision as necessary to protect their interests and investments, some critics worry about the long-term consequences of allowing life forms to be patented.

For example, Sheldon Krimsky, a sociologist at Tufts University, points out that "the right to patent microorganisms, including human cells, will foster manipulation of human genes that could breach an ethical threshold for our society." And Jonathan King, an MIT biologist, believes that "forms of life, be they microbial, plant, or animal, are too important to be allowed to be in private ownership."

In 1985, the Board of Patent Appeals ruled that genetically engineered plants were patentable. At that time the Board was still undecided about what to allow with regard to animal patents. However, in the spring of 1987, the U.S. Patent and Trademark Office ruled that all animals, except for human beings, may be patented. This decision provoked a storm of protests from various political, farm, and religious groups. However, in 1988 the first animal, a mouse carrying a human oncogene, was patented. Almost a hundred animal patents have been applied for since then.

Most scientists and biotechnology company administrators believe that patenting of genetically engineered animals is both necessary and appropriate. As one company spokesperson put it, "If we developed a mouse that was susceptible to the AIDS virus, we certainly would want a patent on it." Obviously, such an experimental animal would be invaluable for AIDS research and for vaccine development. Other people are concerned, however, about where to draw the line on animal patents. Should patents be issued for human cell lines, as well as for human tissues and organs? Should the construction of any and all kinds of animals be patentable?

It is now possible to genetically engineer animals such as cows, for example, so that they can produce human proteins that can be purified from their milk. Such cows become merely biological factories for drugs. Some people view this as a logical extension of existing animal-breeding practices; others are aghast at the prospects of constructing bizarre creatures and the possibilities for animal suffering. What are your views?

READINGS: "Patenting Modified Life Forms." The case for: C. E. Lipsey and C. P. Einaudi. The case against: J. King. *Environment*. July–August, 1982. D. Dickson, "Animals of Invention," *Science*. May 6, 1988.

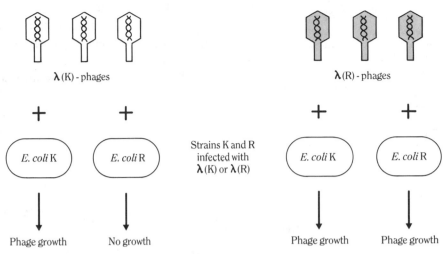

FIGURE 12-1 Pattern of growth of λ-phages in two bacterial strains. Phages grown in one bacterial strain are sometimes unable to grow in another strain. Lambda phages grown in *E. coli* strain K are labeled λ(K); phages grown in *E. coli* strain R are labeled λ(R). When these phages are used to infect the same two bacterial strains, the λ(R)-phages grow in either strain but λ(K)-phages grow only in strain K.

and destroys it. How DNA molecules are destroyed by restriction enzymes is now understood and is the basis for all recombinant DNA experiments and for the genetic engineering of new organisms.

Restriction and Modification Enzymes

When DNA is replicated in bacteria, chemical modification of bases in the newly synthesized

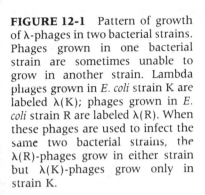

FIGURE 12-2 Modification of DNA by the attachment of methyl (CH₃) groups to DNA. In this example, the modification enzymes recognize the CAT base sequence in either strand and attach a methyl group to adenine. Each species of bacteria has its own unique methylation enzymes and pattern of methylation to protect its own DNA from the restriction enzymes that it produces. However, restriction enzymes from bacteria of different species will destroy the DNA because they do not recognize the pattern of methylation.

DNA occurs. The sites in DNA are recognized by **modification enzymes** that attach the methyl groups to bases. The attachment of methyl groups to particular sites in the DNA protects it from being destroyed by restriction enzymes also present in the cell (Figure 12-2).

The specific methylation pattern of a bacterium's DNA gives it a means of distinguishing its own DNA from that of any other DNA that might happen to enter the cell. For example, bacteria may be infected by different kinds of phages, but all of the phage DNA will itself be methylated according to the pattern characteristic of the particular strain of bacteria in which the phages were grown. Phages will be restricted from growing if the methylation pattern of the phage DNA tells the infected bacterium that the DNA is not the same as that of the new host strain. If bacteria of different strains or species conjugate, or mate, any DNA that is transferred and is recognized as having a different pattern of methylation will be destroyed.

Each kind of microorganism contains restriction enzymes that recognize a different sequence of bases and pattern of methylation (Table 12-2). A restriction enzyme, which

TABLE 12-2   Restriction enzymes can be isolated from many different species of microorganisms, particularly bacteria. Each restriction enzyme recognizes a specific sequence of bases in any DNA (commonly, groups of four or six bases). The enzymes may cut both DNA strands at the same place or the cuts may be staggered, producing small tails of single-stranded DNA. The vertical lines indicate where the DNA is cleaved.

Restriction enzyme	Extracted from	Sequence recognized
Eco RI	Escherichia coli	G|A A T T C C T T A A|G
Hpa I	Haemophilus parainfluenzae	G T T|A A C C A A|T T G
Hae III	Haemophilus aegypticus	G G|C C C C|G G
Msp I[6]	Moraxella species	C|C G G G G C|C
Bam HI	Bacillus amyloliquefaciens	G|G A T C C C C T A G|G
Pst I	Providencia stuartii	C T G C A|G G|A C GT C

breaks DNA molecules only at certain sites, will attack any DNA *in vitro* and always cleave it at the same recognition sites (Figure 12-3). Most restriction enzymes are named for the microorganism from which they were isolated. The restriction enzyme first discovered was named Eco R1 because it is coded for by a gene on the R-plasmid of *E. coli* (hence, Eco R1). Several hundred different restriction enzymes have been isolated from different microorganisms so far. These restriction enzymes are commercially available and are used to construct novel DNA molecules and to make restriction maps of the DNA from various organisms (discussed later).

Restriction Maps of DNA

DNA molecules contain different recognition sites for each of the restriction enzymes. The longer the DNA molecule, the more restriction sites that occur by chance. Cleaving DNA molecules with a restriction enzyme produces a characteristic number of DNA fragments that can be separated on an agarose gel and made visible by straining the separated DNA fragments. Figure 12-4 shows the patterns of DNA fragments produced when λ-phage DNA is

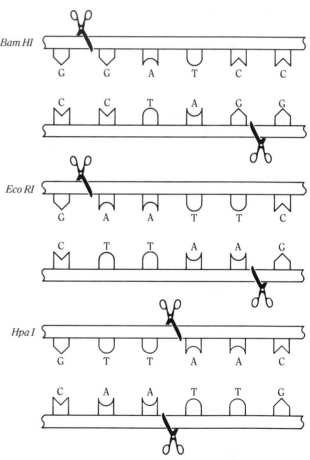

FIGURE 12-3   Restriction enzymes recognize specific short sequences of bases (usually between four and six bases). The restriction enzymes can make staggered cuts, leaving short, single-stranded tails as shown for *Bam HI* and *Eco RI*. Or the cuts can be directly opposite one another on each strand (*Hpa I*), producing DNA fragments with blunt (flush) ends.

cleaved by two different restriction enzymes, Eco R1 and Hind III. The patterns differ because the enzymes cut the DNA in different places, creating fragments of different sizes.

Characterizing a DNA molecule by its pattern of DNA restriction fragments is what is meant by a **restriction map**. By analysis of the number and sizes of the various DNA fragments, the restriction sites can be positioned along the DNA molecule or chromosome (Figure 12-5). The greater the number of different restriction enzymes that are used, the more sites that can be mapped and the more detailed and precise the map becomes. The process is analogous to placing cities, towns, and villages on a road map. If the map includes only cities, it is a crude map because there are few points of reference. If it includes all towns and villages as well, there are many more points of reference and the map is more useful. An enormous advance was made in 1987, when a complete restriction map of the human genome was achieved by a biotechnology company.

Restriction maps of human chromosomes are particularly useful for diagnostic purposes. Quite often, a mutation that produces a hereditary disease will also alter a restriction-enzyme site in the DNA because one base pair in the recognition sequence will be changed. This restriction-site alteration will be inherited along with the defective gene. By analysis of the pattern of restriction fragments in DNA extracted from fetal cells obtained by amniocentesis, it is now possible to predict if a fetus is carrying a particular hereditary defect. It is also sometimes possible to determine which members of a family are carriers of the defective gene by making restriction maps of DNA extracted from a sample of their cells. More will be said about the specific human diseases that can be analyzed in this way in Chapter 16.

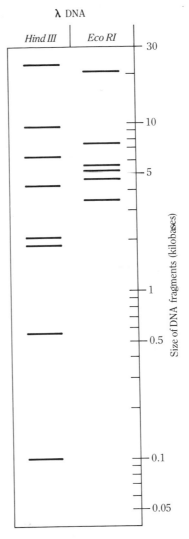

FIGURE 12-4 Separation of DNA fragments of different sizes by gel electrophoresis. Two different restriction enzymes were used to cut the λ-phage DNA in two different samples. Each sample produces a different pattern of fragments. Large DNA fragments move slowly through the gel and are near the top; small ones move fast and are near the bottom.

As soon as the mechanisms of DNA restriction enzymes were understood, many scientists realized that it should be possible to join together pieces of DNA from completely different organisms. The next step was to construct

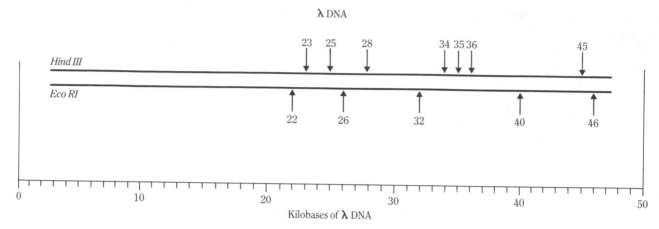

FIGURE 12-5 A restriction map of the λ-phage DNA constructed from the enzyme-digest patterns shown in Figure 12-4. Because the complete base sequence of the lambda genome is known, the exact positions of the restriction sites can be deduced from the fragment sizes determined on the gel.

cloning vehicles—small self-replicating plasmids that carry along and replicate pieces of foreign DNA that have been inserted into their circular DNA molecules. The plasmids replicate in bacteria and billions of copies of the foreign gene are produced as the bacteria grow.

Plasmids as Cloning Vehicles

A vehicle is something used to transport an object from one place to another. **Cloning vehicles** (also referred to as *cloning vectors*) are plasmids or viruses (Figure 12-6) whose DNA has been altered in such a way that foreign DNA fragments produced by restriction enzymes can be inserted into them. When these cloning vehicles are introduced into cells, they replicate both their own DNA and the foreign genes that they carry. The foreign DNA is thus cloned, given that identical copies are made each time the cloning vehicle replicates. Usually, the microorganisms used to replicate cloned genes are themselves genetically altered so that they no

longer produce restriction or modification enzymes, which would alter the foreign DNA that has been introduced.

An outline of a typical cloning experiment is shown in Figure 12-7. The first step is the isolation and cleavage of the DNA into fragments, using a particular restriction enzyme. Some restriction enzymes, such as Eco RI, produce complementary single-stranded ends on the DNA fragments. Next, a plasmid (cloning vehicle) is chosen into which the foreign DNA is to be inserted. Cloning vehicles are available that contain single restriction sites for a number of different restriction enzymes. The plasmid DNA is cut open with the restriction enzyme previously used, and fragments of the foreign DNA are inserted. The plasmids then contain an assortment of foreign DNA fragments. The enzyme ligase is used to seal the ends of the DNAs together.

The recombinant plasmids are mixed with bacteria and the DNA is taken up by the cells and replicated. At this point, there is no way of knowing which bacterium carries the foreign DNA fragment of interest. A variety of techniques are used to isolate and identify cells

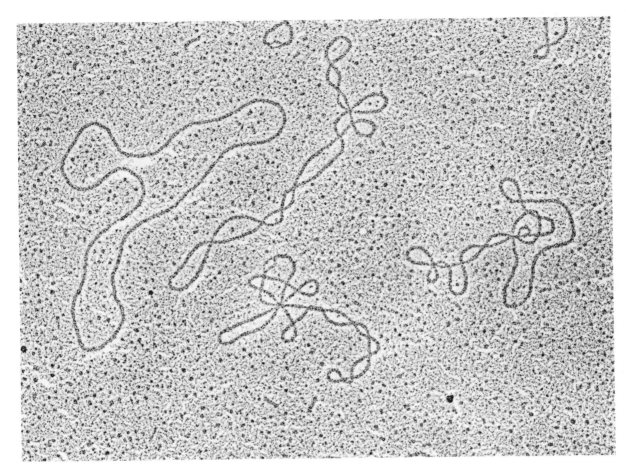

FIGURE 12-6 An electron micrograph of several small, circular cloning vehicles—plasmids that replicate their own DNA and any foreign DNA inserted into them. Plasmids exist in coiled or uncoiled circular forms.

carrying the foreign gene. Generally, the bacteria are spread onto a solid medium to grow, and the plasmid DNA carried in individual bacterial colonies is analyzed.

Suppose that we want to clone the genes that are responsible for making microorganisms resistant to destruction by the antibiotic penicillin. First, we isolate DNA from a naturally occurring penicillin-resistant microorganism. Then we cleave the foreign DNA into pieces with a restriction enzyme such as Eco R1. Next, we open up the circular plasmid DNA by exposing plasmids to the same Eco RI enzyme. Then we randomly insert and seal the foreign DNA fragments into the plasmids using the enzyme ligase. Some plasmids, by chance, will contain the gene or genes that code for penicillin resistance.

To select from the mixture of plasmid-containing bacteria only those few that contain the plasmids carrying the penicillin-resistance genes, we spread the entire mixture of bacteria

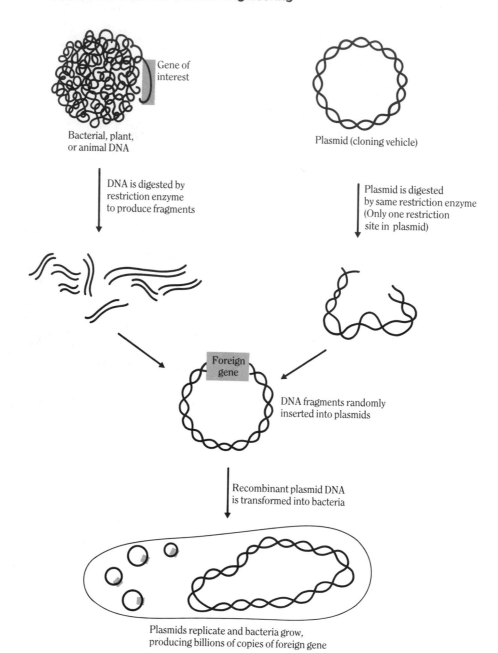

Gene of interest

Bacterial, plant, or animal DNA

Plasmid (cloning vehicle)

DNA is digested by restriction enzyme to produce fragments

Plasmid is digested by same restriction enzyme (Only one restriction site in plasmid)

Foreign gene

DNA fragments randomly inserted into plasmids

Recombinant plasmid DNA is transformed into bacteria

Plasmids replicate and bacteria grow, producing billions of copies of foreign gene

FIGURE 12-7 A schematic outline of a typical recombinant-DNA experiment. A DNA fragment from a foreign organism is inserted into a plasmid and cloned into a bacterium.

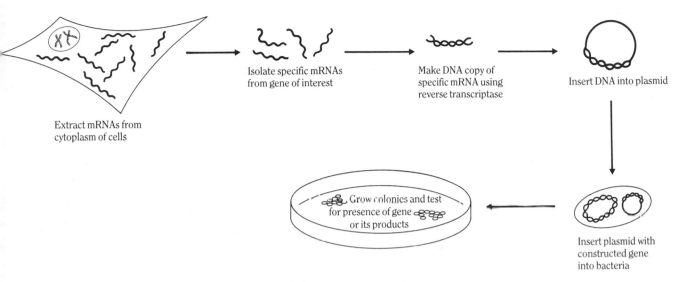

Extract mRNAs from
cytoplasm of cells

Isolate specific mRNAs
from gene of interest

Make DNA copy of
specific mRNA using
reverse transcriptase

Insert DNA into plasmid

Grow colonies and test
for presence of gene
or its products

Insert plasmid with
constructed gene
into bacteria

FIGURE 12-8   Synthesis of genes from mRNAs. If a particular mRNA can be isolated and purified in sufficient amounts, the enzyme reverse transcriptase can synthesize a piece of DNA whose base sequence is complementary to the base sequence of the mRNA. This enzymatically constructed gene can then be cloned in a plasmid in the usual manner.

onto a solid medium that contains penicillin in addition to the nutrients necessary for bacterial growth. Of the millions of bacteria spread on the plates, virtually all will be destroyed by the antibiotic, except for the one in a million or so that is resistant to penicillin. We then allow the antibiotic-resistant bacteria to grow into colonies. Each one of the millions of bacteria in these colonies should contain the penicillin-resistance gene. In this way, the gene for penicillin resistance has been cloned and can be studied in detail.

Other Techniques for Cloning Genes

The techniques just described permit the construction of gene libraries and the cloning of specific genes such as the penicillin-resistance gene. However, unlike the penicillin-resistance gene, most genes or pieces of DNA cannot be selected. For example, no convenient method exists for selecting bacteria that contain the genes coding for human growth hormone or those coding for the human antiviral substance interferon. Thus, other methods for cloning and selection of desired genes had to be devised.

Two other methods of cloning genes have been developed. The first requires the isolation of the mRNA transcribed from a specific gene. If sufficient amounts of the specific mRNA can be purified, the information in the sequence of bases in the mRNA can be converted into a DNA copy by using the enzyme reverse transcriptase (Figure 12-8). Once the DNA has been synthesized, it can be inserted into a plasmid, which is then cloned in the usual manner. Because all of the pieces of DNA are identical (because they were all copied from the same mRNA), all of the bacterial clones will be the same and no selection procedure is necessary.

Sometimes it is not even possible to isolate the particular mRNA of interest from the other mRNAs in cells. Often, however, the polypeptide or protein product can be isolated. If sufficient quantities of the protein can be purified, the complete amino acid sequence of the protein can be determined by chemical techniques. For example, the amino acid sequence for the human hormone somatostatin has been determined in this fashion. Using the genetic code, it is possible to determine the bases in the DNA that code for the particular amino acid sequence in somatostatin. However, because the genetic code is degenerate, more than one codon can be chosen to specify any amino acid other than tryptophan and methionine. Fortunately, only one of the possible codons for an amino acid is usually preferred in *E. coli*; so the most frequently used bacterial codon is the one chosen. Once the sequence of bases has been determined from the genetic code, the gene is chemically synthesized and cloned in the usual manner.

To summarize, three basic methods are used in recombinant DNA technology for cloning genes of interest: (1) isolating and cloning the piece of DNA itself, (2) cloning DNA that has been synthesized from specific mRNAs, and (3) cloning DNA that has been chemically synthesized by knowing the amino acid sequence of the protein.

Many problems and stumbling blocks must be overcome before the product of a foreign gene can be manufactured in microorganisms. First, the correct regulatory sites must be positioned at the beginning of the cloned DNA fragment to ensure correct transcription and translation of the gene. However, even if the gene is correctly transcribed and translated, cells may recognize the products of the foreign gene as

foreign proteins and destroy them before they can be isolated. And the recently discovered fact that eukaryotic DNA is organized as split genes with introns and exons means that genes cloned directly from human DNA cannot be translated successfully in microorganisms because they do not have the necessary RNA splicing enzymes that are present in animal cells.

Despite these difficulties, the genetic engineering of products, plants, and animals has advanced at a phenomenal pace. Many problems had to be overcome to successfully produce the human protein drugs listed in Table 12-1. In addition to developing drugs through the use of recombinant DNA techniques, the biotechnology industry has succeeded in cloning and expressing foreign genes in both plants and animals. However, a new technique called PCR (polymerase chain reaction) now makes the cloning of genes obsolete for many purposes (see Box 12-2).

Constructing Gene Libraries

Gene-splicing techniques are now being used to store all of the genetic information for a specific organism in a "library" of bacteria. In this way, the genes of species that are in danger of extinction can be preserved. Suppose, for example, that we want to preserve all of the genetic information carried in the cells of a condor—a species of bird that is on the verge of extinction in California because so few members remain. Here is how it can be done.

DNA is extracted from condor cells and is cleaved into small pieces by one or more restriction enzymes. (To get pieces of DNA that are small enough to clone, cleavage by several enzymes is usually required.) The small condor

Amplifying Genes
The PCR Technique

BOX 12-2

A powerful new technology called polymerase chain reaction (PCR) now allows genetic engineers and other scientists to amplify specific sequences of DNA and thus obtain large quantities of genes of interest. The amplified DNA can then be analyzed or used for a variety of purposes. For example, the DNA extracted from a single human sperm can be examined by PCR to determine if the sperm carried a defective hemoglobin gene or a virus capable of causing disease. DNA extracted from a single blood or hair cell can be amplified and specific base sequences compared with DNA obtained from cells of crime victims or crime suspects. Even minute samples of DNA from the tissues of extinct plants or animals now can be amplified and compared with the genetic information in cells of existing organisms.

The PCR technique is used to amplify a specific DNA sequence according to the following procedure. First, two short, single-stranded DNA primers from twenty to thirty bases in length are synthesized. The sequences of these primers are constructed so that they are complementary to base sequences located at the ends of a particular gene of interest in the DNA (see the accompanying illustration). Next, an unusual DNA polymerase that is stable at very high temperatures

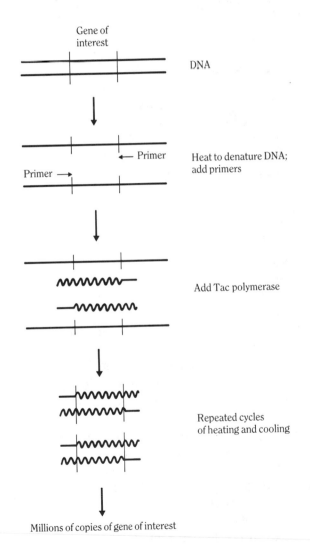

Gene of interest

DNA

Heat to denature DNA; add primers

← Primer

Primer →

Add Tac polymerase

Repeated cycles of heating and cooling

Millions of copies of gene of interest

(70°–80°C) is added to the DNA reaction mixture. This Tac DNA polymerase is isolated from the bacterium *Thermus acquaticus*, which grows in hot springs. The advantage of using a heat-stable DNA polymerase is that repeated cycles of heating and cooling can be used to separate DNA strands without destroying the DNA polymerase activity. Thus, the enzyme (which is quite ex-pensive) can be used over and over in repeated cycles of gene amplification.

After twenty cycles of DNA synthesis, the original gene of interest has increased from one copy to about ten million copies. The amplified DNA can now be extracted and used for clinical, or other, purposes. The PCR technique has been automated so that the entire series of reactions takes place in about an hour. Automation also means that the clinical evaluation of many human genetic disorders can be made rapidly and reliably.

READINGS: Jean L. Marx, "Multiplying Genes by Leaps and Bounds," *Science*, June 10, 1988. R.K. Saiki, D.H. Gelfand, S. Stoffel, *et al.*, "Primer-directed Enzymatic Amplification of DNA with a Thermostable DNA Polymerase," *Science*, 239 (1988).

DNA fragments are then inserted into plasmid DNA. The mixture of plasmids carrying different fragments of condor DNA is reintroduced into bacteria, so that each bacterium contains a plasmid with some of the condor's genes. Among the billions of bacteria, all of the condor's genetic information is preserved.

Because billions of bacteria can be frozen in a few drops of liquid and stored almost indefinitely, bacterial libraries are a means of preserving the genetic information of any species. Fewer than 100,000 bacteria clones are needed to preserve the entire human genome (all of the genetic information in a sperm or egg), and such libraries have already been constructed.

In a conventional library, the information is catalogued and physically organized so that it is readily accessible—the books and periodicals are not just randomly piled together. But, in a bacterial library, the genes are randomly inserted into the bacteria, which are indistinguishable one from another. Thus, to find a particular gene in a bacterial gene library, we must first devise a procedure that will enable us to identify and isolate the few bacteria containing the gene of interest from all the rest. Clever selection procedures have indeed been devised for a number of genes, but often a large number of bacterial clones must be examined to find the desired piece of DNA—an arduous task.

Genetic-Engineering Applications

The applications of genetic engineering are not limited to improving the manufacture of drugs and other chemicals. Recombinant DNA research is also being used to develop microorganisms that can economically convert agricultural waste into fuels such as hydrogen, ethanol, and methanol. Certain microorganisms play critical roles in the manufacture of beverages and foods—including beer, wine, bread, pickles, cheese, yogurt, tofu, and soy sauce—and so it is likely that the fermentations carried out by these microorganisms will be made more efficient and more productive by recombinant DNA technology.

Many scientists believe that the greatest practical benefits of genetic engineering will be in agriculture. Genes will eventually be moved from one kind of plant into another to improve virtually any desired crop characteristic, such as disease resistance, flavor, or nutritional value. The cloning vehicle that seems to have the

FIGURE 12-9 Crown gall tumor on a turnip that was infected with bacteria carrying Ti plasmids. (Courtesy of C. I. Kado, University of California, Davis.)

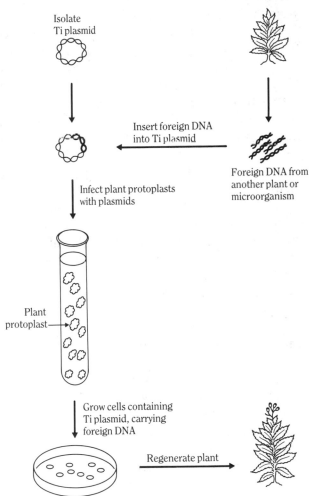

Isolate
Ti plasmid

Insert foreign DNA
into Ti plasmid

Foreign DNA from
another plant or
microorganism

Infect plant protoplasts
with plasmids

Plant
protoplast

Grow cells containing
Ti plasmid, carrying
foreign DNA

Regenerate plant

FIGURE 12-10 Construction of new kinds of plants through the use of recombinant DNA technology. The Ti plasmid from *Agrobacterium tumefaciens* is used as a cloning vehicle for plant cells. Foreign DNA containing genes from other plants can be inserted into Ti plasmids, which are subsequently introduced into plant protoplast cells (cells from which the cell walls have been removed). Cells containing the Ti plasmid carrying foreign DNA are grown, and in some cases the foreign DNA will integrate into the cell's chromosomes. These cells can then be used to regenerate the entire plant.

greatest potential for engineering new plant varieties is the Ti (tumor inducing) plasmid.

This plasmid was originally isolated from *Agrobacterium tumefaciens*, a bacterium that causes tumors on many varieties of plants (Figure 12-9). The Ti plasmid has been genetically altered so that it no longer causes tumors and so that foreign DNA can be inserted at specific restriction-enzyme recognition sites. This altered plasmid can be used to transfer desirable genes into the chromosomes of plant cells. Entire plants can, in some instances, be regenerated from individual plant cells (Figure 12-10).

This capability holds out the promise of constructing new species of plants from single cells whose DNA contains genetic information from one or more other species. As mentioned earlier, these plants are transgenic because they contain genes from another species.

To date, more than a dozen species of plants have been successfully regenerated from single cells. Tobacco was the first and is still the easiest plant to grow from single cells. When tobacco plants have been regenerated from single cells containing Ti plasmid vehicles, the foreign DNA has been found to be present in all of the tissues in some transgenic plants.

A dramatic example of what can be accomplished in plants is demonstrated by the insertion of the luciferase gene from fireflies into tobacco plants. The luciferase gene produces an enzyme that is responsible for the ability of fireflies to give off flashes of light. The luciferase gene was cloned into the Ti plasmid and inserted into tobacco cells. These cells were then grown until eventually complete tobacco plants had been regenerated. Lo and behold, the tobacco plants now glow in the dark when they are sprayed with the chemical that is converted by luciferase into flashes of light.

Aside from the novelty of this experiment, it has practical value. The luciferase-gene activity can be used as a signal to detect other useful genes that scientists want to introduce into plants. Some desirable traits are disease resistance, tolerance to salty soil, increased oil content, and herbicide resistance. If the luciferase gene and another gene are closely linked on the same fragment of cloned DNA, then any cell or plant that "lights up" should contain the second gene also. Thus, glowing-in-the-dark becomes a plant trait that can be used to monitor the presence and activities of other genes.

As any grower of plants knows, gardens have to be weeded. Weeds can be either physically re-

TABLE 12-3 Common herbicides and their modes of action in plants. Most herbicides interfere with an essential enzyme function in plants.

Herbicide	Enzyme affected	Mode of action
Dimethazone	Geranyldiphosphate transferase	Not known
Glyphosate (Roundup)	5-Enolpyruvyl 3-phosphoshikimate synthase	Enzyme inhibitor
Imidazolinone	Acetolactate synthase	Enzyme inhibitor
Phosphinothricin	Glutamine synthase	Enzyme inhibitor
Sulfonyl urea	Acetolactate synthase	Not known

moved or killed by spraying with chemicals called herbicides. If a field of vegetables is sprayed with a herbicide, it not only kills the weeds, but also kills or damages the crop. Genetic engineering can help here also. Herbicides kill plants generally by inhibiting a specific plant enzyme (Table 12-3). It is relatively easy to isolate herbicide-resistance genes in plants or in microorganisms. If these genes are introduced into specific crops, it should become possible to spray a field with the herbicide and selectively kill the weeds without harming the crop plants. One drawback is that it will result in more herbicides in the environment. Genes that confer resistance to Roundup—one of the most widely used herbicides—have already been successfully cloned into tobacco and tomato plants. Although Roundup breaks down rapidly in the environment, many other pesticides and herbicides do not.

Hazards of Recombinant DNA Technology

In February of 1975, scientists from all over the world gathered at Asilomar, California, for a meeting that was without precedent in the history of science. The scientists met to discuss and adopt guidelines under which all further recombinant DNA research should proceed. For almost a year before the Asilomar meeting, these scientists had observed a self-imposed moratorium on recombinant DNA research until safety procedures could be evaluated and agreed upon. Most of them were eager to get on with the experiments that all of them knew would yield exciting information on how genetic information is stored in DNA and how genes are regulated and expressed in different kinds of cells.

After much discussion, all participants at Asilomar agreed upon guidelines that were formally adopted by the National Institutes of Health in Washington, which provides much of the funding used to support basic biological research in the United States. Paul Berg, a Nobel laureate and one of the originators of recombinant DNA techniques, summed up the scientists' concern at that time: "We are placed in an area of biology with many unknowns; indeed, the greatest risk may well be our ignorance. And it is this ignorance which compels us to pause, reflect, and assess the magnitude of the potential risks associated with this line of research."

Since 1975, an enormous amount has been learned about recombinant DNA molecules and about microorganisms that carry plasmids with foreign DNA. Furthermore, public and scientific concern about the hazards has abated considerably. Even though vast numbers of recombinant DNA experiments have been performed through the years, no adverse effects on the environment or on human health have been reported (see Box 12-3). Guidelines for research are still in effect, but the rules have been greatly simplified, and most of the original restrictions have been eased or eliminated.

Most scientists will admit that the release of genetically engineered plants or bacteria into the environment might entail *some* risk; however, society has always been willing to accept reasonable risks when the benefits exceed them by far. For example, automobiles are not illegal in the United States even though 50,000 persons are killed annually in auto accidents. And the live polio vaccine produces paralytic polio in 1 per 11 million doses. It is a fact of life that there is no such thing as "zero risk."

We should realize that it is not always man-made products that cause harm. Among epidemics caused by microorganisms, AIDS is but the latest of many that have swept through human populations for centuries. Cultivated crops and domesticated herds are routinely decimated by natural pathogens such as the corn blight that ruined the corn crop in the 1970s. Newcastle Disease Virus forces destruction of millions of chickens annually. Mites destroy honey bees, and citrus canker destroys millions of orange trees. Some recombinant DNA organism or microorganism *may* cause environmental harm after its release, but no amount of testing will tell us what the danger may be.

However, certain environmental hazards can be anticipated and regulated. Extensive use of herbicide-resistant plants will necessarily encourage heavier use of herbicides to control weeds in fields of resistant crops. People need to ask how much herbicide use is safe and necessary. Plants that are altered so that they are resistant to attack by insects may cause other kinds of problems that need to be evaluated. For example, disease resistance in plants invariably results from higher levels of toxic chemicals synthesized in the cells of the plants. If people are to eat considerable amounts of genetically engineered, insect-resistant plants, we must be sure that the chemicals that they contain are not toxic or harmful.

The ethical and environmental issues of biotechnology tend to polarize people into opposing

"Ice-minus" Bacteria

BOX 12-3

Protection of crops from frost damage is both highly desirable and potentially quite lucrative for companies that can market an effective frost-preventive agent. It turns out that plants could withstand temperatures at freezing and below if it were not for the bacteria that cover the surfaces of their leaves.

Experiments show that a common species of bacteria (*Pseudo-monas syringae*) found on leaves and stems is responsible for frost damage to plants. These bacteria have proteins on the surfaces of their cells that catalyze the formation of ice crystals even when the air temperature is several degrees above freezing. If it were not for the presence of these bacteria, plant cells would not freeze and fruits and vegetables would not be damaged by frost.

Recognizing the problem, Steven Lindow, a professor at the University of California, Berkeley, genetically engineered a strain of *P. syringae* in which the ice-forming gene was deleted and the ice-forming protein was missing. So-called ice-minus bacteria were now available. Lindow and a biotechnology company wanted to spray a field of plants with ice-minus bacteria just before a freeze to see if frost damage could be prevented. If the laboratory-engineered bacteria could be sprayed over the plants and covered enough of the leaves' surfaces, the plant cells might not freeze. However, the proposed experiment initiated an acrimonious legal debate over whether to permit the deliberate release into the environment of genetically modified microorganisms, specifically those altered by recombinant DNA techniques.

The experiment became even more controversial when it was discovered that a California biotechnology company, Advanced Genetic Sciences, had inoculated some plants on the roof of their research facility without government authorization. After several years of litigation and arguments over the safety of releasing ice-minus bacteria, court approval was finally given. On April 24, 1987, a small patch of strawberries was sprayed with the controversial bacteria developed at Advanced Genetic Sciences. As everyone had expected and predicted, no ecological disaster occurred, but a few environmental activists sneaked in at night and pulled up the strawberry plants.

The lessons of the ice-minus controversy have been important. Rules governing the release of organisms have been clarified. Environmental impact reports must be filed. And, perhaps most important of all, companies have learned that they must

Spraying strawberries in a plot near Brentwood, California, with ice-minus bacteria. Even though the bacteria are not harmful or dangerous, regulations require this "space age" protective clothing. (Courtesy of the *San Francisco Chronicle*.)

communicate more honestly and effectively with the public in the future. Most scientists and companies agree that continued surveillance of the release of genetically engineered organisms is appropriate, both to alleviate public fears and to avoid, insofar as possible, any damage to the environment.

READINGS: J. Doyle, *Altered Harvest*. Viking, 1985. J. L. Marx, "Assessing the Risks of Microbial Release," *Science*. September 18, 1987. D. Russell, "Rush to Market: Biotechnology and Agriculture," *The Amicus Journal*. Winter, 1987.

camps. The most-visible and effective opponent of biotechnology has been Jeremy Rifkin who is the director of an activist organization he founded called the Foundation on Economic Trends. Even Rifkin does not believe that it is possible to halt biotechnology research and product development. He has been effective, however, in slowing things down. Although his views are considered radical, and some would even argue irrational, he has provoked a serious examination of some of the consequences of the advances in biotechnology among scientists and government officials. Rifkin says, "Before we pell-mell rush in to violate species walls and re-engineer biological traits in genetic codes, we ought to take a closer look at it, shouldn't we? We ought to at least ask what is the worthiness of doing this?"

Like most other major technological achievements (the automobile, nuclear energy, computers), those of biotechnology hold great promise for benefiting humankind in many different ways. And, as in all other technologies, the potential for abuse and harm is ever present. This is the most-compelling reason why each one of us needs to understand the genetic and biological principles that underlie genetic engineering. Only then can we, as individuals, make informed judgments about what is best for society and what is consistent with our personal beliefs.

Summary

Biotechnology is the application of modern genetic knowledge to the production of biological products, including not only vaccines, food supplements, diagnostic tools, monoclonal antibodies, and human proteins, but also transgenic plants and animals. It is an outgrowth of recombinant DNA research, which refers to the construction of novel DNA molecules whose genetic information is often derived from different species. These DNAs are inserted into cells and organisms. The DNA is said to be cloned.

In early recombinant DNA research, studies of bacteria infected by phages led to the discovery of restriction enzymes (those that attack DNA at specific sites and destroy it by cutting it into fragments) and modification enzymes (those that attach chemical groups to particular bases in DNA, which prevents its destruction). The potential for joining fragments of DNA from different organisms was soon realized and self-replicating plasmids carrying DNA fragments were used as cloning vehicles.

Three alternative methods are used for cloning a gene or genes: (1) isolation of the piece of DNA itself; (2) isolation of a specific mRNA from which its corresponding DNA fragment can be synthesized; and (3) the chemical synthesis of DNA corresponding to the sequence of amino acids in the protein that it encodes.

Applications of genetic engineering are wide ranging: the manufacture of drugs and other chemical substances, the development of microorganisms for the profitable conversion of agricultural waste into fuels, and the improvement of microorganisms used in the fermentation of foods and beverages. Agriculture will benefit from genetically engineered crops and new plant varieties.

Key Words

bacterial strain   Bacteria of the same species that differ from one another in a specific way that can be measured.

biotechnology   The application of scientific knowledge by industries that produce biological products such as food supplements, enzymes, drugs, and so forth. Some companies use recombinant DNA techniques.

cloning vehicle (vector)   Self-replicating molecules such as plasmids or viruses that carry segments of foreign (cloned) DNA.

gene cloning   The insertion of a particular gene into bacteria or other microorganisms, where it multiplies.

gene library   Millions of bacteria containing all of the genetic information from another organism whose DNA has been cloned and propagated in the bacteria.

genetic engineering   The construction and utilization of novel DNA molecules that have been engineered by recombinant DNA techniques.

modification enzymes   Enzymes that recognize specific base sequences and attach methyl (CH_3) groups to DNA at specific sites.

recombinant DNA   The construction and cloning of novel combinations of DNA molecules not found in nature.

restriction enzymes   Enzymes that recognize specific base sequences and cut DNA molecules at specific sites.

restriction map   A genetic map constructed from the localization of restriction-enzyme sites in DNA molecules.

transgenic plant or animal   An organism into which DNA from another species has been stably inserted.

Additional Reading

Commoner, B. "Bringing Up Biotechnology." *Science for the People*, March–April, 1987.

Davis, B.D. "Bacterial Domestication: Underlying Assumptions." *Science*, March 13, 1987.

Drlica, K. *Understanding DNA and Gene Cloning*. Wiley, 1984.

Hall, S.S. "One Potato Patch That Is Making Genetic History." *Smithsonian*, August, 1987.

Mench, J.A., and A. von Tienhoven. "Farm Animal Welfare." *American Scientist*, November–December, 1986.

Pimentel, D. "Down on the Farm: Genetic Engineering Meets Ecology." *Technology Review*, January, 1987.

Sharples, F.E. "Regulation of Products from Biotechnology." *Science*, March 13, 1987.

Tiedje, J.M., R.K. Colwell, Y.L. Grossman, R.E. Hodson, R.E. Lenski, R. Mack, and P.J. Regal. "The Planned Introduction of Genetically Engineered Organisms: Ecological Considerations and Recommendations." *Ecology*, April, 1989.

Torrey, J.G. "The Development of Plant Biotechnology." *American Scientist*, July-August, 1985.

Yoxen, E. *The Gene Business*. Oxford University Press, 1983.

Study Questions

1   What kind of enzymes are absolutely essential for any recombinant DNA experiment?

2   What are three different methods used to clone a gene of interest?

3   Which virus led to the discovery of restriction enzymes?

4   What is meant by a restriction map?

5   Why are human genes not expressed if human DNA is cloned into bacteria?

6   What is the usefulness of having a gene library?

7   What is meant by the term "ice-minus" bacteria?

Essay Topics

1   Give your views on the safety of releasing genetically engineered bacteria, plants, or animals into the environment.

2   Explain how a human gene such as the one coding for insulin was cloned into bacteria.

3   Discuss whether companies should be allowed to patent any kind of organism that they construct.

Human Reproduction

Sex Determination and Sexual Development

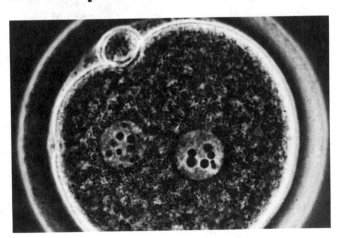

Does science promise happiness? I do not believe so. It promises truth, and it is questionable whether one can ever be happy with the truth.

EMILE ZOLA, *author*

THROUGH THE CENTURIES, philosophers and natural historians have observed that the principal goal of all life is its preservation and continuance. What this means is that the act of reproduction is the most-important behavioral and biological function of any organism. All organisms, from the simplest bacteria to the most-complex animals, propagate themselves and reproduce their likeness. Microorganisms and some plants can reproduce asexually (without mating), but virtually all animals rely on sexual matings to reproduce. Sexuality is the most-essential biological condition for the continuance of life. Thus, it is not surprising that sexual development is one of the most fundamental of all biological processes.

In infancy, we automatically begin to explore the sexual parts of our bodies. In early childhood, we begin to notice that boys and girls are physically different and behave differently. In adolescence, we pass through the psychological stress of establishing our sexual identities and of learning how to relate to persons of the opposite sex. In adolescence, too, our bodies complete the process of sexual differentiation, and we mature into men and women who are capable of participating in the endless cycle of human reproduction.

For many species of animals, especially birds, the observable physical differences between males and females are dramatic; generally, the male is the larger and more-colorful of the pair. **Sexual dimorphism** is the term for this distinctly different physical appearance of males and females of a species. Darwin was very aware of sexual dimorphism among the different species of animals and wrote an entire book on the various mechanisms of sexual selection and their relevance to evolution. It is clear from Darwin's work that he viewed sexual selection as being only slightly less important than natural selection in creating new types of individuals and species of organisms.

Despite the enormous effort that has been made in the study of **sexual differentation**— how males become males and females become females—only the general scheme of the biological processes that produce males and females is understood even today. Sexual differences, sexual preferences, and mating mechanisms are thought to play important roles in the survival and evolution of organisms, but the evidence supporting these quite-reasonable ideas is still inconclusive. Even the question of whether males choose the females with whom they mate or whether females choose males is a difficult one to answer from studies of animal species. Human sexual behaviors are even more complex than those of other animals because much sexual behavior— and the attitudes underlying it—is learned, not instinctive.

Some plants and a few animals have developed strategies for reproduction that do not require matings between individuals of the opposite sex. About half of all flowering plants are hermaphrodites; that is, an individual plant (such as the pea plants in Mendel's experiments) has both male and female reproductive organs. Hermaphrodites are much less common among animal species; hermaphroditism is not a normal developmental pattern in human beings, although, on rare occasions, human hermaphrodites, or pseudohermaphrodites, have been observed. Many female plants and some female animals can reproduce without being fertilized by male gametes. The process by

Parthenogenesis
Could It Happen in Human Beings?

BOX 13-1

Some species of plants and animals are able to reproduce without fertilization of the egg by sperm—a process known as *parthenogenesis*. Reproduction by parthenogenesis occurs only in females; therefore, all parthenogenetic progeny also are female. And, because all the daughters are genetically identical with their mother, they are clones.

Parthenogenesis is rare in animal species. It has been observed among some species of insects, fishes, salamanders, and lizards, but it has never been reliably demonstrated for any species of mammal. Nevertheless, numerous claims of human parthenogenesis, or "virgin birth," have been made through the years.

Within the past few years, a careful study was undertaken of parthenogenesis in a species of whiptail lizards that live in New Mexico (see the accompanying drawing). These lizards can now be raised in the laboratory, and occasionally a female that is raised separately from male lizards will give rise to daughters who, in turn, grow up and produce more daughters without any male contact. In nature, the whiptail lizards normally reproduce sexually and the female's eggs are fertilized by male sperm. However, in whiptail lizards, and in certain other species of lizards, the ability for certain females to reproduce by parthenogenesis arose as a consequence of one or more mutations in some individuals. The mutations alter genes that affect the meiotic process in such a way that the egg retains the diploid number of chromosomes instead of having the haploid number. These diploid eggs are able to develop without being fertilized by sperm because they have received all of the necessary chromosomes from their mother.

In environments where male lizards are scarce, parthenogenetic females have a distinct reproductive advantage over sexually reproducing lizards. When males are scarce, parthenogenesis allows females to reproduce, maintaining the survival of the population. Although parthenogenesis has not been observed in mammals, the chance of its happening as a result of mutations cannot be ruled out. If you read one day that women have found a way to reproduce by parthenogenesis, it may be time to prepare for the best of all possible worlds—or the worst, depending on your sex.

READING: David Crews, "Courtship in Unisexual Lizards: A Model for Brain Evolution," *Scientific American*. December, 1987.

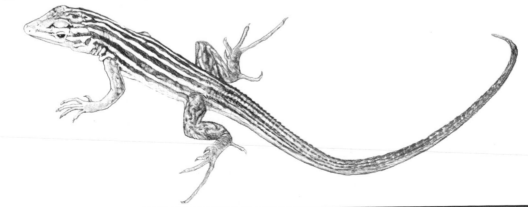

which this is accomplished is called **partheno-genesis**; such females produce only female offspring (see Box 13-1).

The biological mechanisms of sex determination and sexual reproduction vary from species to species, but for most organisms male and female characteristics are genetically determined by the sex chromosomes. These chromosomes, however, only start the overall process of sexual differentiation resulting in males and females. This developmental process is also affected by specific genes on other chromosomes and by the production of sex hormones at the appropriate time in each organism's development. For human beings, the psychological and social environment in which the newborn infant is raised also affects its sexual development.

This chapter discusses the genetic basis for sexual differentiation and the functions of the sex chromosomes in sexual development. Most of us hold strong beliefs about what we regard as "normal" sexual behaviors, sexual appearance, and reproduction. Yet, as you will discover in this chapter, external sex characteristics can be misleading indicators of what sex chromosomes people carry and even of their reproductive capability as males or females. Also under consideration are the basic processes of human reproduction, including spontaneous abortion and *in vitro* fertilization.

Discovery of Sex Chromosomes

Until the beginning of this century, nothing was known about how plants and animals produce either male or female offspring. Since the time of the early Greeks, people had held to the notion that a miniature being, called a **homunculus** ("little man"), existed inside the head of each sperm. It was thought that, after a sperm

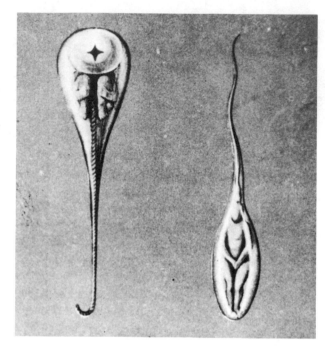

FIGURE 13-1   The homunculus idea. For more than two millennia, it was believed that little persons called homunculi (sing. homunculus) were carried in sperm. This drawing is a seventeenth-century interpretation. (Courtesy of Dorsey Stuart, University of Hawaii.)

was deposited in the woman's womb, the homunculus in the sperm head increased in size until it was finally born (Figure 13-1).

Sexual characteristics are among the most easily recognized of all traits, yet, before Mendel conceived of the idea of particulate hereditary factors (genes), no one had ever been able to formulate a genetic basis for sex determination because all views of heredity were based on the incorrect ideas of pangenesis (discussed in Box 1-1). Following the rediscovery of Mendel's ideas at the beginning of this century, scientists began to look for hereditary factors that would explain sex determination and that would be analogous to the ones Mendel had shown to exist for other traits.

At the turn of the century, it was discovered that male and female characteristics are determined not by single genes but rather by a pair of sex chromosomes. Microscopic examination of chromosomes in insect reproductive cells revealed that the cells of some individuals contained a single unusual chromosome; it was named "X" because its function at that time was unknown. Other reproductive cells did not have the X chromosome but had a much-smaller chromosome that was called "Y." Eventually, it became apparent that the reproductive cells of all animals contain either a large X chromosome or a small Y chromosome in addition to the normal number of autosomes. Examinations of sperm from various species of animals ultimately produced the hypothesis that cells contain specific sex chromosomes that determine the sexual phenotype of the individual.

In most species of plants and animals, male somatic cells contain both X and Y chromosomes, and male reproductive cells produce sperm bearing either an X or a Y chromosome. Female somatic cells contain two X chromosomes and female reproductive cells produce eggs that all bear one X chromosome (Figure 13-2). Individual organisms that produce gametes of only one type with respect to the sex chromosomes are called the **homogametic sex**; those that produce gametes with two types of sex chromosomes are called the **heterogametic sex**. In human beings, females are the homogametic sex and males are heterogametic.

Variation in Sex Determination

Although in many species of plants and in all species of mammals (including human beings) females have two X chromosomes and males have an X and a Y, other forms of sex determi-

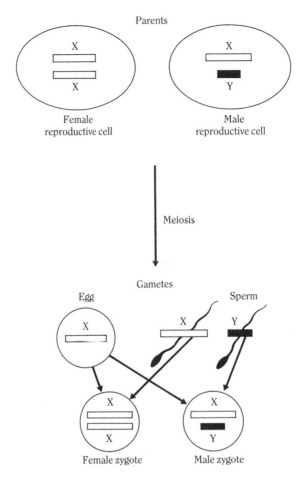

FIGURE 13-2 Sex-chromosome constitution in male and female human beings. Sperm normally contain either an X or Y chromosome; normal eggs contain one X chromosome.

nation are found in nature (Figure 13-3). In many species of birds, the males and females are determined by X and Y chromosomes, but the chromosomal constitutions of the sexes are reversed when compared with mammals. In chickens, for example, roosters are the homogametic sex, whereas hens are the heterogametic sex. Different letters are assigned to the sex chromosomes of birds to distinguish them

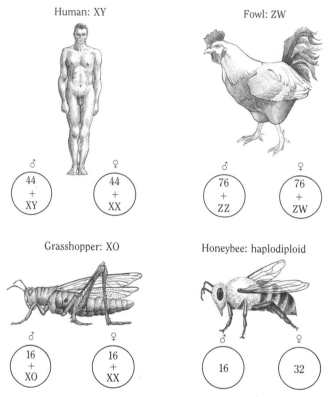

FIGURE 13-3 Variation in sex determination among species of animals. In human beings, males are the heterogametic sex (XY) and females are the homogametic sex (XX). Chickens are the reverse: hens are XY (usually the letters ZW are used instead of XY) and roosters are XX. Grasshoppers have no Y chromosomes; males have one X chromosome and females have two. The sex of honeybees is determined by whether an individual bee has the haploid or the diploid number of chromosomes.

from the X and Y chromosomes of mammals. Roosters (males) are ZZ and hens (females) are ZW with respect to their sex chromosomes (Z = X, W = Y).

In some insect species, including grasshoppers, the cells have no Y chromosome and sex is determined by the number of X chromosomes. Thus, female grasshoppers have two X chromosomes, whereas males have only one. Honeybees streamline the process of sex determination even further; their cells do not contain any specific sex chromosome. Instead, all fertilized eggs of honeybees are diploid and automatically become females; any eggs that remain unfertilized are haploid and develop into males.

Among animals whose sex is determined by the usual complement of XX and XY chromosomes, the sex and fertility of an individual is often affected by the presence of more or less than the normal number of sex chromosomes. For example, a single X chromosome (XO) results in a sterile human female, whereas a fruit fly with only one X chromosome becomes a male. And, in mice, a single X chromosome results in the development of a fertile female (Table 13-1). The presence of an extra X chromosome in the cells of male human beings and of male mice (XXY) results in sterility, but an extra X chromosome in fruit flies (XXX or XXY) produces fertile females. This brief survey

TABLE 13-1   The effect on reproduction of abnormal sex-chromosome constitution in three different species of animals.

Sex-chromosome constitution	Human beings	Mice	Drosophila
XX	Normal female	Normal female	Normal female
XY	Normal male	Normal male	Normal male
XXY	Sterile male	Sterile male	Fertile female
XYY	Fertile male	Semisterile male	Fertile male
XO	Sterile female	Fertile female	Sterile male

of the diverse effects of the number of sex chromosomes in different animal species shows that male and female characteristics and an organism's fertility are determined in quite different ways by the complement of X and Y chromosomes.

Abnormal Human Sex Determination

Determining a person's sex usually poses no special problem: we do not need to know anything about the number of sex chromosomes to decide whether a person is male or female. Whenever we find ourselves in a new group of people, we unconsciously take note of the sex of those around us on the basis of their appearance or behavior. Indeed, the first piece of information that our minds instinctively seek about another person is whether that person is male or female. Generally, we infer this information from the person's dress, hairstyle, presence or absence of facial hair, and body build.

Chapter 3 described the effect of an abnormal number of sex chromosomes in either men or women on their development and reproductive capacities. Recall that women with only a single X chromosome (Turner syndrome) are sterile, as are men with an extra X chromosome (Klinefelter syndrome). Sexual abnormalities result not only from chromosomal abnormalities, but also from abnormalities in sex hormones or in their receptors on cells. One of the most remarkable of the many kinds of human sexual abnormalities is **androgen insensitivity syndrome** (also called *testicular feminizing syndrome*), in which a person's outward physical sexual appearance is completely female despite the fact that she is genetically male (XY) in all of her somatic cells. Such people generally become aware of their unusual condition in adolescence, when they fail to begin to menstruate, or even later, when they discover that they are sterile. They have internal testes, which quite often produce functional sperm. However, the sperm cannot be transported to any external sexual organ because the sperm duct has not developed and the external sex organs are female.

It is now known that androgen insensitivity syndrome is caused by a mutant allele of a gene located on the X chromosome. The product of this gene, which is normally present on the surface of male cells, allows sexually undifferentiated cells to respond to male hormones (called **androgens**) during development of the embryo. If the cells are unable to respond to male hormones, embryo development continues as female.

It should be emphasized, particularly in view of the widespread misguided notions of male superiority in our society, that all human embryos will develop as females, irrespective of whether the sex-chromosome constitution is XX or XY, unless the appropriate male hormones are produced and function properly at early stages of development (Figure 13-4).

Another uncommon abnormality in human sex determination can be caused by a hormone imbalance that occurs early in development. Some embryos begin with a normal female karyotype (XX) but develop as males and are classified as boys at birth. It turns out that the developing female embryo can be masculinized by a genetic defect that blocks production of the hormone cortisol in the embryo's adrenal glands, causing them to release male hormones (androgens) instead. Thus, we can see from these examples of abnormal sexual phenotypes that human sex determination results from a series of complex interactions between sex chromosomes, autosomal genes, and hormones.

FIGURE 13-4   Because of the biblical description, artists traditionally show Eve being formed from Adam's rib. Could it have been the other way around? (Courtesy of Archive Alinari, Florence, Italy.)

Male-Female Differences

All babies are sexed at birth and, except for the most-obvious abnormalities, are classified as either male or female. But can any person be regarded as 100 percent male or 100 percent female? It seems reasonable to suppose that, within these two absolute sex categories, there are small differences in sex determination. All normal men may possess testes and male genitals and produce viable sperm, but they may differ in hormone levels, sperm production, and other aspects of maleness. So, too, small differences in female sexual development may produce women with different female hormone levels and other differences in feminine characteristics.

In addition to biological differences in sex determination, a person's sex-role identity is influenced by the environment in which he or she is raised. Our sexual development and preferences depend on the particular sexual behaviors and acts that are encouraged or discouraged by the society in which we live. In recent years, the unfounded belief in female mental inferiority and physical weakness in western societies has been actively opposed by organized groups of both men and women (see Box 13-2). Although it is true that there are biological differences between men and women, it certainly has not been shown that sex-determined differences affect any person's overall capacities.

The H-Y Antigen and Male Development

While studing the acceptance and rejection of various skin grafts in highly inbred strains of mice in the 1950s, Ernst Eichwald, a scientist at the University of Utah, noticed that the grafts were always successful between any pair of inbred strains of mice *except* when the donor mouse was male and the acceptor mouse was female. From this observation, it was eventually discovered that male mice have a protein on the surface of their cells that is recognized by antibodies circulating in the blood of female mice and that causes the females to reject the male tissue (tissue rejection and the immune system are discussed in Chapter 17).

This male-specific protein is now called the **H-Y antigen** (H-Y stands for histocompatibility-Y chromosome) and, for many years, was thought to be the protein that con-

Sexism
Prejudice Against Women

Throughout history, women have been regarded as fundamentally different from men—and "different" has almost invariably meant inferior—both in physical and in mental capabilities. Women have been defined by men as "the weaker sex": physically weaker, psychologically more unstable, and in general biologically inferior. In addition to being physically smaller (hence, weaker), women have been regarded as having more "irritable" nervous systems, which accounts for their more-frequent "nervous breakdowns" and higher incidence of insanity. In the nineteenth century, brain size was assumed to measure intelligence, and the smaller brain size of women, as well as their overall lack of intellectual accomplishments, was "proof" of their mental inferiority. After all, men argued, how many great artists, musicians, scientists, or philosophers have been women?

Even the astute reasoning of Charles Darwin (who is regarded by many as exemplifying a man of particular brilliance) led him to biased and unfounded conclusions on the issue of female intelligence. In his book *The Descent of Man and Selection in Relation to Sex*, published in 1871, Darwin wrote:

The chief distinction in the intellectual powers of the two sexes is shown by man's

BOX 13-2

attaining to a higher eminence, in whatever he takes up than can woman—whether requiring deep thought, reason, or imagination, or merely the use of the senses and hands.

Even Darwin was unable (or unwilling) to consider that social and cultural prejudices and conditions may have prevented women from "attaining to a higher eminence" throughout history.

Controversy over female intellectual ability is still very much in evidence. In 1980, psychologists Camilla P. Benbow and Julian C. Stanley published a study on the mathematical abilities of seventh- and eighth-grade boys and girls. Both groups had taken the same number of courses in mathematics, yet the boys achieved better grades on the mathematical sections of scholastic aptitude tests. From this evidence, Benbow and Stanley concluded that boys are biologically (genetically) more talented in mathematics than are girls. The conclusions of these psychologists were questioned for many reasons, not the least being that the social and cultural exposures that boys and girls have are very different. For example, how many girls receive calculators or computers as presents?

Lest any reader think that women's groups make much ado about nothing, it is worth quoting what a respected male scientist, Gustave Le Bon, a founder of social psychology and a contemporary of Darwin, wrote in 1879:

All psychologists who have studied the intelligence of women . . . recognize today that they represent the most inferior forms of human evolution and that they are closer to children and savages than to an adult civilized man. They excel in fickleness, inconstancy, absence of thought and logic, and incapacity to reason. Without doubt there exist some distinguished women, very superior to the average man, but they are as exceptional as the birth of any monstrosity, as for example, of a gorilla with two heads.

Even though this grossly sexist statement was made more than a hundred years ago, less-blatant views are often expressed by contemporary psychologists and scientists in language that is scarcely more rational. Sexism is only one example of the misuse of biological facts to justify prejudice against a particular group of people.

READINGS: Stephen Jay Gould, "The Brain Appraisers," *Science Digest*. September, 1981. Jon Beckwith and John Durkin, "Girls, Boys and Math," *Science for the People*. September–October, 1981. C. Benbow and J. C. Stanley, "Sex Differences in Math Reasoning," *Science News*, 119, 147 (1981).

trols the formation of testes in both mouse and human male embryos. The H-Y antigen is thought to be coded for by a gene or genes on the Y chromosome, but its definite assignment to the Y chromosome is still uncertain. If the H-Y antigen is eventually located on the Y chromosome, it will be the first definite assignment of a gene to that chromosome. (A gene for hairy ears in men is often cited as residing on the Y chromosome, but that assignment has been questioned.)

Recent research suggests that the role of the Y chromosome in determining maleness is not yet well understood. For example, some persons have a sex-chromosome complement that is inconsistent with their sex. Six phenotypically normal but infertile women were described, and all of them have XY chromosomes. However, the Y chromosome is smaller than usual, suggesting that a piece had been deleted.

When cells were examined from men who produced no sperm, it was found that all of them had two X chromosomes. Further analysis showed that they failed to develop as females because a small piece of the Y chromosome had translocated onto one of their X chromosomes. Although the Y chromosome must participate in some way, no genes have been identified that determine maleness with certainty.

Hormones and Sexual Development

The complete development of a human male or female is conveniently described as taking place in three stages. The first step in sex determination begins with the chromosome constitution of the zygote. As already noted, two X chromosomes usually lead to the development of a female, and X and Y chromosomes usually lead to the development of a male. Chromosomes determine **primary sex differentiation**: the

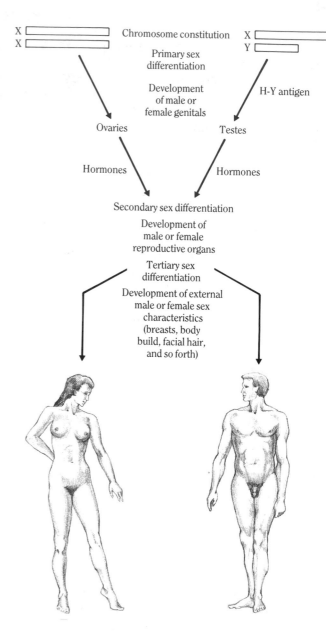

FIGURE 13-5 Stages in the sexual development of males and females. In human beings, XX chromosome constitution determines primary female characteristics; XY constitution determines primary male characteristics. Hormones then direct the development of secondary and tertiary sex characteristics.

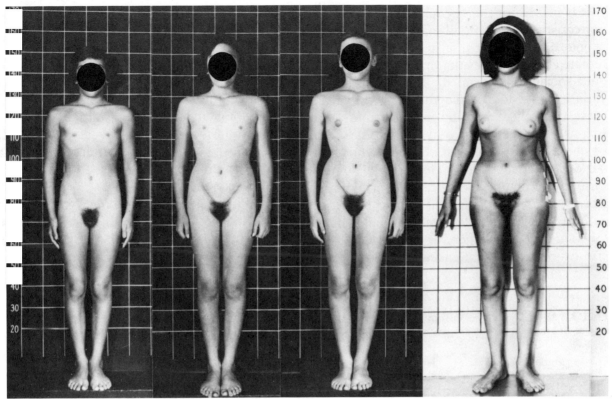

FIGURE 13-6 Medical transformation of a male (XY) into a female. The person, shown at ages eleven, twelve, thirteen, and nineteen, was raised as a male until age eleven, when it was decided to change her secondary sexual characteristics. At ages eleven and nineteen, she received surgical feminization; in the intervening years she was given hormone therapy. (Courtesy of Johns Hopkins School of Medicine and Johns Hopkins University Press.)

formation of female ovaries or male testes, which in the adult will produce eggs or sperm, respectively. However, once the embryonic gonads have formed (usually about the sixth week after fertilization), the internal sex organs themselves produce hormones that influence further sexual differentiation during development of the embryo and that continue to affect sexual development after birth. The hormones produced by the gonads affect other tissues and direct what is called **secondary sex differentiation**: the formation of male and female reproductive organs. **Tertiary sex differentiation** occurs at puberty and includes deepening of the voice in males, development of breasts in females, and growth of pubic hair in both sexes (Figure 13-5).

Regardless of a person's chromosome consitution or even the presence of ovaries or testes, secondary and tertiary sex characteristics can be modified or reversed by surgery and hormone therapy. Some people are born with abnormal secondary sex characteristics that may have been caused by abnormal chromosome constitution, by gene defects, or by hormone imbalances. If such persons are identified, modern medical therapies and psychological counseling applied at the appropriate time in development can help them cope with their aberrant sexual phenotypes (Figure 13-6).

It is important to remember that nature does not construct perfect "all male" or "all female" people. Each person is a mixture of male and female characteristics that are both biologically and socially determined. Thus, considerable variation in sexual development is possible, and these differences may be manifested in adult sexual behaviors. In recent years, our society has become more tolerant of people whose sexual characteristics and preferences fall outside formerly acceptable biological and social norms. However, many people still adhere to narrow definitions of what is acceptable sexual behavior for men and for women, as well as what constitutes normal sexual appearance.

Human Gametes

Plants and animals normally produce new individuals by the fusion of male and female gametes to form a *zygote*, a fertilized egg that is the first cell of a new individual. The time scale of development in different animal species varies enormously: a new fruit fly larva develops from a fertilized egg in about thirteen hours, whereas a new human being requires about nine months of development (Figure 13-7).

In human beings, each male gamete contains the haploid number of chromosomes in a sperm cell that is only about 0.06 mm in length. Each sperm carries either an X or a Y chromosome in

FIGURE 13-7   The stages of human development from zygote to birth.

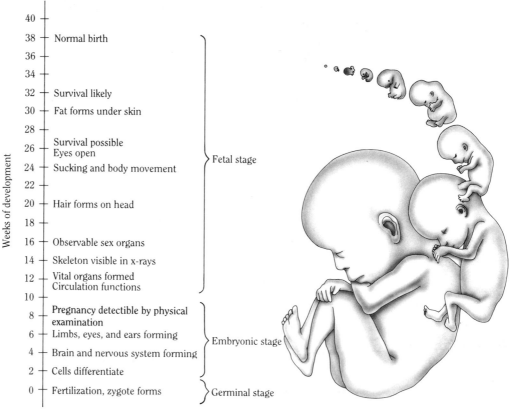

Weeks of development

40
38 — Normal birth
36
34
32 — Survival likely
30 — Fat forms under skin
28
26 — Survival possible / Eyes open
24 — Sucking and body movement
22
20 — Hair forms on head
18
16 — Observable sex organs
14 — Skeleton visible in x-rays
12 — Vital organs formed / Circulation functions
10

} Fetal stage

8 — Pregnancy detectible by physical examination
6 — Limbs, eyes, and ears forming
4 — Brain and nervous system forming
2 — Cells differentiate

} Embryonic stage

0 — Fertilization, zygote forms

} Germinal stage

addition to the twenty-two autosomes. About 300 million sperm, synthesized in the male testes, are released in each ejaculation. Many sperm will attach to an egg; however, one and only one sperm will penetrate and fertilize the egg.

Male chromosomes are contained in the nucleus of the sperm. The nucleus is located beneath the **acrosomal cap**, which contains proteins that recognize the egg's surface and help the sperm attach to the egg and penetrate it (Figure 13-8). Mitochondria, located in the tail close to where it joins the head, supply the energy that the sperm needs to swim vigorously through the vagina and uterus en route to an egg traveling toward it down the Fallopian tube. Depending on conditions in the vagina and uterus, the journey of the millions of sperm may take hours or even days before they reach the egg and one of them fertilizes it.

Human eggs (*ova*), which are the largest of all human cells (about 0.1 mm in diameter), are produced in the ovaries. At birth, a woman's ovaries have already produced all of the eggs she will have during her lifetime, and all of the eggs will have undergone the first meiotic division. At puberty, when a woman begins to menstruate, the first eggs are released in a process that will be repeated each month for thirty to forty years.

The human egg is much more complex in structure than the sperm. In addition to having the haploid number of chromosomes consisting of an X chromosome and 22 autosomes in the nucleus, an egg has a complex internal structure that provides all of the cellular components necessary for the early stages of development. The outer surface of the egg is surrounded by a gelatinous substance called the **zona pellucida**, which acts as a barrier against the egg's being fertilized by sperm from other species. The zona pellucida is also thought to provide a

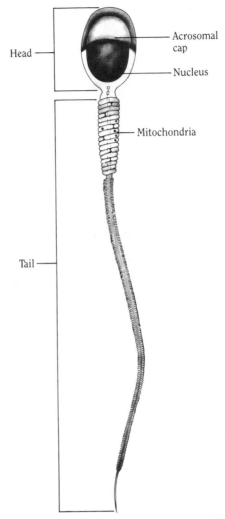

FIGURE 13-8 Diagram of a human sperm. The acrosomal cap consists of proteins that help the sperm attach to an egg and penetrate it. The nucleus contains the haploid set of chromosomes. The mitochondria supply the energy that allows the sperm to swim.

mechanism that allows only a single sperm to penetrate the egg.

When a sperm finally becomes attached to the surface and penetrates an egg, the sperm's tail is left behind on the egg's inner surface, where it disintegrates. The head of the sperm is

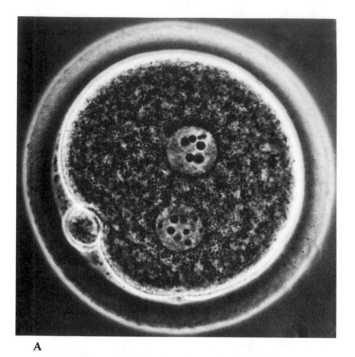

A

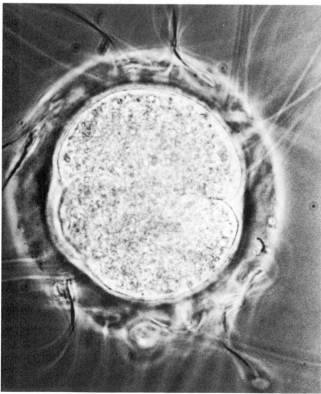

B

absorbed into the egg and begins to swell before releasing its nucleus. A human sperm can attach to an egg and be swallowed in less than a minute. Within 20 minutes, the sperm and egg nuclei have fused, the diploid chromosome number has been restored, and **fertilization** has been accomplished (Figure 13-9A). After the egg has been fertilized, development proceeds very rapidly. DNA synthesis begins almost immediately, and within a few hours the zygote undergoes its first mitosis and division into two cells (Figure 13-9B).

Perhaps what is most remarkable about fertilization and embryonic development is that the process functions correctly so much of the time. However, there are many ways in which reproduction can go awry. During the production of hundreds of millions of sperm (many billions over a man's lifetime), the synthesis of abnormal sperm is not uncommon. Sperm with two tails or two heads are often observed. If the proportion of abnormal sperm in a man's ejaculate is too high, he may be infertile.

Reproductive abnormalities in women also can prevent fertilization and cause infertility. After sperm have been deposited in a woman's vagina (either by intercourse or by artificial insemination), they undergo chemical changes called **capacitation** as they swim through the secretions in the vagina and uterus and into the Fallopian tubes, where fertilization takes place. The secretions in a woman's reproductive system may fail to capacitate sperm, or they may even inactivate them. Another possible cause of female infertility is defective Fallopian tubes.

FIGURE 13-9
A. A fertilized egg showing both the egg's nucleus and the sperm's nucleus. The structure at the lower left is a polar body, one of the meiotic products that does not become an egg.
B. A fertilized egg about to undergo the first cell division. The nuclei from the egg and sperm have fused and the chromosomes are no longer visible. (Courtesy of Jane Rogers, Vanderbilt University.)

After it is released from the ovary, the egg must travel down a Fallopian tube to become fertilized; then it must attach to the wall of the uterus. If the tubes are blocked as a consequence of disease, injury, or malformation, fertilization may be prevented. Another reason for infertility is that the uterus may not have been adequately prepared by hormones for the egg's attachment.

In Vitro Fertilization and Surrogate Motherhood

The first successful *in vitro* fertilization (IVF) of a human egg was achieved in 1969 by three English scientists. Nine years later, two of them, Richard G. Edwards and Patrick C. Steptoe, implanted the egg that became Louise Joy Brown, the first baby to be born from an *in vitro* fertilized human egg. Human *in vitro* fertilization has been attempted many times since then, but many more of the attempts fail than succeed because the procedure is difficult. First, several eggs must be surgically removed from a woman's ovaries without damaging the eggs. Next, the eggs must be fertilized *in vitro* and cell division must be initiated. Then, a single fertilized egg must be selected and implanted in the woman's uterus and a pregnancy established. The implantation is the most-difficult procedure to control because the uterus must be prepared for pregnancy by hormones that are given the woman before implantation. The dividing egg (now called a **conceptus**) is implanted at the four-to-eight-cell stage of cell division (Figure 13-10). Usually, several fertilized eggs are implanted to increase the likelihood of a pregnancy. However, this also increases the chance of multiple births, such as twins or triplets.

The success rates for pregnancy and live birth by IVF range from about 15 to 30 percent. Thus,

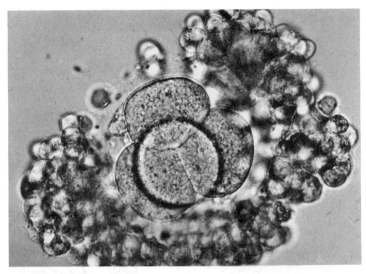

FIGURE 13-10   An *in vitro* fertilized egg that has undergone several cell divisions. It is now ready for implantation into a woman's uterus. The woman has been given hormones to assist in establishing a pregnancy. (Courtesy of S. Fishel, University of Nottingham.)

many women will have to try repeatedly before becoming pregnant by IVF. And some women never become pregnant even after repeated attempts. Despite the relatively low success rate, more than two hundred IVF clinics are now operating in the United States.

The increased ability to freeze human embryos at the four-to-eight-cell stage for later implantation has given rise to unusual legal and ethical problems. In one case, a woman whose fertilized eggs were frozen died in a plane crash, along with her husband. The question then is: Who "owns" the frozen concepti? In another case, successful pregnancies were established about a year apart in a woman using her own frozen embryos. The babies turned out to be identical twins—but one twin is a year older than the other!

Another option for infertile couples is to contract for a surrogate mother to bear a baby for them. Advertisements such as the following one are typical:

> WANTED: surrogate mother. Infertile married couple willing to pay fee to white single woman to bear child for them. Conception to be achieved through artificial insemination supervised by medical doctors. Child to be given to married couple through adoption court. All expenses paid plus fee. All replies confidential.

Such advertisements point up the willingness of many infertile couples to have another woman bear their child. The husband's sperm is used to artificially inseminate a fertile woman who agrees to deliver a child to the couple, generally for a fee. Although it has not yet been reported, it is technically possible for the *in vitro* fertilized egg derived from one couple to be implanted in another woman, who could go through pregnancy and give birth to the couple's child. A woman might make a successful career of bearing other couples' children for a fee. Some people are morally outraged by the prospect of made-to-order babies and surrogate motherhood; others view such prospects as beneficial and logical outgrowths of the recent advances in reproductive technology.

Chromosomal Abnormalities and Spontaneous Abortions

Although most of the controversy over abortion in the United States is concerned with medically assisted termination of pregnancy, the fact is that most human abortions occur spontaneously in pregnant women. As mentioned in Chapter 3, it is estimated that as many as 15 percent of all confirmed pregnancies spontaneously terminate *in utero*. When cells from spontaneously aborted embryos are examined, at

TABLE 13-2 Chromosomal abnormalities observed in cells from spontaneously aborted human embryos.

Chromosomal abnormality	Frequency (percent)
Trisomy 16	7–8
Trisomy 13, 18, or 21	4–5
All other trisomies	13–16
Extra or missing X or Y	9–10
Triploidy	6–7
Tetraploidy	2–3

least half of them have observable chromosomal abnormalities of various kinds (Table 13-2). The remainder also may have genetic defects, such as small translocations or inversions, which cannot be detected by examining chromosomes under the microscope. The frequency of chromosomal abnormalities indicates how often mistakes occur in meiosis. The high frequency of spontaneous abortions also shows that nature has evolved mechanisms to recognize and eliminate genetically abnormal embryos. As Table 13-2 shows, some chromosomes are more likely to segregate improperly in meiosis than are others. Most embryos with an extra chromosome do not survive to term; one of the few exceptions is an extra chromosome 21, which results in Down syndrome.

From about 5,000 to 10,000 IVF babies have been born worldwide since Louise Brown was delivered in 1978. Studies of these pregnancies show that the frequency of spontaneous abortions is about twice as high among IVF women as it is among women who conceive through intercourse. Part of the difficulty with IVF pregnancies may be due to the hormonal manipulations that the women undergo because most women who elect IVF are older than the age at which most women become pregnant. However, it should be noted that the overall health and intelligence of children resulting from IVF pregnancies is not significantly different from that of the general population of children.

We have just seen that, if either the egg or the sperm has an abnormal number of chromosomes, the embryo may abort spontaneously early in development. During embryonic development, cells must undergo hundreds of cell divisions, and each time the cell's chromosomes must undergo mitosis and the cells must divide with precision. Thus, genetic errors may arise in chromosomes during development of the embryo, causing it to abort. Finally, during the many months of embryonic development, genes must be switched off and on and expressed in a precise sequence if abnormal development is not to result. That so many of us are born biologically and genetically normal testifies to the many failsafe mechanisms that nature has developed. Despite these mechanisms, hereditary diseases and defects do occur with predictable frequencies. The various kinds of hereditary disorders are discussed in the next chapter.

Summary

Sexual differentiation in most diploid organisms is genetically determined by the sex chromosomes. In human beings, female cells normally contain two X chromosomes and male cells have one X and one Y. However, sexual differentiation is also affected by autosomal genes and, in the course of development, by hormones. It can be thought of as a three-stage process: chromosomes determine primary differentiation, the formation of ovaries or testes; sex hormones determine both secondary differentiation, the formation of reproductive organs, and tertiary differentiation, the development of sexual characteristics at puberty. Thus, sexual abnormalities may be caused not only by chromosomal abnormalities, but also by hormonal abnormalities or by abnormalities in the cellular receptors of sex hormones.

In human reproduction, male and female gametes fuse to form a zygote, or fertilized egg. Of the approximately 300 million sperm released in one ejaculation, many will attach to an egg, but only one will penetrate it. Proteins in the acrosomal cap of the sperm enable it to recognize and penetrate the egg's surface. Like sperm, egg cells contain twenty-two autosomes and one sex chromosome, but they are more complex in that they contain all the cellular components necessary for early development. After fertilization, the first cellular division starts the process of about nine months of development from zygote to birth.

In *in vitro* fertilization, eggs are surgically removed from a woman's ovaries and fertilized. After cell division begins, the conceptus is implanted in the uterus, which has been prepared for pregnancy by hormones that have been administered to the woman. The procedure is difficult, as indicated by the rates of success for pregnancy and live births by IVF, which range from 15 to 30 percent.

Most abortions are spontaneous, and examination of the cells of aborted fetuses reveals that at least half of those examined contain chromosomal abnormalities. Such abnormalities may be present in the egg or sperm or they may arise in embryonic cells undergoing mitosis in the course of development.

Key Words

acrosomal cap Proteins surrounding the head of a sperm that facilitate attachment to an egg.

androgen insensitivity syndrome A syndrome characteristic of women who have an XY chromosome constitution.

androgens   A group of male hormones (the major one is testosterone) that are produced early in development and result in the formation of testes.

capacitation   Physiological changes in a sperm as it moves up the reproductive tract that enable it to penetrate an egg.

conceptus   A fetus; the product of conception.

fertilization   The union of male and female gametes to produce the first cell (zygote) of a new organism.

heterogametic sex   The sex of a species that produces gametes with two types of sex chromosomes.

homogametic sex   The sex of a species that produces gametes containing only one type of sex chromosome.

homunculus   A little individual that early Greeks believed was contained in the head of each sperm.

H-Y antigen   A protein that participates either in testes development or in sperm production in males.

in vitro **fertilization**   The formation of a zygote in a test tube by mixing sperm and eggs together.

parthenogenesis   Reproduction by females without fertilization by male gametes. This form of reproduction produces only female offspring.

primary sexual differentiation   Chromosomal determination of male or female development.

secondary sexual differentiation   Hormonal determination of male or female development.

sexual differentiation   Expression of a male or female phenotype through development of sexual organs and characteristics.

sexual dimorphism   The distinctly different physical forms of males and females in a species; for example, male birds are generally larger and more colorful than their female counterparts.

tertiary sexual differentiation   Changes that occur at puberty in male and female human beings.

zona pellucida   The gelatinous outer covering of an egg that allows sperm of the same species to attach and that facilitates fertilization by a single sperm.

Additional Reading

Birnholz, J., and E.E. Farrell. "Ultrasound Images of Human Fetal Development." *American Scientist*, November–December, 1984.

Eberhard, W.G. "Runaway Sexual Selection." *Natural History*, December, 1987.

Edwards, R.G., and P.A. Steptoe. *A Matter of Life*. William Morrow, 1980.

Griffin, J.E., and J.D. Wilson. "Syndromes of Androgen Resistance." *Hospital Practice*, August 15, 1987.

Grobstein, C. "The Early Development of Human Embryos." *The Journal of Medicine and Philosophy*, 10, 213 (1985).

Hassold, T.J. "Chromosomal Abnormalities in Human Reproductive Wastage." *Trends in Genetics*, April, 1986.

Morse, G. "Why Is Sex?" *Science News*, September 8, 1984.

Wassarman, P.M. "The Biology and Chemistry of Fertilization." *Science* 235, 553 (1987).

Study Questions

1   What class of molecules determines a person's primary sexual characteristics? Secondary sexual characteristics?

2   What are the sexual phenotypes of people having the following chromosome constitutions: XO, XX, XXX, XY, XXY, XYY?

3   What is meant by sexual dimorphism?

4   If an animal reproduces by parthenogenesis, what will be the chromosome consititution of its progeny?

5   Other than abnormalities in sex chromosomes, what kinds of molecular changes can cause abnormal sex determination in people?

6   What is the most-common cause of spontaneous abortions?

7   What is a major problem with *in vitro* fertilization among women who conceive by this technique?

Essay Topics

1   Discuss whether or not you would consider having a child by *in vitro* fertilization if you were unable to conceive by intercourse.

2   Discuss whether you think that sexism exists in our society and, if so, describe the form that it takes.

3   Discuss some of the biological mechanisms that contribute to abnormal sexual development in humans.

14

Human Hereditary Diseases

Patterns of Inheritance

The breakthroughs in genetic diagnosis are providing us with far more information that we could have dreamed of 40 or even 30 years ago, but the problem of how to use that information wisely remains — as it always has been — difficult and elusive.

F. CLARKE FRASER, *physician*

ENORMOUS STRIDES have been made in the past century in the prevention and cure of many serious human diseases. Epidemics that formerly decimated human populations have been eliminated in many areas of the world. Bacteria that caused outbreaks of plague, cholera, pneumonia, and typhoid fever have been virtually eliminated in the United States as a result of improved sanitary conditions, better nutrition, and modern medical care. Many significant viral diseases have also been combatted successfully: smallpox is thought to be totally eradicated throughout the world, and polio has nearly disappeared as a result of vaccination. In short, modern medicine has been extremely successful in eliminating or reducing many infectious diseases. However, it has been much less successful in coping with hereditary diseases, either by prevention or by treatment.

Hereditary diseases—more commonly called **genetic diseases**—result from abnormal chromosomes or mutant genes that are passed from one or both parents to their offspring. Although these inherited defects exist in the egg or sperm at the moment of fertilization, the phenotypic effects are not usually observable until birth or, in some instances, until much later in life.

Many people are confused about what causes hereditary diseases and what can be done about them. The questions students frequently ask underscore some of the common fears and misunderstandings: "My sister is mentally retarded. Does that mean that my children have a good chance of being retarded also?" "My aunt and my grandmother both died of breast cancer. The doctor says that I am much more likely to get breast cancer because it runs in my family. Is it true that I've got genes that will give me breast cancer?" "People have told me that diabetes and high blood pressure are genetic diseases, but other people say these diseases are caused by diet. Which is it?" "Everybody in my family is fat. Is there something in my genes that makes me fat, too?" After reading this chapter, you should be able to formulate answers to these questions for yourself.

To emphasize the differences between hereditary diseases or defects and other kinds of diseases, the term *hereditary* will be used in this chapter instead of the term *genetic* because genetic changes may also occur in somatic cells. Recall that germinal mutations are hereditary, whereas somatic mutations are not. **Germinal mutations** occur in the chromosomes of the reproductive (sex) cells, whereas **somatic mutations** occur in all body cells except the sex cells.

For many human diseases and physical defects, it is often difficult to say whether the disease is caused by a person's genes. The assignment of heredity to diseases is complicated by the fact that not all are caused exclusively by mutant genes; many diseases result from complex gene-environment interactions. Moreover, the genetic and environmental contributions to hereditary diseases often vary from person to person.

About 5 percent of all newborn infants have some observable anatomical or physiological abnormality that can be traced to a hereditary defect. However, during pregnancy, exposure to infections, drugs, and other environmental agents also may cause abnormal development of the embryo, resulting in a **congenital defect**—a physical abnormality that is detectable at birth. Congenital defects are not necessarily inherited, although genes may play a role. This

TABLE 14-1 The Apgar Test. This test determines the health status of a newborn right after delivery. If the score is 4 or less, immediate medical attention is indicated.

Sign	Criteria	Score
Heart rate (beats/ minute)	100 or more	2
	Less than 100	1
	Not detectable	0
Respiration (lungs)	Lusty breathing and crying	2
	Shallow or irregular	1
	Not breathing	0
Reflexes	Vigorous response to stimulation	2
	Weak response	1
	No response	0
Muscle tone	Limbs flex spontaneously, resist applied force	2
	Limpness, lack of resistance	1
	Flaccid response	0
Skin color	Pink all over	2
	Partly pink	1
	Bluish or yellowish	0

chapter describes the basic rules that distinguish human hereditary diseases from the many sorts of other congenital defects.

Hereditary Diseases

The first opportunity to examine a person in detail is usually at his or her birth, at which time any physical abnormality is noted. Immediately after birth, the Apgar Test (Table 14-1) is given to each newborn to determine its health status and whether medical attention is required. Also, each newborn is tested for phenylketonuria (PKU) and for other hereditary diseases if there is an indication that the child is at risk for them. Some well-characterized human hereditary diseases, their phenotypic consequences, and their modes of inheritance are listed in Table 14-2.

The most-frequent autosomal chromosome abnormality observed at birth is trisomy 21

TABLE 14-2 Some well-characterized human hereditary disorders and diseases and their phenotypic consequences.

Disease	Major consequences
Chromosome abnormalities	
Down syndrome	Mental retardation
Klinefelter syndrome	Sterility, occasional mental retardation
Turner syndrome	Sterility, lack of sexual development
Autosomal dominant mutations	
Achondroplasia	Dwarfism
Retinoblastoma	Blindness
Porphyria	Abdominal pain, psychosis
Huntington chorea	Nervous-system degeneration
Neurofibromatosis	Growths in nervous system and skin
Polydactyly	Extra fingers and toes
Autosomal recessive mutations	
Cystic fibrosis	Respiratory disorders
Hurler syndrome	Mental retardation, stunted growth
Xeroderma pigmentosum	Skin cancers
Albinism	Lack of pigment in skin and eye
Phenylketonuria	Mental retardation
Progeria	Premature aging; early death
Maple sugar urine disease	Convulsions, mental retardation
Alkaptonuria	Arthritis
Galactosemia	Cataracts, mental retardation
Homocystinuria	Mental retardation
Tay-Sachs syndrome	Neurological deterioration
Sickle-cell disease	Anemia
X-chromosome mutations	
Lesch-Nyhan disease	Mental retardation, self-mutilation
Hemophilia	Failure of blood to clot, bleeding
Duchenne muscular dystrophy	Progressive muscular weakness
Agammaglobulinemia	Defective immune system, infections
Testicular feminizing syndrome	Sterility, lack of male organs

Mongolism
How Prejudice Created a Description for a Hereditary Disease

BOX 14-1

In 1866, a London physician, J. Langdon H. Down, published a report entitled *Observations on an Ethnic Classification of Idiots*. Down had had the opportunity to study a number of institutionalized, retarded children who today would be recognized as being trisomic for chromosome 21. At that time, however, nothing was known about chromosomes or about the biochemistry of hereditary diseases, and Down ascribed the signs and symptoms to tuberculosis in the children's parents. Having observed the physical signs and mental defects of these children for a number of years, Down decided that about 10 percent of all children classified as congenital idiots appeared to be "typical Mongols." He described the characteristics of one such child in these terms:

> The face is flat and broad and destitute of prominence. The cheeks are roundish, and extended laterally. The eyes are obliquely placed. . . . The lips are large and thick with transverse fissures. The tongue is long, thick, and is much roughened. The nose is small. The skin has a slight dirty, yellowish tinge and is deficient in elasticity. . . . There can be no doubt that these ethnic features are the result of deterioration.

Thus did the term *mongolism* come into general use to describe a particular kind of congenital idiocy in children. Moreover, the term *mongolism* implied that such children represented a degenerate ethnic group of human beings, specifically, Orientals.

Down was attempting to scientifically justify a common prejudice of the period: that Caucasians were evolutionarily the most advanced of the human races and that the English, in particular, were biologically far superior to all other people. Down also represented his ethnic classification as being in accord with Darwin's ideas of evolution. He reasoned that, since English parents could give birth to children with mongoloid features, the mentally defective children must have throwbacks to some less-evolved, "degenerate" form of the human species. Down closed his article with what might have been construed at the time as a liberal comment: ". . . the result of degeneracy among mankind, appears to me to furnish some arguments in favor of the unity of the human species."

Down may have been proposing biological unity among individuals or races, but it was obvious which ethnic group he believed to be most highly advanced and which groups he regarded as degenerate. This kind of *scientific racism* (the use of scientific facts or observations to support a prejudice) received widespread acceptance during the early 1900s. Such pseudoscientific nonsense about "inferior races" and a "master race" was ultimately used to justify genocide—the mass murder of religious and ethnic groups by the Nazis during World War II (see Chapter 20).

Some years ago, a group of scientists recommended that the term *Down syndrome* replace the designation of mongolism in all scientific publications. Thus, the term *mongolism*, which carries with it the stigma of racial degeneracy, is no longer used by scientists and others who understand its historical significance.

READING: Stephen Jay Gould, "Dr. Down's Syndrome" in *The Panda's Thumb*. Norton, 1980.

(**Down syndrome**), which affects 1 of every 10,000 to 15,000 children born in the United States (see Box 14-1). The extra chromosome 21 can arise by nondisjunction in either the mother's or the father's sex cells at the time the parental gametes are formed, but the evidence indicates that most of the time the extra chromosome originates in the egg. Down syndrome is not a typical hereditary disease because it is rarely passed on to progeny and does not occur generation after generation in a family lineage. People having Down syndrome, who typically are severely mentally retarded, usually do not mate. However, because the chromosomal abnormality is present in sperm or egg, Down syndrome, like other diseases caused by chromosomal abnormalities, is classified as hereditary.

As noted earlier, **phenylketonuria** (PKU) is a hereditary disease caused by a defect in the gene that codes for an enzyme that normally breaks down the amino acid phenylalanine. In PKU, the enzyme is wholly or partly inactive, so that phenylalanine and its abnormal breakdown products build up in the developing brain, causing mental retardation.

Phenylketonuria is a recessive trait because phenotypically normal parents can have a child who is afflicted with the disease. For PKU to manifest itself, both parents must carry the PKU-determining allele on one of their chromosomes, and both genes must be passed on by chance when the sperm and egg unite to form the zygote. Phenylketonuria can be detected during the first few weeks of life by means of a blood or urine test; such tests are now routine in the United States. Once the disease has been diagnosed, it can be controlled by a low-phenylalanine diet.

Congenital Defects

Down syndrome and PKU are classified as hereditary diseases because the genetic defects were present in the eggs or sperm of the parents before the child was conceived. In contrast, most congenital defects are due to a complex interaction of genetic information and environmental factors. Examples are *cleft lip*, *cleft palate*, and *spina bifida* (cleft spine), which are the results of developmental abnormalities in the formation of the oral cavity and the spine, respectively. Cleft lip has been known to occur in only one of a pair of identical twins; so critical environmental factors other than genes must determine this abnormality.

Spina bifida afflicts 1 of every 1,000 newborns. It occurs when one or more spinal vertebrae fail to close and the spinal cord and related nerves bulge through the cleft, forming an easily damaged, fluid-filled sac. The protruding spinal nerves are vulnerable to paralysis-causing damage and life-threatening infection. Most spina bifida babies also suffer from mental retardation caused by **hydrocephalus**, an abnormal accumulation of fluids in the brain. Today, spinal surgery can repair the exposed nerves and fluid-filled sac, and brain surgery can prevent hydrocephalus-caused mental retardation by the installation of tubes that continually drain the excess fluid. Spina bifida is more prevalent among persons of Northern European (white) ancestry than it is among Jews or blacks, suggesting that it has a genetic, as well as an environmental, component.

Neither cleft lip nor spina bifida can be classified as hereditary diseases, although it is likely that the parents' genetic constitution contributes to the risk of progeny developing these defects. Because both cleft lip and spina bifida are detected at birth, they are classified as congenital defects.

FIGURE 14-1 A victim of the teratogenic drug thalidomide, this armless boy has learned to use his toes and feet as substitutes for hands and fingers. (United Press International.)

Although genes play a role in some congenital defects, many birth defects are caused solely by environmental agents called *teratogens*, as mentioned in Chapter 5. Teratogenic agents that can adversely affect embryonic development include viruses, radiation, and many kinds of drugs. For example, if a woman is infected with German measles (rubella) while she is pregnant, the embryo may develop heart defects or suffer damage to its eyes or ears. Exposing the embryo's cells to radiation such as diagnostic x-rays can cause somatic mutations, which change normal development of tissues or organs. Finally, any of numerous drugs that a pregnant woman may take for one reason or another can affect embryonic development and cause congenital defects.

A particularly tragic example of a drug-induced congenital defect is that caused by **thalidomide**, a tranquilizer that was prescribed in the 1950s and 1960s to alleviate anxiety in expectant mothers. Thalidomide is a potent teratogen in human beings, and thousands of women who took this drug early in pregnancy gave birth to children who had abnormal or missing arms and legs (Figure 14-1). These children are genetically normal and can have normal children. Thanks to the alertness and efforts of Dr. Frances O. Kelsey, a physician at the Food and Drug Administration in Washington, D.C., thalidomide was not prescribed in the United States; virtually all of the affected children were born in Europe, Asia, or Canada.

Alcohol is another teratogen; it can cause serious physical defects, including some degree of mental retardation (Figure 14-2). That is why pregnant women are advised not to drink alcohol, particularly early in pregnancy, when most development takes place. Cortisonelike drugs are teratogenic in mice; they may or may not cause developmental defects in human beings, but pregnant women are advised not to take them. Even as common a substance as the caffeine in coffee, tea, and many soft drinks is teratogenic in some animals, but there is no evidence yet that caffeine causes developmental defects in human embryos.

Because of the uncertainty about which substances are teratogenic, pregnant women are advised to forego taking drugs of any kind, including alcohol, caffeine, aspirin, and tobacco. The early months are the most important for development of the embryo; by five to six months most tissues and organs, including the brain, have been formed, and from then until birth the fetus mainly increases in size. However, this does not mean that damage cannot occur from teratogenic substances during the last months of pregnancy.

What is important to remember is that congenital defects may or may not be due to genetic defects; only developmental abnormalities that are inherited will be passed on to one's children. Keep in mind, however, that developmental ab-

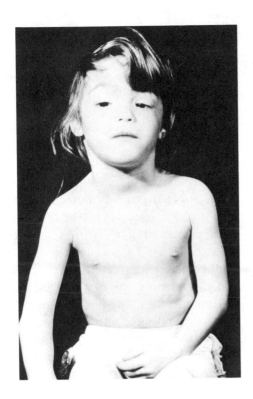

FIGURE 14-2 A child with fetal alcohol syndrome. Such children, born to women who consumed significant amounts of alcohol while pregnant, are undersize and have eyes and facial features that are abnormally proportioned or positioned. Many have heart defects, and most are mentally retarded to some degree. (Courtesy of Kenneth Lyons Jones, M.D.)

normalities may also be caused by mutations in the chromosomes of the embryo's somatic cells. These kinds of mutations are not germinal and cannot be passed on to offspring. Because developmental abnormalities and congenital defects can be caused by either hereditary or environmental factors or a combination of both, it is important to understand the criteria that establish whether a disease or defect is hereditary.

Ways of Establishing Whether a Disease Is Hereditary

Three well-established rules are applied to determine whether a congenital defect or disease is due primarily to abnormal genes that have been passed on from parents—that is, whether the disease has been inherited. If any one of these criteria for inheritance can be demonstrated, we are justified in classifying the disease as hereditary.

The three rules are:

1 A Mendelian pattern of inheritance; that is, a predictable pattern of inheritance (a pedigree) over several generations that obeys Mendel's laws.
2 A visible chromosomal abnormality; that is, loss of a chromosome, presence of an extra chromosome, or rearrangements of chromosome segments from their normal locations, as shown by a cytological examination or karyotype analysis (Figure 14-3).
3 A biochemical defect that can be measured, such as in a defective protein. Abnormal hemoglobin molecules, which cause sickle-cell anemia, are an example.

If a trait, defect, or disease can be characterized by one of these three rules, its hereditary nature is established. Each criterion is proof that a mutation of one sort or another was present in one or both parental gametes and was transmitted to the progeny. However, the fact that a trait does *not* meet one of these criteria does not mean that the trait lacks a genetic component; it simply means that the issue is more complex. As will be seen in the next chapter, there are important ethical, medical, and psychological reasons for requiring conclusive proof before a person is told that he or she has a hereditary defect.

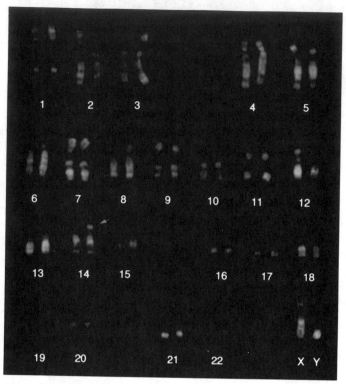

FIGURE 14-3 Karyotype of a child with Down syndrome. Sometimes the extra chromosome 21 becomes attached to another chromosome, in this case to chromosome 14; this occurrence is called a Robertsonian translocation. (In this photograph, chromosomes 19, 20, and 22 are faint but present.) (Courtesy of Irene Uchida.)

demonstration that traits are determined by discrete genetic factors expressed in a dominant or a recessive manner and transmitted in predictable ratios, it became possible to identify hereditary diseases by constructing a **family pedigree**—the pattern of a trait's occurrence in family members from generation to generation. To construct a pedigree, it is necessary to have medical histories of all the family members, or as many as possible, spanning several generations. Some hereditary defects may skip a generation completely because of their particular mode of inheritance, as in male color blindness (Figure 14-5). Thus, the more extensive and detailed the pedigree, the easier it is to accurately determine the mode of inheritance.

In diagraming pedigrees, geneticists use a few universally recognized symbols (Figure 14-6). With the system used in the United States, females are always indicated by circles and males by squares. If a person is affected (has the trait or the disease), the square or circle is specifically marked, usually filled-in. **Carriers** of the defect—that is, people who carry the mutant gene on one chromosome but are not themselves affected because they carry a normal allele on the other chromosome—are indicated

Pedigree Analysis

For centuries before the experiments of Gregor Mendel showed how traits are inherited, breeders of plants and animals kept records of desirable traits (Figure 14-4). After Mendel's

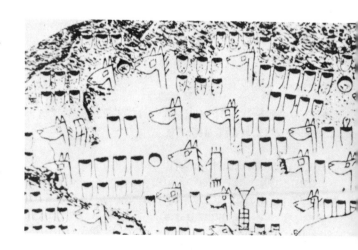

FIGURE 14-4 A carved stone showing the pedigree of horses raised by a breeder about 4,000 years ago. This stone, discovered in Asia, shows that breeders kept records of desirable traits in their animals. (Courtesy of Dorsey Stuart, University of Hawaii.)

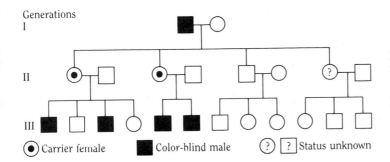

Generations

I

II

III

⊙ Carrier female   ■ Color-blind male   (?) [?] Status unknown

FIGURE 14-5 A family pedigree for color blindness. The Mendelian mode of transmission of a trait from generation to generation establishes that a particular defect or disease is inherited. Color blindness is a sex-linked hereditary defect that shows up almost exclusively in men. Can you figure out why the status of one of the women in the second generation is unknown?

by a different marking, usually by partly filled-in squares or circles. If a sufficiently detailed or complete pedigree can be constructed spanning several generations, a geneticist can determine not only whether the trait is inherited but also whether the gene for the trait is recessive, dominant, or sex linked.

Because the royal families of Europe kept detailed family records and because they tended to have large numbers of offspring, their pedigrees are particularly informative. **Hemophilia**—a sex-linked, recessive trait caused by a gene on the X chromosome—used to be called a royal disease because many male members of the royal families of Europe and Russia suffered from this hereditary affliction. Hemophiliacs, or "bleeders," bleed profusely from the slightest scratch or injury because they lack one or more blood proteins called *clotting factors*. In the past, hemophilia was a life-threatening disease and

FIGURE 14-6 The meaning of the various symbols used in the United States in constructing pedigree diagrams.

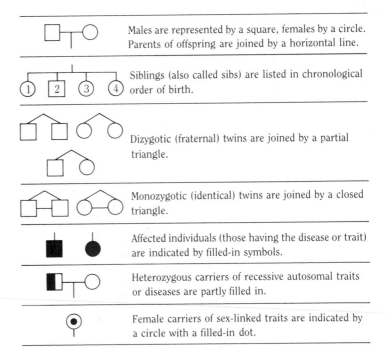

Males are represented by a square, females by a circle. Parents of offspring are joined by a horizontal line.

Siblings (also called sibs) are listed in chronological order of birth.

Dizygotic (fraternal) twins are joined by a partial triangle.

Monozygotic (identical) twins are joined by a closed triangle.

Affected individuals (those having the disease or trait) are indicated by filled-in symbols.

Heterozygous carriers of recessive autosomal traits or diseases are partly filled in.

Female carriers of sex-linked traits are indicated by a circle with a filled-in dot.

FIGURE 14-7 Queen Victoria and her family. Her granddaughter Alexandra (above Victoria and to our left, with neck fur), married to Nicholas II, Tsar of Russia, gave birth to a son, Alexis, who was a "bleeder." Victoria's granddaughter Irene (above Victoria and to our right, with neck fur) married Prince Henry of Prussia, and the mutant gene was transmitted to the German royal family as well (see Figure 14-8). (Courtesy of Photography Collection, The Harry Ransom Humanities Research Center, University of Texas at Austin.)

afflicted males had to be extremely careful not to be cut or bruised. Today, the disease, though still quite serious, can be controlled by transfusions and periodic injections of the missing clotting factor.

Queen Victoria of England (1819–1901) was a carrier of hemophilia, and it is possible that it arose in her family as a result of a mutation that she inherited (Figure 14-7). Be that as it may, among Victoria's nine children, one son was a bleeder and two daughters, Alice of Hesse and Beatrice, were carriers. Alice, in turn, had seven children and passed the mutant gene on to at least two daughters, Irene and Alexandra (Figure 14-8). As a result of intermarriage with the German and Russion royal families, hemophilia was transmitted to other European royal families, and males in succeeding generations were

FIGURE 14-8 A pedigree of the European and Russian royal families showing the transmission of hemophilia. The disease can be traced to Victoria, who was a carrier of the X-linked mutant gene. Two of Victoria's daughters also were carriers.

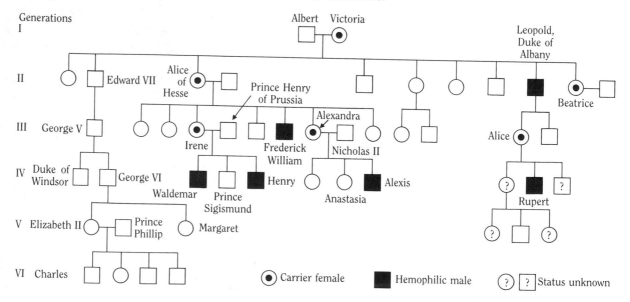

affected. Because Edward VII was not affected, the disease was not transmitted to living descendents of the English royal family.

X-linked Inheritance

Genes that are located on the X chromosome are said to be *sex-linked* (X-linked) because generally only one sex (male) is affected. A female also can have X-linked diseases, but both her X chromosomes must carry the mutant alleles if she is to be affected (assuming that the genes are recessive, which they frequently are for serious hereditary diseases). Female hemophiliacs, for example, are rare, as their absence from royal pedigrees shows. Today, however, a few females who have hemophilia are known, because male hemophiliacs now survive, marry, and have children. Female hemophiliacs must carry two mutant alleles.

Because chromosomes segregate at random into the eggs and sperm in meiosis, it is possible to calculate the probability that a daughter will be a carrier of an X-linked disease or that a son will be affected by an X-linked trait (Figure 14-9). On the average, if the mother is a heterozygous carrier, half of her daughters will be carriers and half of her sons will have the disease. Albert and Queen Victoria had nine children — five girls and four boys, very close to the expected 50–50 sex ratio. Of the five female offspring, two were definite carriers of the defective, hemophilia-causing allele, close to the expected frequency of one-half, although it is not certain that the others were not carriers. However, of the four male offspring, only one — Leopold, Duke of Albany — was affected, whereas two out of four would be expected to be affected. It is important to remember that probabilities become more accurate when the numbers are large; if the number of individuals (or

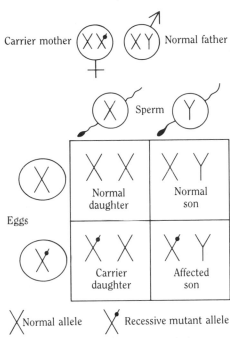

FIGURE 14-9 The pattern of inheritance of an X-linked trait. Both parents are phenotypically normal, but the mother is a carrier of a recessive gene on the X chromosome. All daughters of this couple will be phenotypically normal, but on the average half of them will be carriers. Half of the sons, on the average, will be affected, because there is one chance in two that they will receive the X chromosome carrying the mutant gene from their mother.

events) is small, deviations from predicted numbers will often be observed.

In calculating the probability that a particular allele or group of alleles will be inherited, it may be helpful to think of the segregation of chromosomes as being analogous to the tossing of a coin. Just as a coin always comes up either heads or tails, eggs and sperm always receive either one or the other chromosome of each pair. On the average, tossing a coin a large number of times produces nearly equal numbers of heads and tails. However, it is possible to get runs of either heads or tails, and a run can badly skew the outcome of a limited number of tosses.

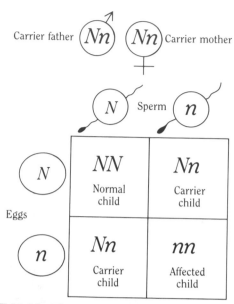

FIGURE 14-10   The pattern of inheritance of an autosomal recessive trait (n). Both parents are phenotypically normal and carriers of the same mutant gene. Half of the children will be heterozygous carriers of the trait like their parents and a quarter of the children will be affected.

So it is with inheritance: all the offspring in a family may get the abnormal chromosome, all may get the normal one, or the frequency may approximate the expected 50–50 ratio.

Autosomal Recessive Inheritance

An abnormal recessive gene is of little or no consequence as long as cells contain a normal allele whose expression masks that of the recessive allele. This is why heterozygous carriers of harmful recessive genes are not themselves affected by the disease or defect. For offspring to be affected, both parents must be carriers of the recessive gene. Because of the random segregation of chromosomes in meiosis, on the average one-fourth of the offspring of two carriers will be affected and one-half will be carriers themselves (Figure 14-10). Some autosomal recessive diseases are listed in Table 14-2.

Autosomal recessive genes are usually recognized in human pedigrees by two distinguishing characteristics: (1) the trait is usually present in only one generation or it skips one or more generations, because as long as people marry outside of their immediate families it is unlikely that two carriers will mate; and (2) when the parents are heterozygous for the recessive gene, among a large number of children the ratio of those not affected to those affected should be 3 to 1. Rarely do pedigrees conform to the predicted pattern, but it is usually possible to distinguish an autosomal recessive trait from an X-linked or autosomal dominant trait.

More than one thousand autosomal recessive human genetic disorders have been described, many of them within the past several years. Most recessive diseases, however, are quite rare. The two most-frequent ones are sickle-cell anemia, which affects about one in every six hundred blacks, and cystic fibrosis, which affects about one in every two thousand whites. **Cystic fibrosis** is due to an abnormal protein that causes the production of a thick mucus in organs, particularly the lungs. All together, about twelve hundred babies are born each year in the United States with either sickle-cell anemia or cystic fibrosis.

The defective gene that causes sickle-cell anemia is known; it is the gene that codes for the β-chain of hemoglobin in red blood cells. Even though the nature of the defective protein is well known, there is still no cure for the disease, nor is one imminent. Recent molecular techniques (described in Chapter 16) have determined the cystic-fibrosis gene to be on chromosome 7. However, neither the gene nor the defective protein that presumably is responsible for cystic fibrosis has been identified. Although progress has been made in understanding the underlying cause of this severe disease, no cure is yet in sight.

Autosomal Dominant Inheritance

To say that a gene is expressed in a dominant fashion means that the phenotype is determined by that gene regardless of what other allele is present. The number of recognized dominant genes that cause serious human hereditary defects is about the same or slightly greater than the number of deleterious recessive genes.

If a deleterious dominant gene were to significantly reduce a person's chances of survival to reproductive age, he or she would not pass that gene on to the next generation. Thus, dominant genes, even in heterzygotes, should disappear in time or at least decrease in frequency in a population. The large number of dominant, disease-causing genes in the human population is explained by the mild symptoms that they cause, by the late onset of symptoms, or by a combination of both factors. Also, most of the disorders caused by dominant genes do not affect a person's reproductive potential.

The most-frequent class of deleterious dominant genes (though still rare) are those that cause **hyperlipidemia** and **hypercholesterolemia**. These disorders elevate the levels of various kinds of lipids and cholesterol in the blood. This, in turn, is thought to contribute to the development of heart and artery disease. Although these dominant hereditary disorders may cause health problems, generally people have already had their families before serious symptoms appear. Like the genes that cause hyperlipidemia and hypercholesterolemia, some of the more-common dominant genes that cause serious diseases have survived in human populations because they are not expressed until later in life.

Huntington disease (also called *Huntington chorea*), a neurological disorder that results in death, is one of the best-known examples of a dominant inherited disease. Woody Guthrie, the famous folk singer, died of this hereditary disease. At this writing, Arlo Guthrie, his son and a famous folk singer in his own right, does not yet know whether he carries the gene for Huntington chorea, nor does he know whether he has passed the gene on to his children. About half the people who have Huntington disease do not show signs of the disease until age forty; others may die of other causes before the gene is ever expressed. Screening for this disease early in life has created serious ethical dilemmas for many families (discussed in Chapter 16).

If only one parent has an autosomal dominant allele for a disease or physical defect, half of the couple's children will be affected if the gene is fully expressed, but this is not always the case (Figure 14-11). For example, *porphyria* is caused by an autosomal dominant allele, but its expression varies from person to person and the presence of the mutant gene does not always cause clinical symptoms (Box 14-2). The symptoms of Huntington disease vary markedly with age; some persons with the mutant gene may even die of other causes before the mutant gene is expressed and symptoms develop. However, because dominant genes are commonly expressed to some extent in all persons who possess them, there are not "carriers" of dominant traits as there are for recessive ones.

A rather common autosomal dominant gene that *is* fully expressed causes **achondroplasia**, or dwarfism, which is characterized by a disproportionate shortening of arms and legs. This hereditary defect causes abnormal growth of cartilage but has no other serious phenotypic effects. Persons with this autosomal dominant gene are fertile; so the mutant gene is passed on to children.

A Hereditary Disease and American History

BOX 14-2

Investigation of royal pedigrees has uncovered another hereditary disease, **porphyria**, which is caused by a biochemical imbalance that sometimes produces insanity and a variety of other symptoms. Some historians think it likely that this hereditary disease, which can be traced through more than 200 years of British royalty (see accompanying pedigree), had important effects on the War of Independence and on American history.

King George III (1738–1820), Queen Victoria's grandfather, was England's monarch during a turbulent period of world history that included both the French Revolution and the American Revolution (1775–1783). During these years in which new governments and countries were created, George III was having serious health problems that affected his mental state and judgment and certainly reduced his ability to manage colonial affairs. In 1765, at age twenty-six, he had his first documented bout of insanity; another more-serious attack occured in 1788, when he was fifty. Altogether, his medical records show five periods of insanity that were described by his physicians as including severe physical symptoms, as well as

delirium and hallucinations. He was treated with accepted medical practices of the period: being shackled to a bed or chair while subjected to forced vomitings, cuppings, blisterings, and leechings. None of his doctors understood what was wrong, but they knew that insanity among members of the English royal family was quite common and that insanity had also affected the royal houses of Stuart, Hanover, and Prussia through the centuries.

It has now been established from medical records, recorded symptoms, and pedigree analysis that George III, as well as some of his ancestors and descendants, was afflicted with a hereditary disease known today as *acute intermittent porphyria*. The symptoms are acute pain in the abdomen and sometimes in the limbs and back, headaches, insomnia, nausea, hallucinations, depression, and delirium. The disease is caused by a dominant gene with variable expression, which simply means that the severity of the disease ranges from mild symptoms that may go unnoticed to full-blown psychosis and even death due to respiratory paralysis.

Porphyrin molecules are present in most body cells and are essential for the functioning of many enzymes. They play a particularly important role in the transport of oxygen to body cells because they are part of the hemoglobin molecule in red blood cells. During the reign of George III, nothing was known about the genetics or biochemistry of hereditary diseases or about porphyrins, and so his insanity was presumed to have a psychological basis. Modern medical research has shown that all the symptoms of this disease, including insanity, are due to a toxic effect from the overproduction of porphyrins, which is affected by diet, alcohol, and other factors.

American history might have been different if George III had not suffered from porphyria. How different we will never know. Nor will we know how Russian history might have been affected if Alexis, heir to the throne, had not suffered from hemophilia. Alexis was cared for by the "mad monk" Rasputin, who played an important role in Russian politics at the time of the Russian revolution.

READING: V. V. Vadakan, "Porphyria: Curse of British Royalty," *Hospital Practice.* September 15, 1987.

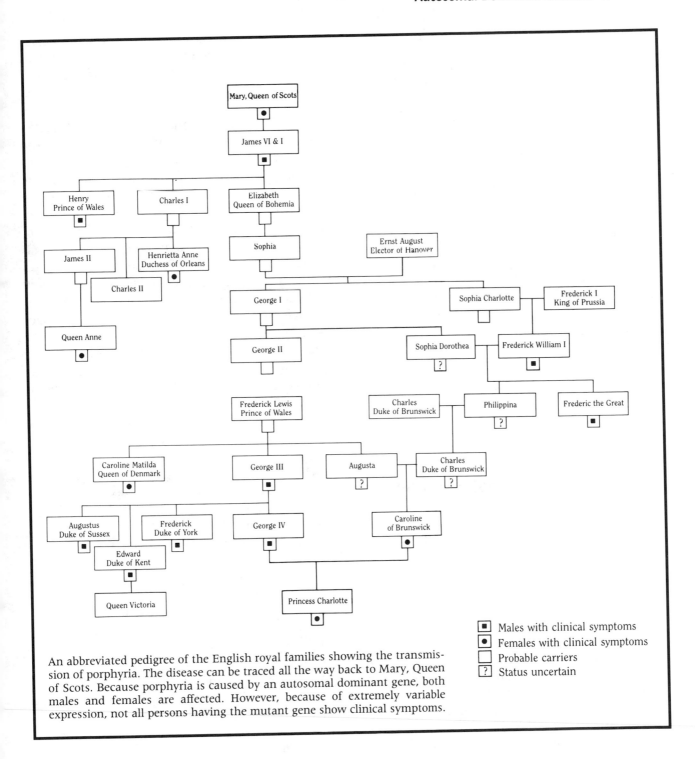

An abbreviated pedigree of the English royal families showing the transmission of porphyria. The disease can be traced all the way back to Mary, Queen of Scots. Because porphyria is caused by an autosomal dominant gene, both males and females are affected. However, because of extremely variable expression, not all persons having the mutant gene show clinical symptoms.

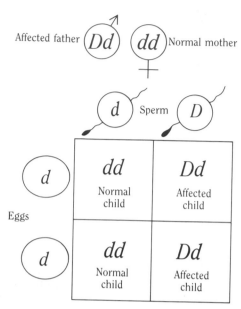

FIGURE 14-11 The pattern of inheritance of an autosomal dominant trait. One parent is affected and the other is phenotypically normal. Half of all children, on the average, will be affected and half will be phenotypically normal. This pattern assumes that the dominant trait is fully expressed in persons carrying the mutant gene.

Polygenic Multifactorial Inheritance

Although hundreds of human diseases and physical defects are caused by X-linked, autosomal recessive, or autosomal dominant mutations, many others have an ambiguous hereditary basis and do not obey any of the genetic rules discussed so far. Diseases such as hypertension (elevated blood pressure), diabetes, schizophrenia, and even allergies and cancers are often described as genetic diseases or as diseases that show familial patterns and thus by inference are assumed to have some hereditary component. These particular human diseases, as well as such normal traits as stature, weight, skin and hair color, intelligence, and muscular-

ity are described as **polygenic multifactorial traits**. This means that these traits or diseases are caused by several genes whose numbers, chromosomal locations, and degree of expression are unknown.

For all polygenic traits, environmental factors influence the expression of the genes; hence the term multifactorial. For example, height is influenced by nutrition and disease during the growing period of life, skin color is influenced by exposure to sunlight, and intelligence is influenced by the amount of stimulation, attention, and love received in the developing years. Even congenital disorders such as the more-common forms of cleft lip and cleft palate are strongly influenced by environmental factors. It has been shown that a woman's use of the tranquilizer Valium during pregnancy quadruples the chance that her baby will be born with a cleft lip or palate. Characterizing a trait or disease as polygenic multifactorial simply means that it cannot be ascribed to any known genes and that it is difficult to say how much of the trait is due to genetic or to environmental factors.

One of the best-studied polygenic multifactorial traits are fingerprints. There are three basic types of fingerprints (Figure 14-12), and no two persons have the same set. Fingerprints are formed in the embryo by the twelfth week of pregnancy and remain unchanged throughout life. Their patterns are determined by an unknown number of genes interacting with the intrauterine environment. Strange as it may seem, even the fingerprints of identical twins, although quite similar in their general pattern and number of ridges, are not identical, despite the fact that identical twins have identical genes and a common intrauterine environment. In fact, a finger-by-finger comparison of identical twins leads to the conclusion that, in early development, external factors play a major role in the formation of fingerprint patterns.

Ridge count   0   13   17 + 8

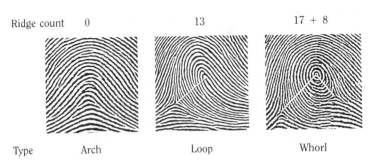

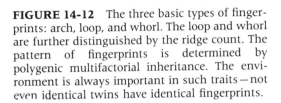

FIGURE 14-12 The three basic types of finger- prints: arch, loop, and whorl. The loop and whorl are further distinguished by the ridge count. The pattern of fingerprints is determined by polygenic multifactorial inheritance. The envi- ronment is always important in such traits—not even identical twins have identical fingerprints.

Type   Arch   Loop   Whorl

Another easily measured polygenic multi- factorial trait is stature, or height. Although researchers have identified numerous genes that can cause dwarfism in human beings, they do not yet know how many or which genes are important in determining normal stature. As mentioned earlier, the environment plays an important role in determining stature. In one study of stature, for example, it was shown that Japanese men and women who migrated from Japan to Hawaii had children who were, on the average, from 2 to 4 inches taller than they were (Table 14-3). It is clear that the genes in this population of Japanese had not changed in one generation; therefore, the increase in stature must have been due to environmental factors. It is interesting to note that the increase in stature was complete in males in one generation, whereas females continued to increase in height for at least two generations.

Presumably the factor that changed most sig- nificantly in these migrating families was their diet. Even in Japan, the stature of persons born after World War II increased as a result of increased amounts of milk in the diet. Other studies have shown that nutrition affects not only physical stature but also many other growth characteristics, including the growth of brain cells.

What can be said about the inheritance of such diseases as hypertension, diabetes, schizo- phrenia, and allergies? Can these diseases be

regarded as hereditary in the same sense as those listed in Table 14-2? Clearly, the answer is no. Most of them can be prevented or controlled by appropriate changes in the affected person's environment. For example, blood pressure can often be brought under control by weight loss and diet; similarly, diabetes can be controlled in many instances by weight loss, reduced sugar consumption, and, if necessary, insulin injections.

Most people regard hereditary diseases as be- ing incurable—a view that to a considerable degree is correct. That is why it is important for people to understand the rules that establish whether or not a disease is hereditary. It is worth remembering that families share environ- ments as well as genes; in the polygenic inher- itance of diseases or traits in particular, the environment may well be the decisive factor. We cannot speak of the inheritance of polygenic

TABLE 14-3 The effect of environmental factors (diet) on the stature of Japanese-Americans in Hawaii.

| | Stature (height in inches) | |
Generation	Women	Men
First	58.1	61.4
Second	59.8	65.6
Third	61.0	65.6

Source: J. W. Froelich, "Migration and the Plasticity of Physique in the Japanese-Americans of Hawaii," *American Journal of Physical Anthropology*, 32, 429 (1970).

diseases or traits with the same degree of assurance that we can speak of traits and diseases determined by single genes with a well-defined mode of inheritance.

In recent years, it has become increasingly popular to characterize abnormal human behaviors and problems as "genetic" disorders. For example, health problems such as alcoholism, obesity, phobias, and even suicide have been ascribed to biological abnormalities caused by "genetic" susceptibility or "genetic" predisposition in people. Clearly, the use of the term genetic in this way is quite vague and imprecise. It certainly does not fit with any of the more-rigorous definitions of hereditary disease presented in this chapter.

It may well be that certain destructive human behaviors have an underlying biological basis. However, no genes have been identified for any of the behavioral disorders just mentioned. Thus, it is not yet appropriate to refer to these conditions as hereditary diseases. Moreover, all of them can be modified or even cured by counseling or psychological therapy—therapies that do not change genes.

Modern medical techniques have become increasingly powerful in detecting genetic abnormalities in developing embryos, as well as in carriers. The ability to diagnose genetic defects has created ethical dilemmas and made many medical decisions complex and difficult. The new techniques and the issues that they raise in diagnosing and treating hereditary diseases are discussed in the next two chapters.

Summary

Hereditary diseases are caused by abnormal chromosomes or mutant genes that are passed from one or both parents to their offspring. Congenital defects are physical abnormalities detectable at birth—defects that may or may not be inherited. They are usually due to complex interactions between genes and environment.

Any one of the following criteria can be used to classify a disease as being hereditary: (1) a Mendelian pattern of inheritance; (2) a visible chromosomal abnormality, as shown by cytological examination or karyotype analysis; or (3) a biochemical defect that can be measured.

A family pedigree is the pattern of a trait's occurrence in family members from generation to generation. Pedigrees show people who have a given trait and those who carry it but are not affected. From a sufficiently detailed pedigree spanning several generations, a geneticist can determine not only whether a trait is inherited, but also whether the gene for that trait is dominant, recessive, or sex linked.

Most sex-linked (X-linked) diseases are caused by recessive alleles and are inherited by male offspring. For a daughter to inherit such a disease, both of her X chromosomes must carry the recessive allele. One-half of the male offspring of a mother who is heterozygous for an X-linked trait can be expected to inherit that trait, and one-half of her female offsrping can be expected to be carriers.

For offspring to inherit an autosomal recessive trait, both parents must be carriers, and the ratio of those not affected to those affected should be three to one; one-half of the offspring of two carriers will be carriers themselves.

A dominant allele determines phenotype regardless of what other allele is present. So, there are no carriers of dominant traits. If one parent has an autosomal dominant allele that causes a disease or physical defect, one-half of the offspring will be affected if the gene is fully expressed.

Polygenic multifactorial traits are the expression of several genes whose number, chromosomal locations, and degree of expression are unknown. The expression of genes for all such traits is affected by environmental factors.

Key Words

achondroplasia An autosomal dominant disorder causing dwarfism due to retarded growth of long bones.

carrier An individual who is heterozygous for a recessive allele.

congenital defect Any abnormality detectable in a newborn at birth.

cystic fibrosis An autosomal recessive disorder characterized by abnormal mucus production in the lungs and other organs.

family pedigree The pattern of a trait's occurrence in family members from generation to generation.

germinal mutation One that arises in the chromosomes of reproductive (sex) cells.

hemophilia A recessive, X-linked disorder that causes excessive bleeding due to lack of an essential clotting factor.

hereditary (genetic) disorder or disease A disorder or disease due to a defect in a gene or chromosome inherited from one or both parents.

Huntington disease (Huntington chorea) An autosomal dominant disease that is expressed late in life, producing neurological symptoms and death.

hydrocephalus A congenital defect caused by excess cerebrospinal fluid in the brain.

hypercholesterolemia Elevated levels of cholesterol in the blood.

hyperlipidemia Elevated levels of lipids in the blood.

polygenic multifactorial inheritance Any trait caused by an undetermined number of genes and environmental factors, all of which interact together.

porphyria An autosomal dominant disorder caused by abnormal metabolism of porphyrin molecules.

somatic mutation Any mutation arising in a somatic cell.

spina bifida A congenital defect caused by abnormal development of the spinal column.

Additional Reading

Edlin, G.J. "Inappropriate Use of Genetic Terminology in Medical Research: A Public Health Issue." *Perspectives in Biology and Medicine*, Autumn, 1987.

Grady, D. "The Ticking of a Time Bomb in the Genes." *Discover*, June, 1987.

Nicol, S.E., and I.I. Gottesman. "Clues to the Genetics and Neurobiology of Schizophrenia." *American Scientist*, July–August, 1983.

Patterson, D. "The Causes of Down Syndrome." *Scientific American*, August, 1987.

Valle, D. "Genetic Disease: An Overview of Current Therapy." *Hospital Practice*, July 15, 1987.

Vehar, G.A., and R.M. Lawn. "The Cloning of Factor VIII and Genetics of Hemophilia." *Hospital Practice*, May 15, 1986.

Study Questions

1 What is meant by a congenital defect? Are all congenital defects due to heredity?

2 What kind of genetic defects are responsible for Down syndrome? For PKU?

3 Why should pregnant women drink little or no alcohol?

4 What are the three rules that allow one to say that a disorder is due to a hereditary defect?

5 What are the two most-common autosomal recessive human disorders? What are the two most-common autosomal dominant ones?

6 How does a polygenic-multifactorial disorder differ from an X-linked disorder?

7 From which parent do offspring inherit an X-linked disorder most often?

Essay Topics

1 Discuss whether you would want to know at your present age that you carry the dominant gene for Huntington disease.

2 Explain what can be learned from a family-pedigree analysis.

3 Discuss how you would react if you had a child with a severe genetic handicap.

15

Human Hereditary Diseases

Screening and Counseling

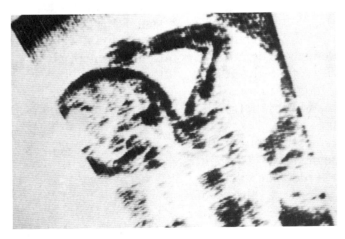

Knowledge about one's genetic makeup is one of the most personal and private pieces of information one can possess about oneself. It vies with private thoughts, fantasies, and dreams as being the essence of self-identity.

MARGERY W. SHAW, *lawyer*

MORE THAN A MILLION persons are hospitalized each year in the United States for a hereditary or congenital disease. It is estimated that at least one of ten Americans suffers at some time in his or her life from a serious disease that has a significant hereditary component. Among people with severe mental retardation—those who are incapable of functioning in society—many have an observable chromosomal abnormality, and most of the rest have either hereditary or developmental defects. As mentioned earlier, about five thousand children are born with Down syndrome in the United States each year, with an average life expectancy of about forty-five years. A majority of these people will sooner or later be relinquished by their families into foster homes or institutions.

Despite nature's ability to eliminate defective fetuses before and at birth, about 5 percent of all live newborns have some congenital defect, half of which are hereditary. With recent advances in medical care, many thousands of infants now survive and grow up with handicaps that would have been fatal a few years ago.

This rather depressing presentation of the numbers, kinds, and consequences of congenital defects emphasizes the burden placed on the feelings of families and on the resources of society. Because hereditary diseases cannot be cured, only ameliorated, most research focuses on their detection and prevention. Amniocentesis, a medical technique now widely used to detect and diagnose defects early in pregnancy, is of particular benefit to women or couples who are at risk for bearing children with genetic defects. Genetic counseling both before and after amniocentesis can help high-risk prospective parents decide whether they want to abort the pregnancy, and it can help them to cope with their child if it is eventually born with a defect. Genetic counseling can also help them prevent having other children who may also be genetically handicapped. This chapter takes a look at what new medical techniques and genetic counseling programs can accomplish and discusses the ethical and moral problems that arise in regard to amniocentesis and abortion.

Determining Who Is at Risk

The first problem in trying to reduce the number of babies born with serious genetic diseases or defects is to determine which couples, and more particularly which pregnant women, are at a higher than average risk for giving birth to a genetically defective child. Certain factors are known to increase the risk sufficiently for some form of genetic counseling and prenatal testing to be advisable:

1 The prospective mother is thirty-five years of age or older or the father is older than fifty. It is known that chromosomal abnormalities occur more frequently with increasing maternal age and to a lesser degree with increasing paternal age.
2 The couple has had a child with either a chromosomal abnormality or a neural-tube defect or has had such an abnormality in a close relative. Such couples have an increased risk of producing another child with a defect.
3 The prospective mother is known to be or is suspected of being a carrier of a deleterious X-linked trait. It can be predicted that about half of the male children of mothers who are carriers will be affected.

4 Both prospective parents are known to be carriers of serious recessive autosomal genes or one parent has a defective autosomal dominant gene. A high proportion of the children of such couples will be affected.

5 Polygenic-multifactorial traits are present in an earlier offspring or close relative. In such cases, the risk of having an affected child is increased.

Because it is impossible to genetically screen or provide genetic counseling for everyone who decides to have children, these five criteria help determine who is most likely to benefit.

The AFP Test for Neural-Tube Defects

Among the most-serious congenital defects—which may be due to hereditary factors to various degrees—are those in which the neural tube (the brain and spinal cord) fails to develop properly in the embryo. If embryonic development of the nervous system is defective, usually other bodily and intellectual functions are also impaired. Approximately one in every 500 children born in the United States has a **neural-tube defect**, of which there are three basic types: anencephaly, hydrocephalus, and spina bifida.

Anencephaly results from abnormal development of the brain; most babies with this defect are stillborn or they die soon after birth. As mentioned in Chapter 14, *hydrocephalus* is caused by the accumulation of excess cerebrospinal fluid in the brain. If the pressure exerted on the brain by this fluid is not relieved, brain damage and mental retardation result. The most-common form of neural-tube defect (also considered in Chapter 14) is *spina bifida*, in which one or more vertebrae do not close completely during development. This congenital defect has a variety of consequences, including partial or

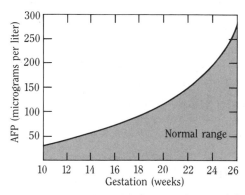

FIGURE 15-1 Normal alpha-fetoprotein (AFP) levels in pregnancy. Neural-tube defects in a developing embryo are often signaled by abnormal levels of AFP circulating in a pregnant woman's blood. AFP levels that fall outside the normal range indicate that further diagnostic tests are needed.

complete paralysis from the waist down and, in severe cases, mental retardation.

Anencephaly and spina bifida can be detected in most cases by examining a pregnant woman's blood for the presence of a particular protein called **alpha-fetoprotein (AFP)**. During the development of the embryo, AFP is present in high levels in the uterus and enters the woman's bloodstream through the **placenta**, the organ that connects her blood supply with that of the fetus. The level of AFP in the woman's blood increases during pregnancy but usually remains within certain normal limits (Figure 15-1). If the midpregnancy (16–18 weeks) AFP levels are found to be abnormally high on repeated blood tests, the embryo may have some kind of neural-tube defect and further tests are recommended. However, several other factors—both normal and abnormal—can complicate the interpretation of high levels of AFP, including the development of twins. Moreover, because the level of AFP is determined by the length of development of the embryo, the time of conception must be known with some degree of accuracy; otherwise, a correct interpretation of the AFP level may be impossible.

The use of the AFP blood test for detecting neural-tube and other developmental defects is still controversial because the test's degree of accuracy leaves something to be desired. About fifty of every thousand pregnant women test positively for high AFP levels. A second blood test eliminates about thirty of those testing positively. If the remaining twenty or so women agree to further tests, such as ultrasound scanning and amniocentesis, ultimately only one or two embryos of the original thousand will be shown to have neural-tube defects.

The AFP test is so simple that many pregnant women elect to have it done. If the AFP levels are normal, they can be reasonably sure that their babies are developing normally. However, for those women who have abnormally high AFP levels, deciding whether to undergo further tests can be distressing. For those who do seek reassurance, further prenatal tests can determine whether the embryo is normal, usually in time to abort it if the parents wish to do so. The decision of whether to abort a pregnancy entails many moral questions and ethical judgments that are not easy for prospective parents to deal with (see Box 15-1).

Ultrasound Scanning and Fetoscopy

Until recently, the only way to visualize an embryo in a womb was by radiological examination (x-rays). However it is now generally recognized that x-rays are potentially harmful to the developing embryo because such radiation may damage tissue and may induce mutations or other chromosomal abnormalities. Thus, pregnant women are now advised not to undergo abdominal x-ray examination unless the benefits to the fetus or the woman clearly outweigh the risks. Fortunately, two other tech-

FIGURE 15-2   Image of a fetus obtained by ultrasound scanning. Such ultrasound scans reveal the position of the fetus and may indicate certain physical abnormalities. (Courtesy of Willard Centerwall, University of California, Davis, Medical Center.)

niques are now available for visual examination of the fetus.

Ultrasound scanning has become an accepted and (generally) safe technique for visualizing the fetus. Ultrasound scanning, or *sonography* as it is sometimes called, employs high-frequency sound waves to observe the structural features of the embryo's body. The sound waves penetrate the womb and are reflected differently from embryonic tissues that differ in composition and density. The reflected sound waves are displayed on a screen, and the pattern of the picture is interpreted by a physician (Figure 15-2).

Ultrasound scans can be used to detect multiple fetuses and to determine the orientation of the fetus in the womb. They can also ascertain the location of the placenta, which is particularly important if amniocentesis is to be performed. And they can gauge the embryo's head

The Abortion Controversy
When Does Human Life Begin?

BOX 15-1

Killing another human being without just cause is a crime in all civilized societies. Once a live baby is born, everyone agrees that it is alive and that anything that is knowingly done to stop its life constitutes murder. But determining when (or if) an unborn fetus is alive is another issue entirely. Recent advances in medical care for premature babies, in amniocentesis, and in *in vitro* fertilization have created thorny medical, legal, and moral questions in this regard. The most-vexing problem concerns abortion, which is inextricably linked to the question of when life begins.

Opinions differ enormously on this question. Some people argue that human eggs and sperm are alive and are already the basis for determining the beginning of life because the fertilized egg will eventually develop into a human being. Other people insist that life begins at the moment of conception—when the sperm penetrates the egg—because at that moment a human being begins to develop. Yet about half of all human conceptions abort spontaneously; are the pregnant women in whom this happens guilty of involuntary manslaughter?

Some people argue that human life begins when the person is born and able to survive on its own. Still others point out that even normal infants cannot survive without years of care. But infants born prematurely—after only six or seven months of development instead of the usual nine—can now be kept alive by heroic medical efforts, and most "preemies" survive into adulthood and live healthy, normal lives. Are "preemies" not to be considered living human beings?

There are even some people who are convinced that there is no such thing as a time when human life begins, because individual human life is part of a continuous process. They hold that the act of reproduction is merely a stage in an endless cycle of human lives. Some scientists point out, in what could be construed as support for this view, that for human life to "begin" at some moment it logically must have "stopped" at some other moment; so the whole issue of the beginning of human life is a red herring.

After all the arguments and facts have been heard, the legal issues of abortion remain unresolved. In the 1970s, abortion laws in the United States were liberalized so that most women who desired to abort a fetus were able to do so legally and with medical assistance. Now, strong pressure is being exerted by certain groups to revise the laws so that all abortions will be illegal regardless of the circumstances. The legal issue of abortion has been well stated by Brian G. Zack of the Rutgers Medical School:

> The issue is at what stage of development shall the entity destined to acquire the attributes of a human being be vested with the rights and protections accorded that status. It is to the moral codes of the people that the law must turn for guidance in this matter, not to the arbitrary definitions of science. . . . Science may never make moral judgments; the law must.

Ultimately, any legal definition of when life begins must be provided through the political process, which is to say that (in a democratic society, at any rate) it must be a matter of social consensus. In a pluralistic society, a consensus on the abortion issue is agonizingly difficult to reach, and it will never be unanimous.

READINGS: Brian G. Zack, "Abortion and the Limitations of Science," editorial in *Science*. July 17, 1981. John D. Biggers, "When Does Life Begin?" *The Sciences*. December, 1981.

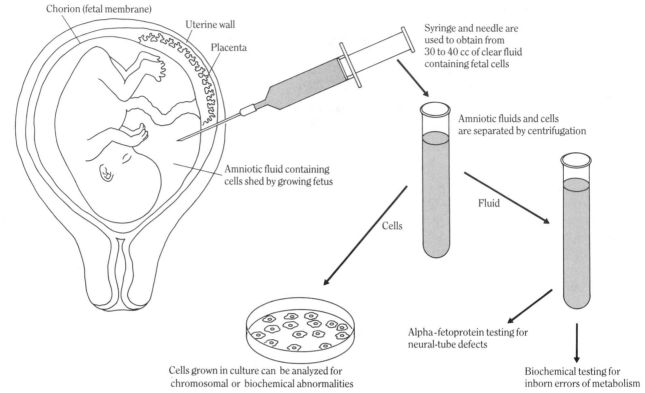

Chorion (fetal membrane)

Uterine wall

Placenta

Syringe and needle are used to obtain from 30 to 40 cc of clear fluid containing fetal cells

Amniotic fluids and cells are separated by centrifugation

Amniotic fluid containing cells shed by growing fetus

Fluid

Cells

Alpha-fetoprotein testing for neural-tube defects

Cells grown in culture can be analyzed for chromosomal or biochemical abnormalities

Biochemical testing for inborn errors of metabolism

FIGURE 15-3 Amniocentesis. This procedure is used to test for hereditary or developmental defects in the fetus. At about the fifteenth week of pregnancy, a sample of amniotic fluid containing fetal cells is removed. The cells are cultured *in vitro* to increase their quantity. The cultured cells are then used for a karyotype (chromosome) analysis and for any biochemical or enzyme tests that are deemed necessary.

size, thereby providing an independent means of determining the age of the embryo and of ascertaining normal or abnormal brain development, such as hydrocephalus. Late in pregnancy, the fetus can also be sexed.

Fetoscopy is another diagnostic technique used to visually examine the developing embryo, but it is more hazardous to the mother and fetus because it requires an instrument (the fetoscope) to be inserted through a small surgical opening in the abdomen and into the amniotic sac. However, developmental defects that would go unnoticed by ultrasound examinations, such as deformed limbs or extra toes or

fingers, can be seen with the fetoscope. If necessary, this instrument can also be used to remove a sample of fetal blood that can then be analyzed for blood diseases or abnormal cellular enzyme levels. Because of the danger of harming the fetus and inducing a miscarriage, fetoscopy is used only for serious conditions that cannot be diagnosed by simpler techniques.

Amniocentesis and Chorionic Villus Sampling

The medical technique that has revolutionized prenatal genetic counseling in recent years is

amniocentesis. As a result of this technique, couples at risk for a hereditary defect are willing to conceive, given that the genetic constitution of the fetus can be determined. Many more normal babies are delivered than fetuses aborted because of the findings obtained from amniocentesis. In this procedure, which is now widely used in the analysis of hereditary diseases and developmental defects, a long needle and syringe are used to remove a small quantity of fluid from the amniotic sac (Figure 15-3), thereby making the amniotic fluid and the fetal cells that it contains available for genetic and biochemical tests. The extracted amniotic fluid and cells are analyzed for abnormal enzyme levels. The fetal cells can be cultured *in vitro*, and their chromosome constitution can be analyzed to determine whether any chromosomes are abnormal. Amniocentesis has been developed to the point at which it is quite safe for both the pregnant woman and the fetus, but there is some risk: in about one of two hundred amniocentesis procedures, the fetus is lost.

Amniocentesis is usually performed about the fifteenth week of pregnancy, the earliest time at which to safely and consistently obtain a sufficient amount of fluid and cells. At this stage of fetal development, there is still enough time to culture the cells and analyze the results (from three to four weeks) and to safely abort the pregnancy, should that option be elected by the prospective parents if the fetus is abnormal. Even earlier analysis would be better, however, and researchers are striving to improve the technique of amniocentesis so that the results can be obtained more quickly.

Amniocentesis is recommended only if there is reason to suspect that the embryo is at risk for a particular hereditary disease or developmental defect, although in some countries it is available to anyone who requests it. Before amniocentesis or any of the other prenatal tests are performed,

---

INFORMED CONSENT FOR AMNIOCENTESIS

I have been informed that my physician has recommended that a chromosome and/or biochemical test be performed on me to provide further information about my pregnancy. I understand that the cells of the fetus required as a basis for such an analysis are obtained in a procedure called transabdominal amniocentesis which involves penetration of my abdominal and uterine walls by a needle and the withdrawal of fluid from the sac in which my unborn child is contained. This fluid is called amniotic fluid.

The reasons for this procedure as well as the limitations and complications, which are listed below, have been explained to me:

1. That although transabdominal amniocentesis is a proven technique which has been used extensively and hazard to me or the fetus is considered to be small, it cannot be guaranteed that the procedure will not cause damage to me or the fetus including infection, bleeding or initiate premature labor possibly resulting in spontaneous abortion.

2. That any particular attempt to obtain amniotic fluid by transabdominal amniocentesis may be unsuccessful and repeat amniocentesis may be required

3. That any attempt to obtain a viable tissue culture from the cells or any particular sample of amniotic fluid may be unsuccessful or the chromosome preparations and/or biochemical analyses may be of poor quality and unusable.

4. That although the likelihood of a misinterpretation of the chromosome analysis and/or biochemical analyses in this case is considered to be extremely small, a complete and correct diagnosis of the condition of the fetus based on the karyotypes or biochemical analyses obtained, cannot be guaranteed.

5. The results provided of normal chromosomes or normal biochemical status of the fetus does not eliminate the possibility that the child may have birth defects and/or mental retardation because of other disorders.

6. In the case of presently undiagnosed twins, the results provided pertain only to one of the twins.

I acknowledge that I have had an opportunity to ask questions and become fully informed about this procedure.

I request that _____ M.D., and/or his associates, assistants of his choice and personnel assigned by the hospital or medical group attempt to perform a chromosome and/or biochemical test (transabdominal amniocentesis) on me.

Date: _____ Patient: _____

Date: _____ Witness: _____

FIGURE 15-4 A typical consent form for amniocentesis. By signing the form, the patient attests that she has had counseling and understands the risks and limitations of the procedure.

---

the prospective parents must be told about the risks of the procedures and about the various options. The reasons for performing amniocentesis, the slight risks that it entails, and stipulations to the effect that the success of the cell and chromosome analyses cannot be guaranteed are spelled out in a consent form that must be signed by the patient (Figure 15-4). If the analyses reveal a genetic defect, the prospective parents receive further genetic counseling so that they will have the best-possible informational basis on which to decide whether to terminate or to continue the pregnancy.

In recent years, another technique called **chorionic villus sampling** has been developed for obtaining embryonic tissue and cells. In this procedure, a sample of cells is obtained by inserting an instrument through the vagina to extract a tiny piece of tissue from the chorion (the membrane surrounding the embryo—see Figure 15-3), which consists only of embryonic cells. These cells can be analyzed for chromosomal and biochemical defects by use of the same procedures employed to analyze cells obtained by amniocentesis. An advantage of chorionic villus sampling is that it can be performed earlier in pregnancy, between the eighth and tenth weeks. The results are thus obtained earlier, which benefits the prospective parents psychologically, and the mother biologically, should they opt for an abortion. However, the risks to the embryo are still being evaluated. It is not yet certain that this procedure is as safe or as informative as amniocentesis.

Genetic Counseling

Just what is **genetic counseling**? The American Society of Human Genetics has formulated this definition:

> Genetic counseling is a communication process concerning the risks of occurrence of a genetic disorder in a family. It involves an attempt to help the person or family comprehend the medical facts, appreciate the hereditary nature and recurrence risks in specific relatives, understand the options dealing with the risk, choose the most appropriate course of action, and make the best possible adjustment.

Because genetic counseling may lead to the abortion of a fetus, counselors must undergo considerable training to be effective. They must also be sensitive and tactful in conveying information to their clients and apprising them of the various options.

Giving advice or making recommendations that affect the life of another person inevitably necessitates moral decisions—decisions that are influenced by the counselor's own views, beliefs, prejudices, and convictions. For example, a counselor's religious beliefs will influence the advice given, as will his or her sensitivity to the fears, social and cultural attitudes, and religious beliefs of the prospective parents.

An ethical dilemma that arises in genetic counseling is that of deciding whether a genetically defective fetus should be aborted. The advice and recommendations that are conveyed to the prospective parents vary with the genetic counselor's own views on the questions. Ideally, the personal views of the counselor should not influence the individuals' or the families' decisions. Clients should feel free to arrive at their own decision after careful consideration of all medical and legal options that have been explained to them.

Even among specialists who have carefully considered the issues, views differ. Joseph Fletcher, a bioethicist formerly at the National Institutes of Health, has stated:

> I would say that it is always unjust and therefore unethical or, if you like, immoral to knowingly and deliberately victimize innocent third parties or the innocent others. And I think that to deliberately and knowingly bring a diseased defective child to the world injures society, very probably injures the family, and certainly injures the individual who is born in that condition.

This forthright statement expresses the viewpoint of many people but is disputed by others

The Baby Doe Controversy

BOX 15-2

In 1982, at a hospital in Indiana, a baby boy was born with severe congenital abnormalities. The newborn had Down syndrome and a severe malformation of the esophagus so that he was unable to eat. The decision was made not to attempt any heroic surgical or other medical measures to sustain the infant because of the serious nature of the problems. The infant was allowed to die of starvation and, as a result, a long, bitter legal controversy ensued that became known as the "Baby Doe" case.

The issue is one of passive euthanasia (a painless death for incurables) in hospital neonatal intensive care units. Babies are sometimes delivered who have severe congenital abnormalities due to genetic or developmental factors or both. Advances in medical techology have made it possible to prolong the life of even severely deformed newborns. Traditionally, physicians and parents have decided when to attempt to prolong the life of a congenitally abnormal newborn and when to let nature take its course. In 1983, the U.S. government intervened in the birthing process by issuing regulations that defined nontreatment of congenital abnormalities as discrimination against the handicapped. Hospitals and physicians would be subject to penalties including the loss of all federal funds if they failed to treat all handicapped newborns no matter how severe the defects.

The U.S. government's ruling was overturned in court decisions in 1983 and 1984. The U.S. Supreme Court also voted against the government in 1986 in a decision that stated that the government had no legal right "to give unsolicited advice either to parents, to hospitals, or to State officials who are faced with difficult treatment decisions concerning handicapped children." Although no longer bound by the legal consequences of the "Baby Doe" rules, pediatricians are still in conflict over whether to treat or not. Those who advocate nontreatment are said to be making moral judgments based on their own value systems; those who advocate indiscriminate treatment are accused of being infatuated with high-technology medicine and a lack of moral judgment.

While this battle over medical care rages in the neonatal intensive care units, infant mortality remains higher in the United States than in any other industrialized country. And it is twice as high among black Americans as it is among whites. Some pediatricians believe that we are not paying enough attention to prenatal care for pregnant women, given that the primary cause of infant mortality is low birth weight. High-technology medicine and the financial rewards of treating the severely handicapped have replaced the more-mundane rewards of prenatal care and counseling and of delivering healthy, full-term babies. As with many other issues, our society will have to establish its priorities and decide how it wants to use financial and medical resources that are limited.

READING: John Lantos, "Baby Doe Five Years Later," *New England Journal of Medicine.* August 13, 1987.

(See Box 15-2). First, it does not take into account the various degrees of disability that the child might have. Some defects are much more debilitating and handicapping than others. For example, children with hemophilia now usually survive to adulthood with medical assistance, and some lead normal lives. (A particular tragedy of the AIDS epidemic in the United

States and elsewhere is that a majority of hemophiliacs have become infected with the AIDS virus. The pooled human blood from which the anticlotting factor that is used to treat hemophiliacs was purified carried the AIDS virus. The contamination was not discovered for several years; by that time, most had been infected.)

Hereditary diseases that cause severe and irreversible mental retardation are usually accompanied by serious emotional problems. They also create other burdens, including financial ones, for the rest of the family—burdens that can cause emotional problems for other family members and can damage family life in other ways. Still, even if the disorder is Down syndrome, some parents elect to have the genetically abnormal child even though they know from the results of amniocentesis that the child will be mentally retarded and suffer from the other disabilities of the syndrome. For some persons, the anguish caused by abortion would be greater than the anguish of caring for a defective child. Because many young people will eventually face these issues in their own lives, it is important that they come to grips with their thoughts and feelings about them.

Although genetic counselors strive to be objective, the counseling process is often, subtly and unintentionally, colored by the counselors' own views and prejudices. For example, prospective parents who carry recessive genes can be told that any child that they bear will have one chance in four of being abnormal or they can be told that the odds are three to one that the child will be normal. Both statements express the same truth about the probabilities, but the prospective parents might well interpret the two statements differently. A counselor with strong antiabortion beliefs may counsel a couple in such a way that an abortion will not seem advisable, whereas a counselor with different attitudes and beliefs may advise clients in such a way that abortion seems to be the only sensible way to cope with the abnormal embryo. So, although genetic counseling begins with objective calculations of degree of risk (which may approach certainty) that an abnormal fetus is present, from that point on subjective elements—value judgments—inevitably come into play and inevitably influence what the counselor tells the clients.

Genetic Screening

Genetic counseling is essentially a communications process that informs prospective parents about the nature of genetic disorders, about the risk of their having a genetically defective child, and about the options available to them in dealing with that risk. Or it can help them cope with the care of an existing genetically handicapped child. **Genetic screening**, in contrast, is a routine diagnostic procedure devised to detect those who are carriers of, or who are themselves affected by, a hereditary disease. Genetic screening applies to populations rather than to individuals.

The most-widespread application of genetic screening in the United States is for phenylketonuria (PKU). All hospitals in the United States screen newborn babies for PKU by a blood test called the Guthrie test. A drop of the infant's blood is checked for the presence of excess phenylalanine, one of the twenty amino acids. If a PKU infant is detected shortly after birth, the serious effects of the hereditary disease can be prevented by providing the infant with a special diet very low in phenylalanine. About one in fifteen-to-twenty thousand newborns is found to have PKU. Because the PKU test is simple, reliable, and inexpensive, it is practicable to screen an entire population—millions of newborns—for this hereditary disease.

TABLE 15-1 Newborn genetic-screening program in New York State from 1965 to 1984: a summary of cases.

Hereditary disease	Testing begun	Live births (millions)	Cases	Incidence
Phenylketo-nuria	1965	5.35	315	1 per 17,000
Galactosemia	1968	3.28	55	1 per 59,000
Maple syrup urine disease	1968	3.28	11	1 per 298,000
Histidinemia	1975	2.42	27	1 per 89,000
Homocysti-nuria	1975	2.42	5	1 per 484,000
Sickle-cell anemia	1975	2.35	947	1 per 2,500
Hypothy-roidism	1978	1.47	419	1 per 3,500

SOURCE: Adapted from T. P. Carter and A. M. Willey, eds., *Genetic Disease: Screening and Management*. Alan R. Liss, 1986.

Although numerous other inherited diseases affect the metabolism of amino acids, sugars, or nucleic acids and thus could be detected by genetic-screening programs, this is not done to any great degree. New York is one of the few states that screen newborns for hereditary diseases other than PKU (Table 15-1). However, the costs and follow-up efforts of screening millions of babies is generally too expensive for most states or countries. Some metabolic disorders, although they produce serious diseases, are so rare that it is simply too costly to screen all newborns for those very few among them who may have inherited the diseases. For other detectable metabolic diseases, there is no existing effective treatment; so there is no value in early detection. Before instituting any large-scale genetic-screening programs, authorities must evaluate the costs of the programs, the benefits to society, and the benefits for the affected persons.

Ethical Issues in Genetic Screening

Aside from the logistic and economic problems of widespread screening for hereditary diseases, questions have been raised concerning the right of any government or society to impose genetic screening for a particular trait. Are personal freedoms and individual rights violated by mandatory screening programs? Advocates of such programs have sometimes been accused of racism or prejudice because the frequency of certain hereditary diseases tends to be higher in particular ethnic groups (see Table 15-2). For example, about three-fourths of all cases of Tay-Sachs disease are found among Ashkenazic Jews, β-thalassemia in the United States is primarily a disease of Italian-Americans and Greek-Americans, and sickle-cell anemia is almost exclusively a disease of black Americans and certain populations of Africans. Singling out a particular hereditary disease or a

TABLE 15-2 The probabilities of hereditary diseases in particular ethnic groups or nationalities.

Hereditary disorder	Ethnic group with highest risk	Probability that a person is a carrier	Probability that carrier's child will inherit the disease
Sickle-cell anemia	Black-Americans	1 in 10	About 1 in 400
Beta-thalassemia	Italian-Americans Greek-Americans	1 in 10	About 1 in 400
Tay-Sachs disease	Jews (Ashkenazic)	1 in 30	About 1 in 4,000
Adult lactose intolerance	Orientals Blacks	Almost all Most	Nearly 100% About 7 in 10
Phenylketonuria	Predominantly Caucasian	1 in 80	About 1 in 20,000
Cystic fibrosis	Predominantly Caucasian	1 in 25	About 1 in 2,500
Mediterranean fever	Armenians	1 in 45	About 1 in 8,000

particular ethnic group for mandatory genetic screening is often interpreted as governmental interference with the right of people to have children with whatever partner they choose.

It should be pointed out that, although screening tests are available for the three hereditary diseases just mentioned, such tests have never been mandatory. And, even for the disorders for which screening tests are not mandatory, no restrictions have been placed on the breeding practices of the affected persons or the carriers who are detected. However, the concern of some people is that someday such infringements on personal liberties will come about. The concern of others is that no adequate checks are available to curb the births of predictably genetically defective infants.

Accusations of prejudice in genetic-screening programs and of interference with a person's right to bear children are often dealt with by comparing these programs with the way in which society deals with communicable diseases (those transmitted from person to person). Few people question the right of county, state, or federal governments to prevent epidemics of serious infectious diseases such as

plague, cholera, typhoid fever, or malaria by mandating sanitation measures, pasteurization of milk products, and immunizations. Most people would agree that the spread of sexually transmitted diseases is also a matter of public concern and that it would be desirable to reduce the incidence of syphilis, gonorrhea, AIDS, and genital herpes infections. Throughout history, societies have used quarantine and other measures to control the spread of infectious diseases in populations.

Why do some people accept and support efforts to eradicate infectious diseases yet oppose proposals to employ similar measures to reduce the incidence of hereditary diseases? One reason is that, despite feeling that laws mandating genetic screening for serious hereditary diseases are useful and beneficial, they worry that such laws may eventually lead to screening for less-serious traits and even to the acceptance of the ideas of eugenics, which advocates the breeding of people with desirable traits. Eugenics is a notion that has fostered an ugly history of racism and persecution and that, in its worst manifestations, has spawned the horrors of genocide (discussed in Chapter 20).

TABLE 15-3 Annual incidence and prevalence of some hereditary disorders in the United States.

Disorder	Approximate annual incidence	Approximate prevalence	Genetic cause	Detected by
Down syndrome	5,100	44,000	Chromosomal abnormality	Amniocentesis: chromosome analysis
Muscular dystrophy	Unknown	200,000	Autosomal recessive	Symptoms at onset
Spina bifida or hydrocephalus or both	6,200	53,000	Polygenic-multifactorial	Amniocentesis; prenatal x-ray; ultrasound; maternal blood test; examination at birth
Cleft lip or cleft palate or both	4,300	71,000	Polygenic-multifactorial	Examination at birth
Cystic fibrosis	2,000	10,000	Autosomal recessive	Sweat and blood tests; amniocentesis
Sickle-cell anemia	1,200	16,000	Autosomal recessive	Blood test; amniocentesis
Hemophilia	1,200	12,400	Sex-linked recessive	Blood test
Phenylketonuria (PKU)	310	3,100	Autosomal recessive	Blood test at birth
Tay-Sachs disease	30	100	Autosomal recessive	Blood and tear tests; amniocentesis
Thalassemia	70	1,000	Autosomal recessive	Blood test
Galactosemia	70	500	Autosomal recessive	Blood and urine tests; amniocentesis

SOURCE: The National Foundation for Birth Defects.

Genetic Screening and Down Syndrome: An Example

A better understanding of the controversies surrounding genetic counseling and genetic screening can be gained by examining one example: Down syndrome. This syndrome causes a number of physical defects and abnormalities, the most serious of which is mental retardation. As mentioned earlier, about five thousand Down-syndrome babies are born each year in the United States, along with thousands of others with hereditary handicaps (Table 15-3). Down syndrome is found in all populations, but women thirty-five years of age or older are particularly at risk for bearing a Down child. In fact, it is now virtually obligatory for physicians to inform any pregnant women in that age group of her increased risk of bearing a child with Down syndrome and of the availability of prenatal detection by amniocentesis. A physician who prefers not to have anything to do with prenatal testing that might lead to abortion must refer the at-risk woman to other persons who can supply the information. Failure to do so opens up the possibility of a financially catastrophic lawsuit if the woman subsequently has a child with Down syndrome.

No law compels a pregnant woman past thirty-five to undergo amniocentesis, however, and many couples simply prefer to take their chances. Then, too, many people feel that what they do with their bodies and their childbearing is their business, not the government's, and so it is unlikely that laws will be enacted that will make genetic screening or amniocentesis obligatory. Thousands of Down babies continue to be born each year—births that could be prevented by genetic counseling, amniocentesis, and abortion. Perhaps, as more people become aware of the serious consequences of hereditary diseases and learn that defects can be detected early enough to take remedial action, attitudes regarding genetic-screening programs and genetic counseling will change. Preventive measures are the only real hope in reducing the number of hereditary diseases; it appears that only a few of these diseases may ever be curable.

Prospects for Gene Therapy

Human hereditary diseases are not curable today. All that medicine can offer at this time is treatment of the symptoms of those genetic diseases that can be managed by administering drugs, by supplying the missing gene product, or by removing an abnormal gene product. For example, PKU is treated by limiting the phenylalanine in the diet. Bleeding in hemophiliacs is controlled by periodic injections of the missing clotting factor. Porphyria is controlled by avoiding certain foods and drugs.

To *cure* a hereditary disease, the abnormal genes would have to be replaced by normal ones in all of the patient's cells, including sperm or eggs if the disease is not to be passed on to offspring. It is impossible to do this by any existing medical techniques. However, it soon may be possible to correct the symptoms of a few, selected hereditary diseases by **somatic gene therapy**. In a general sense, this therapy refers to the insertion of a normal gene into the cells of an affected person who has inherited a nonfunctional gene at that locus. If the normal gene functions and the protein product is able to reach the organs in which it is required, the symptoms of the disease should disappear.

Two aspects of somatic gene therapy are important. First, the hereditary defect cannot be corrected because neither sperm nor eggs are genetically altered. The technological and ethical problems of **germ cell therapy** are so formidable that no attempts to alter genes in human sperm or eggs are expected to be made in the foreseeable future. The second important aspect of somatic gene therapy is that the disease is not cured in that the defective genes in the somatic cells are not removed. Rather, the symptoms of the disease are reduced or eliminated by introducing normal genes that supply the missing gene product. So, in the end, somatic gene therapy really is a remedy in the same sense as the administration of insulin to a diabetic or of clotting factor to a hemophiliac.

There are several technical approaches to accomplishing somatic gene therapy. First, the normal gene must be cloned into a plasmid or virus to obtain many copies. The gene must also be associated with regulatory elements that will allow it to be expressed in human cells at an appropriate level. (This is a problem that has not yet been satisfactorily solved). Next, the gene must be inserted into the affected person's cells by use of viruses or by physical means. This can be most easily accomplished in bone marrow cells. A sample of bone marrow cells is removed from the patient and the cloned, normal genes are inserted into the genetically defective bone marrow cells. These cells are then reinjected into the patient's bone marrow. If the cells are able to grow and reproduce and if the normal gene is expressed, the defect should be corrected and the disease symptoms should disappear.

At present, only three extremely rare hereditary diseases are thought to be good candidates for somatic gene therapy—adenosine deaminase (ADA) and purine nucleotide phosphorylase (PNP) deficiencies, both of which cause severe immune-system diseases, and hypoxanthine-guanine phosphoribosyl transferase (IIPRT) deficiency, which causes Lesch-Nyhan disease. In each of these three diseases, it is thought that the production of even a small amount of normal enzyme in bone-marrow cells should be beneficial, particularly because somatic and germ cell therapy experiments on rodents have been successful to some degree. However, because of the potential abuses and dangers associated with human gene therapy experiments, a number of regulatory safeguards have been instituted in the United States.

Ethics of Human Gene Therapy

In 1980, an American physician at the University of California, Los Angeles, Martin Cline, attempted somatic gene therapy on two patients with thalassemia (defective hemoglobin genes), one in Italy and another in Israel. He removed their bone marrow cells, treated the cells with DNA containing normal hemoglobin genes, and reinjected the modified bone marrow cells. Dr. Cline went ahead with his experiments before receiving authorization from the university and government committees that review human medical experiments. When knowledge of Cline's experiments became public, a hue and cry ensued. He was censured by the National Institutes of Health and his research grants were revoked. (The patients apparently were neither helped nor harmed by the experiment.)

In the aftermath of this case, much public discussion took place about the ethics and advisability of human gene therapy experiments. Almost everyone agrees that it is ethical to insert genes into a patient for the purpose of alleviating pain and suffering caused by a genetic defect. However, most people also believe that there should be a reasonable chance of success and, therefore, the feasability of somatic gene therapy should first be demonstrated in laboratory animals. In other words, can the genes be inserted into a person safely and effectively? Will they function normally and not otherwise harm the patient?

Only a tiny fraction of the bone marrow cells that are removed are of the type that can be expected to grow and reproduce when reinjected, and so the insertion of normal genes into cells must be highly efficient. At present, the use of genetically engineered viruses is the most-effective means of inserting genes into cells. So one concern is that the viruses themselves should do no harm. To correct most kinds of genetic diseases other than those that result in defective red or white blood cells, it will be necessary to target the genes to specific organs, and techniques for doing this are not yet available.

Although it is likely that somatic gene therapy will be approved and attempted in human patients in the future, there seems to be less public enthusiasm for other kinds of gene therapy. As mentioned earlier, germ cell therapy on human sperm or eggs is not likely to be attempted soon, if ever. And the ethics of "enhancement" gene therapy or eugenic gene therapy are even more controversial. *Enhancement gene therapy* refers to inserting genes into a normal person to "enhance" a particular characteristic. For example, a child of normal height might be treated with genes that produce growth hormone to enable him or her to grow taller or faster. *Eugenic gene therapy* refers to the remote possibility of altering complex human characteristics such as intelligence or personality. At present, the genetic basis for these polygenic-multifactorial traits are completely unknown. Also, the social pressure against any form of eugenics is very strong (discussed in Chapter 20).

The basic mechanisms underlying some monogenic diseases are rapidly becoming better understood. The defective genes causing some of the common hereditary diseases have been, or soon will be, identified and normal replacements for them cloned. However, despite recent advances in the detection and diagnosis of hereditary diseases, cures are not yet in sight. The most-effective "treatments" for years to come will be prevention. The next chapter describes some of the recent advances that have been made in detecting human hereditary diseases.

Summary

Hereditary diseases cannot be cured, though some can be treated medically. Congenital defects are found in about 5 percent of all live newborns; about one-half are hereditary. Medical techniques are available for the prenatal detection of many hereditary diseases; for example, amniocentesis is widely used to detect and diagnose defects early in pregnancy.

Factors that are known to increase the risk of giving birth to a genetically defective child are: (1) the prospective mother is thirty-five years of age or older; (2) either the parents have already had a child having a chromosomal abnormality or a neural-tube defect or a close relative has had the condition; (3) the mother is a carrier of a deleterious X-linked trait; (4) both parents are carriers of an autosomal recessive deleterious allele or one parent has an autosomal dominant deleterious gene; or (5) an earlier offspring or close relative has a polygenic-multifactorial trait.

Anencephaly, hydrocephalus, and spina bifida are neural-tube defects—those in which the brain or spinal cord fails to develop normally in the embryo. Anencephaly and spina bifida can be detected by examining a pregnant woman's blood for the presence of alpha-fetoprotein; abnormally high levels of this protein at certain stages of pregnancy indicate the possibility of a genetically defective fetus. Further testing, such as ultrasound scanning or amniocentesis, is usually required to confirm the diagnosis.

Ultrasound scanning and fetoscopy have generally replaced the use of x-rays for the visual examination of a fetus. Of the two techniques, ultrasound scanning is safer. Amniocentesis is used to extract amniotic fluid and the fetal cells that it contains for genetic and biochemical testing. A newer technique, chorionic villus sampling, extracts embryonic tissue from the membrane surrounding the embryo. The cells thus obtained are examined for chromosomal and biochemical defects, as they are in amniocentesis.

The object of genetic counseling is to inform prospective parents about genetic disorders, about the risk of their having a genetically defective child, and about the medical and legal options available to them, such as abortion, in dealing with their genetic risks.

Genetic screening is a procedure used to detect people who are carriers of a hereditary disease, as well as those who have it, in a population. Screening of newborns for diseases such as phenylketonuria allows early medical treatment that prevents the serious effects of the hereditary disorder.

Key Words

alpha-fetoprotein (AFP)   A protein produced during embryonic development; its level in the mother's blood can be used to detect neural-tube defects.

amniocentesis   A procedure in which a sample of amnionic fluid and fetal cells is removed from the uterus and examined for genetic or biochemical defects.

anencephaly   A congenital defect in which a newborn is missing all or a major part of the brain.

chorionic villus sampling   A technique for sampling embryonic cells between the eighth and tenth week of pregnancy to look for genetic defects.

fetoscopy   Insertion of an instrument (the fetoscope) into the amniotic sac to visualize the fetus.

genetic counseling   A process of communicating the risks of having a defective child to prospective parents.

genetic screening   Procedures that detect individual members of a population who carry a defective gene.

germ cell therapy   Correction of a mutant gene in sperm or eggs.

neural-tube defect   A defect in development of the brain or spinal cord that can result in anencephaly, hydrocephalus, or spina bifida.

placenta   The organ in pregnant women (and other mammals) that regulates the exchange of nutrients and chemicals between the mother's blood and the fetus.

somatic gene therapy   Correction of a genetic disease by inserting a normal gene into somatic cells of an individual.

ultrasound scanning   Production of a visible image of a fetus in the uterus by using sound waves.

Additional Reading

Anderson, W.F. "Prospects for Human Gene Therapy." *Science*, October 24, 1984.

Cline, M. "Gene Therapy: Current Status." *The American Journal of Medicine*, August, 1987.

Fuchs, F. "Genetic Amniocentesis." *Scientific American*, June, 1980.

Ledley, F.D. "Somatic Disease Therapy for Human Disease: A Problem of Eugenics?" *Trends in Genetics*, April, 1987.

Mitchell, M.L., and H.L. Levy. "The Current Status of Newborn Screening." *Hospital Practice*, July, 1982.

Motulsky, A.G. "Impact of Genetic Manipulation on Society and Medicine." *Science*, January 14, 1983.

Patterson, D. "The Causes of Down Syndrome." *Scientific American*, August, 1987.

Rathman, D.J. "Ethics and Human Experimentation." *New England Journal of Medicine*, November 5, 1987.

Wertz, D.C., and J.C. Fletcher. "Communicating Genetic Risks." *Science, Technology, and Human Values*, 12, 60 (1987).

Sources of information on medical facilities with genetic specialists:

The National Foundation
1275 Mamaroneck Avenue
White Plains, NY 10605

National Gene Foundation
250 West 57th Street
New York, NY 10019

Study Questions

1   Which congenital defects can be detected by measuring the levels of AFP in the blood of a pregnant woman?

2   What are three procedures used in prenatal diagnosis of fetal defects?

3   Is Down syndrome likely to be corrected by somatic gene therapy or germ cell therapy in the near future? Explain.

4   What hereditary disorder is screened for in all newborns?

5   Which genetic disease is more common in the United States—sickle-cell anemia or phenylketonuria?

6   What three genetic disorders are considered to be good candidates for somatic gene therapy?

7   What is the difference between genetic screening and genetic counseling?

Essay Topics

1   Discuss the issues that would concern you most if you were receiving genetic counseling.

2   What are your views on the Baby Doe controversy?

3   Give your views on the abortion issue with respect to genetically defective fetuses.

16

Human Hereditary Diseases

Mapping and Probing

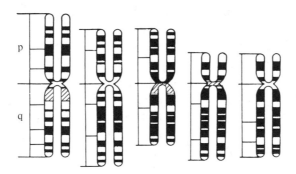

The end of man is knowledge, but there is one thing he can't know. He can't know whether knowledge will save him or kill him. He will be killed, all right, but he can't know whether he is killed because of the knowledge which he has got or because of the knowledge which he hasn't got and which, if he had it, would save him.

ROBERT PENN WARREN, *author*

ULTIMATELY, we study nature to better understand ourselves. When we perform genetic experiments with yeasts, bacteria, flies, or plants, we hope that what is learned from these organisms will not only be informative, but also eventually contribute to a better understanding of human inheritance. Until quite recently, the problem of identifying disease-causing genes in people or of constructing a map of the human genome seemed insurmountable. Unlike flies or pea plants, persons who have particular traits cannot be mated in the name of research. Also, human matings do not produce large numbers of progeny, and the time from one generation to the next is about twenty years.

The first human gene to be mapped to a specific chromosome was that for color blindness. In 1911, Edmund B. Wilson of Columbia University in New York deduced from its pattern of inheritance in families that this gene for color recognition must be on the X chromosome. Eventually, the genes responsible for hemophilia and Duchenne muscular dystrophy also were mapped to the X chromosome by analysis of family pedigrees. However, it was not until 1968 that a human gene was mapped to a specific autosome. This was the Duffy blood group gene and it was mapped to chromosome 1. This assignment was made by studying the linkage between the Duffy gene and another gene for a trait that could be followed in a family pedigree. If one gene has been mapped to a particular chromosome and another gene is found to be linked to it, then both genes must be on the same chromosome. In subsequent years, a few other human genes were mapped to autosomes by linkage studies, but it was not until the development of somatic-cell hybrids

and, more recently, the techniques of restriction fragment length polymorphism (RFLP) that mapping of human genes became more or less routine. By 1987, about a thousand genes had been assigned to a specific one of the twenty-three human chromosomes. This chapter describes how human genes are mapped and the effect that mapping has had on the prenatal detection of genetic diseases, on therapies for inherited diseases, and even on criminal investigations.

Identifying Genes and Chromosomes

As mentioned briefly in Chapter 2, human chromosomes can be displayed in a karyotype. This permits the detection of such gross chromosomal abnormalities as missing or extra chromosomes. As techniques for analyzing chromosomes improved, it became possible to detect even smaller chromosomal abnormalities, such as inversions, deletions, and translocations of pieces. To discuss the architecture of human chromosomes, an agreed-upon nomenclature was needed.

Figure 16-1 is a diagram of human chromosome 1. Chromosomes are numbered according to their size, shape, and banding patterns, which depend on the dyes that are used to stain the chromosomes. Each chromosome is assigned a short arm (p) and a long arm (q). The point of constriction along the chromosome is where the centromere is located. Each arm is broken down into regions and bands. The long arm (q) of chromosome 1, for example, has four regions and thirteen bands; the short arm (p) has three regions and eleven bands. Regions are assigned on the basis of the most-consistent, observable feature of each chromosome. The dark bands are produced by staining with either Quinacrine or Giemsa dyes. Both give essentially the same pattern of light and dark bands. A complete diagrammatic representation of a human karyotype is shown in Figure 16-2. With this system, any human gene that has been mapped can be designated by a chromosome number, an arm, a region, and a band.

The locations of some genes are known with great accuracy; others can be assigned only an approximate position. For example, the mutation that causes PKU is due to synthesis of an inactive enzyme, phenylalanine hydroxylase. The gene is located at 1q24. This means that it is on the long arm of chromosome 1 in region 2, band 4. The color-blindness gene mentioned earlier is at Xq28. It is on the long arm of the X chromosome in region 2, band 8. The exact location of the gene that causes Duchenne muscular dystrophy is still uncertain—its location is given as Xp12-p21. This means that it is on the short arm of the X chromosome somewhere between region 1 band 2 and region 2 band 1. Now that we have some idea of how geneticists describe the location of genes on human chromosomes, we can examine some of the methods used to construct the human genetic map.

A variety of methods have been used to assign genes to chromosomes and to map them to specific regions and bands. Some methods analyze family pedigrees and can determine linkages of two related traits. Others use the relation of chromosomal abnormalities, such as inversions, deletions, or translocations, to specific diseases or traits. However, these methods are tedious and relatively unproductive. Recently developed cellular and molecular techniques have produced enormous advances in the mapping of human genes.

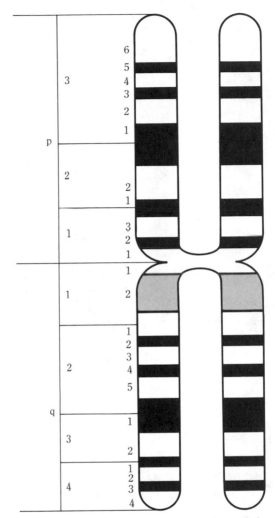

FIGURE 16-1 A schematic representation of human chromosome 1. The dyes most commonly used for distinguishing bands in a chromosome are Quinacrine (Q-bands) and Giemsa (G-bands). Both usually give the same banding patterns. The regions numbered along each arm refer to the most-distinctive bands either by size or intensity of staining or its lack; these regions are further subdivided and assigned numbers according to light- and dark-stained areas. The letters "p" and "q" refer to short and long arms, respectively.

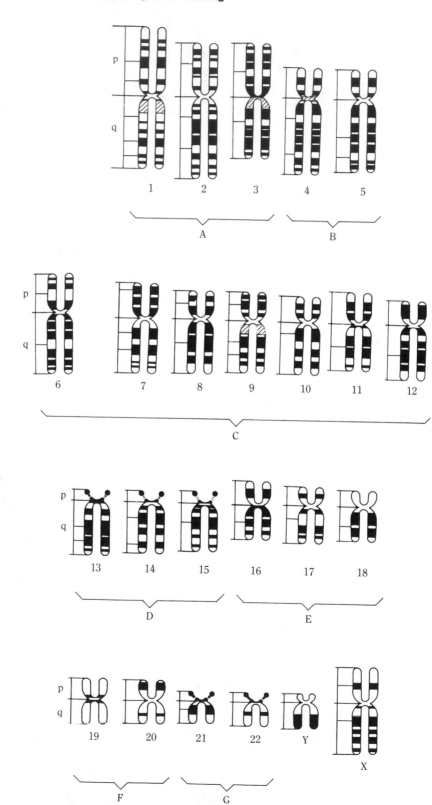

Mapping Genes by the Use of Somatic-Cell Hybrids

Although somatic cells reproduce *in vitro* until their density prevents further growth, they never fuse with one another. Each somatic cell remains isolated from the others by its membrane. About thirty years ago, scientists began to look for ways of making somatic cells fuse with one another while they were growing *in vitro*. They realized that if this could be accomplished—especially with cells that were derived from different species of animals—it would become possible to map genes to particular chromosomes and to learn something about how genes are regulated and expressed.

In 1960, scientists in France discovered that an influenzalike virus, the *Sendai virus*, would facilitate fusion of two different lines of mouse cells. When two cells fuse, first their cytoplasms unite within a common membrane and eventually the nuclei of the cells fuse to form a single nucleus initially containing all of the chromosomes of both cells. This is not a stable condition for most fused cells, and some chromosomes are eventually lost. As will be seen later, because of the progressive loss of chromosomes during cell division in *in vitro* cultures, it is possible to map genes to particular chromosomes using mouse-human cell hybrids.

The first step in cell fusion is to treat the cells with polyethylene glycol. This chemical does not harm the cells but does damage the outer membranes of the cells so that the two cells fuse together and become one. At this stage, the fused cell is called a **heterokaryon** because it contains two distinct nuclei (visible under a microscope) and two different sets of genetic information (Figure 16-3A).

As heterokaryons grow, the nuclei fuse together in some of them, forming true **somatic-cell hybrids**. Such a hybrid is a single cell with one nucleus that contains chromosomes derived from each of the two original cells. By the time hybrid cells are isolated, they have already undergone several mitotic and cellular divisions in which some chromosomes fail to segregate into the daughter cells (Figure 16-3B).

After the original discovery that different mouse cells could form somatic-cell hybrids, researchers developed techniques for fusing cells and for selecting the hybrid cells from different species. By 1967, scientists in the United States were successful in fusing cells from mice with human cells. Once this was accomplished, the mapping of certain human genes to a particular human chromosome became a fairly routine matter.

Mapping by Genetic Complementation

One way to select the relatively uncommon mouse-human hybrid cells from a mixture of fused and unfused cells in an *in vitro* culture is by the technique of **genetic complementation**. This procedure generally begins with two cell lines, each of which lacks the capacity to synthesize some product that is essential for the growth of the cells *in vitro*. In this case, the cells will grow *in vitro* only if the medium is supplemented with the missing gene products.

Consider, for example, a line of mouse cells that cannot synthesize compound "A" (an amino acid, vitamin, or base) and a line of human cells that cannot synthesize compound

◀ **FIGURE 16-2** A schematic representation of all twenty-three human chromosomes (twenty-two autosomes plus X and Y). A karyotype is analyzed by assigning chromosomes to seven major groups (A–G) based on the size of the chromosome and the position of its centromere. Then each chromosome is assigned a long arm, a short arm, numbered regions, and numbered bands, as shown in Figure 16-1. The numbered bands are shown only for chromosome 1.

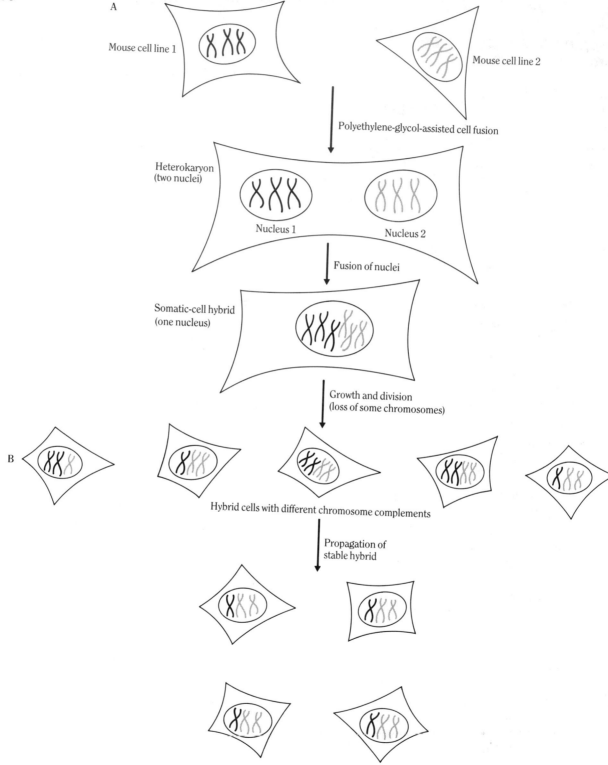

A

Mouse cell line 1

Mouse cell line 2

Polyethylene-glycol-assisted cell fusion

Heterokaryon
(two nuclei)

Nucleus 1

Nucleus 2

Fusion of nuclei

Somatic-cell hybrid
(one nucleus)

Growth and division
(loss of some chromosomes)

B

Hybrid cells with different chromosome complements

Propagation of
stable hybrid

"B" (a different amino acid, vitamin, or base). The mouse cells can grow *in vitro* if compound A is added to the medium, whereas the human cells can grow if supplied with compound B. If a mixture of mouse and human cells is inoculated into a medium that is deficient in both compounds, neither kind of cell will grow or divide. However, if the mouse and human cells are fused before inoculation into the medium, any somatic mouse-human hybrid cells that contain the genes for synthesizing both compounds A and B will be able to grow (Figure 16-4). The hybrid cells grow because functional genes complement one another and both products are synthesized in the hybrid cell. If only a few (ideally, only one) human chromosomes are in the hybrid cell, then the human gene that complements the mutant mouse gene must reside in one of the remaining human chromosomes.

Because mouse-human hybrid cells are chromosomally unstable, human chromosomes are lost as the hybrid cells continue to grow and divide in the *in vitro* culture. By correlating the loss of a particular protein from the hybrid cells (which can be done by biochemical techniques) with the loss of a particular human chromosome, the human gene responsible for the protein can be assigned to that lost chromosome (Figure 16-5). The lost chromosome is identified by a karyotype analysis of the hybrid cells before and after loss of the protein is observed.

◀**FIGURE 16-3** Fusion of mouse cells *in vitro* from different established lines.
A. When two cells fuse, they first form a heterokaryon—a single cell with two nuclei containing part or all of the genetic information of the original cells. Further growth allows the nuclei to fuse, producing a somatic-cell hybrid. B. In the course of growth and cell division of various hybrid cells, chromosomes are lost from one or both original cell lines. If the somatic-cell hybrid becomes genetically stable, millions of genetically identical cells can be derived from the original hybrid cell.

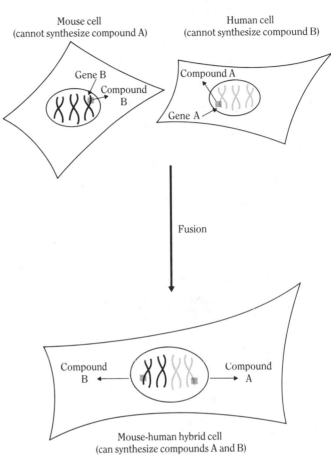

FIGURE 16-4 Genetic complementation. Fusion of cells from different species facilitates selection of somatic hybrid cells. In this example, the line of mouse cells chosen cannot synthesize compound A, which is essential for cell growth *in vitro*, but it does have a gene that synthesizes compound B. The line of human cells cannot synthesize compound B but is able to synthesize compound A. If the mouse and human cells are fused and grown in a medium that lacks compounds A and B, they are able to grow and multiply because of genetic complementation. The somatic-cell hybrids synthesize both compounds A and B because they contain chromosomes from both cell lines and thus carry gene *A* and gene *B*.

Various techniques that employ mouse-human hybrid somatic cells have been used to map human genes. For genes that produce enzymes whose activities can be measured, the use

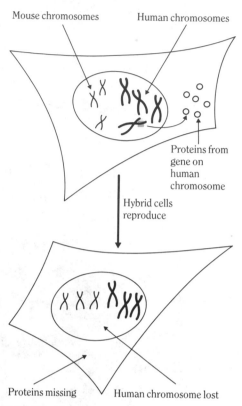

FIGURE 16-5 A technique for mapping human genes. In a mouse-human hybrid line of cells, loss of a human chromosome is correlated with loss of a protein activity. The gene for this protein must therefore be on the chromosome that was lost, which can be determined by a karyotype analysis.

three human chromosomes because, as mentioned earlier, most human chromosomes are spontaneously lost from mouse-human hybrid cells as they undergo cell division. The remainder of the hybrid's chromosomes are derived from the mouse cell. Because these mouse-human hybrid cells are able to grow without proline added to the medium, the remaining human chromosome must carry the genes for proline synthesis. This chromosome can then be stained and identified.

It should be evident from these examples that hybrid somatic cells are a means for mapping certain human genes. Somatic-cell hybrids can also yield information about the expression of human genes that cannot be obtained by other methods. The value of using somatic hybrids for human genetic studies depends on three properties of the somatic hybrid cells: (1) human chromosomes tend to segregate (be lost) from somatic hybrids when cells divide; (2) human chromosomes are readily identifiable by their size, structure, and particular banding patterns; and (3) human proteins can be distinguished from similar proteins produced by other species. Several hundred human genes have been localized to specific chromosomes by using somatic-cell hybrids. Nowadays, however, DNA probes offer a more-powerful technique for mapping human genes to specific locations on chromosomes, particularly those that cause disease.

A Southern Blot

As described in Chapter 12, restriction enzymes recognize specific short sequences of bases in DNA and cleave large DNA molecules into a collection of fragments differing in length. These DNA fragments can be separated by a procedure known as agarose gel electrophoresis. This consists of placing the fragments on a slab

of genetic complementation is quite effective. For example, suppose that the growth of the hybrid cell depends on its ability to synthesize an essential amino acid, such as proline. A mouse somatic cell is selected that is defective in proline synthesis; thus, mouse-human hybrid cells will grow only if the human chromosome that contains the genes responsible for proline synthesis is maintained in the hybrids. After fusing proline-deficient mouse cells with human cells, functional mouse-human hybrid cells are isolated that contain only one of the twenty-of gelatinlike material that is then exposed to an electric current. As the DNA fragments migrate through the gel, the largest ones remain near the top while the smallest ones move more rapidly and migrate to the bottom (Figure 16-6). After the fragments have been separated, the DNA can be visualized by staining the gel with various dyes. Large DNA molecules produce a large number of fragments of all sizes; so the separation appears as a smear of DNA from the top to the bottom of the gel.

Individual fragments can be detected by hybridizing a specific radioactive fragment of DNA—called a **probe**—to the DNA smeared out through the gel. This procedure of probing a mixture of DNA restriction fragments to identify one specific fragment or gene is called a **Southern blot** (Figure 16-7). The first step in carrying out a Southern blot (named for E. M. Southern, of Edinburgh, Scotland, who developed this important technique) is to treat samples of DNA with several different restriction enzymes to cut the DNA into fragments. Each restriction enzyme produces a unique set of DNA fragments because there is a different pattern of restriction sites in the DNA for each enzyme. The DNA fragments from each enzyme digest can be separated by gel electrophoresis, as shown in Figure 16-7A. Note that the size of the fragment carrying the gene of interest depends on the particular restriction enzyme used to cut the DNA.

Next, the DNA fragments are transferred to a thin plastic sheet made of nitrocellulose (Figure 16-7B). This is the "blotting" part of the procedure; DNA binds very tightly to nitrocellulose. The plastic is pressed onto the gel until all of the DNA sticks to the nitrocellulose filter. Specific DNA fragments can be detected on the filter by using a radioactive DNA probe (Figure 16-7C). To enhance understanding of this procedure, consider a real example.

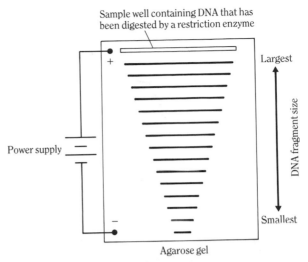

FIGURE 16-6 Separation of DNA fragments on an agarose gel through which an electric current is passed (electrophoresis). Large DNA fragments move through the gel more slowly than do small fragments. The number of fragments depends on the size of the original DNA molecule and the number of restriction-enzyme sites that were cleaved by the particular restriction enzyme used to digest the DNA. Although the DNA fragments are shown as dark bands, remember that DNA is not visible until it is stained with a dye. The DNA fragments at the top are larger than the DNA fragments below them in the gel, though in the gel itself all bands are the same width as that of the sample well at the top.

The gene for the beta chain of human hemoglobin has been cloned into a plasmid, as described in Chapter 12. The purified plasmid DNA carrying the "foreign" β-globin gene is made radioactive, usually by allowing some plasmid DNA to undergo synthesis in the presence of radioactive phosphorus (^{32}P).

The radioactive plasmid (now called the probe) is incubated at high temperature in a salt solution with the nitrocellulose filter. (Remember that the strands of DNA "melt apart" at high temperature because the hydrogen bonds are disrupted.) A radioactive DNA strand of the probe finds the fragment of DNA on the filter that has the complementary base sequence and

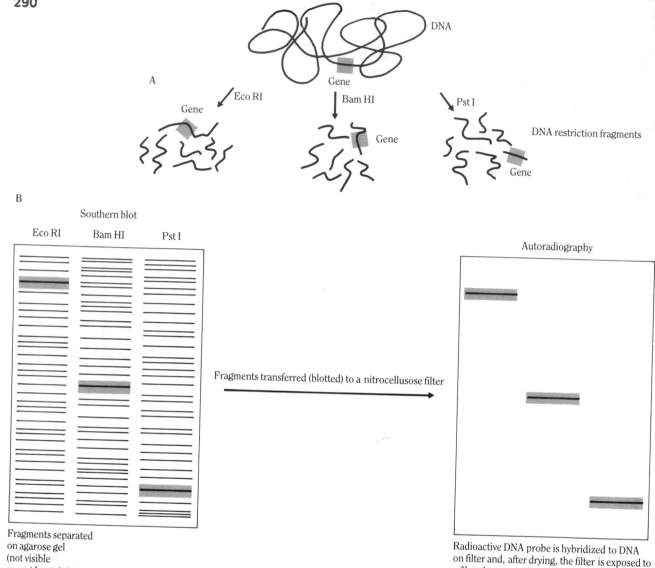

FIGURE 16-7 Schematic diagram of a Southern blot. Samples of DNA are digested with different restriction enzymes—for example, Eco RI, Bam HI, and Pst I. (The abbreviations for all restriction enzymes are derived from the name of the bacterial species from which they are isolated; see Table 12-2). The DNA fragments are separated according to size by placing a sample of the digested DNAs on an agarose gel and passing an electric current through the gel. The separated DNA fragments are transferred (blotted) to a nitrocellulose filter. The gene or base sequence of interest can be identified by hybridizing the DNA trapped on the filter to a radioactive DNA probe.

the two DNA strands hybridize (stick) together. After the hybridization reaction is completed, the filter is dried and covered with a film. The radioactivity on the filter exposes the film, producing a dark band corresponding to the DNA fragment that has bound to the probe. The fragment of DNA carrying the β-globin gene has now been identified.

Probing for Sickle-Cell Anemia

Some probes have been constructed that can distinguish mutant, disease-causing genes in carriers and in fetuses. By chance, the mutation that causes sicke-cell anemia also causes a change in one restriction site that is present in the normal gene. After trying many different restriction enzymes, researchers found one that could be used to distinguish a wild-type β-globin gene from one that carries the sickle-cell mutation (Figure 16-8). Digestion of normal human DNA by a restriction enzyme from the bacterial strain *Microcoleus*, Mst II, produces two bands on a Southern blot when it is probed with a plasmid carrying the β-globin gene. However, the DNA that contains the sickle-cell mutation has lost the middle Mst II restriction site and gives only one band on the Southern blot.

The use of β-globin probes makes it possible to determine if prospective parents are carriers of the sickle-cell mutation. If both parents are carriers, a child is at risk for sickle-cell anemia. Amniocentesis can then be recommended to determine if the fetus is affected or is, like the parents, only a carrier of the sickle-cell trait.

Probing for RFLPs to Detect a Disease

Detection of the mutant β-globin gene in sickle-cell anemia is facilitated because the mutation not only produces an abnormal gene but also changes a restriction-enzyme site within the gene. However, restriction sites that lie outside of, but close to, a mutant gene also can be used to detect persons or fetuses at risk for a hereditary disease. One of the most-dramatic examples is the detection of Huntington disease. As mentioned in an earlier chapter, this disease is caused by an autosomal dominant mutation that is not expressed until middle age. Then, neurological symptoms develop that eventually lead to death, as happened to folk singer Woody Guthrie.

Just as a population of people can differ in the alleles that they carry at a genetic locus (a **polymorphism**), so can they differ in a particular restriction-enzyme site in their DNA (a **restriction fragment length polymorphism**, or RFLP). For years, researchers looked for a means to detect the presence of the mutant Huntington-disease gene. By studying a very large family (more than three thousand people) in Venezuela, many of whose members had Huntington disease, they were eventually able to find an RFLP that could be used to detect the mutant gene in individual members. It was discovered that every person in this Venezuelan family who carries the Huntington gene also carries an additional site for the restriction enzyme Hind III, located near the mutant gene (Figure 16-9A). Thus, a diagnosis of Huntington disease can be made on the basis of a difference in RFLP in the DNA between normal and affected persons in this particular family.

When the study of Huntington disease began, the gene had not been mapped and no one had any idea of what human DNA probe to use to search for the gene. A group of scientists decided to use a brute-force approach. They systematically tested all of the available probes carrying human DNA in the hope of finding one that would show a different pattern of hybridization in a Southern blot analysis between the normal gene and the Huntington gene. Consider the enormity of their task. There are hundreds of different restriction enzymes that could be used in the hope of showing a difference in RFLPs between normal and mutant DNA. And hundreds of human DNA probes were available from the cloning of human DNA fragments, only one of which might possibly carry the sequence complementary to that of the Huntington gene.

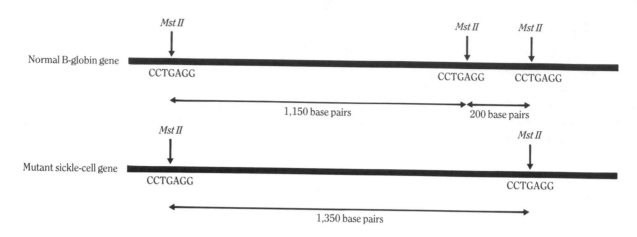

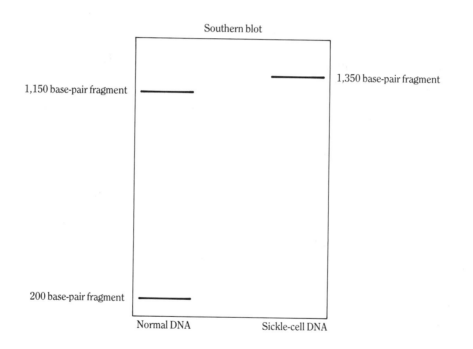

FIGURE 16-8   Detection of the sickle-cell gene in carriers by the use of a Southern blot and a radioactive DNA probe that detects an RFLP that is always linked to sickle-cell DNA. A normal β-globin gene has three sites in DNA that are recognized by the restriction enzyme Mst II. The sickle-cell mutation eliminates one of the three Mst II sites, leaving only two sites. As a result, the Southern blot shows a different RFLP for normal DNA compared with that for DNA from carriers of the sickle-cell mutation.

Luck, again, played a critical role. One of the restriction enzymes chosen for the initial analysis was Hind III and among the first dozen probes used was one called G8. This probe contains the wild-type base sequence at the Huntington locus (now mapped to chromosome 4).

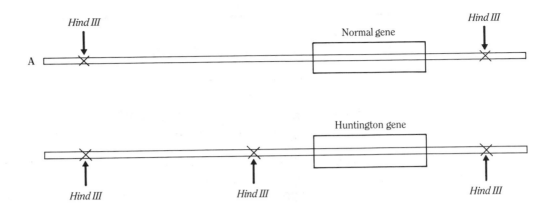

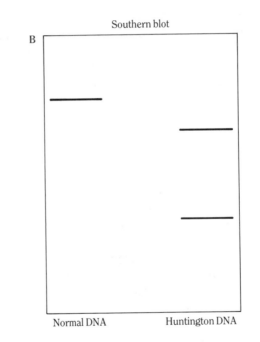

FIGURE 16-9 Detection of the mutant gene in carriers of Huntington disease.
A. A person carrying the mutant gene is identified by the presence of an additional fragment of DNA produced by an extra site for the Hind III restriction enzyme. DNA carrying the normal gene at the Huntington locus does not have this extra restriction site located nearby.
B. DNA extracted from cells of a normal person will give one band after digestion by the Hind III restriction enzyme and exposure to a particular radioactive probe. A carrier of the Huntington gene will have two bands after identical treatment.

The G8 probe is diagnostic for Huntington disease because it detects the extra DNA fragment produced by the Hind III site located near the mutant gene (Figure 16-9B). Many Huntington families have now been screened by the use of Hind III DNA digest and the G8 probe. The altered restriction site (and abnormal RFLP) detected by the G8 probe is always linked to the Huntington gene and thus can be used for the diagnosis of carriers. An uncertainty of about 5 percent exists in the diagnosis because of chiasma formation and a possible recombination event between the extra Hind III site and the mutant gene during meiosis, when eggs or sperm are produced. If recombination does occur, the extra Hind III restriction-enzyme site and the Huntington allele become separated from one another.

DNA Fingerprints
Probing for Criminals

In 1985, Alec J. Jeffreys, a professor of genetics at the University of Leicester in England, made a series of remarkable discoveries. While analyzing a fragment of DNA that had been isolated from a human myoglobin gene, Jeffreys and his colleagues noticed that the DNA contained a sequence of thirty-three bases that were repeated over and over—more than twenty times in their DNA fragment. They were curious to know how often this particular repeating sequence, or others similar to it, occurred in the human genome. They cloned their original repeating DNA sequence, and, using this as a probe, they hybridized it to a library of human DNA fragments that had been cloned into lambda bacteriophages (refer to Chapter 12 for a discussion of DNA libraries).

They found to their surprise (and delight) that these repeating sequences occur quite frequently throughout the human genome. When they analyzed DNA from different people, they found that the pattern of these sequences—called *hypervariable regions*—is unique and

distinctive for every person. It turns out that, like genes, these hypervariable sequences are inherited in a Mendelian fashion. Identical twins have identical patterns of hypervariable regions, but otherwise no two people are the same (see the accompanying illustration).

When DNA from a person is analyzed and probed by a Southern blot (see text), a unique set of bands is observed that is called a "genetic fingerprint." It is analogous to a person's fingerprints used for identification but is far more accurate and informative. This technique has already been used to resolve paternity and maternity disputes. For example, a boy from Kenya was denied permission to immigrate to England because the authorities did not believe that the woman living in England was the boy's mother as she claimed. Dr. Jeffreys did a fingerprint analysis of DNA extracted from cells from the woman and the boy. The pattern of the hypervariable regions

showed conclusively (the accuracy was greater than one in a billion) that the woman was the boy's biological mother. Entry was granted.

The uses of the Jeffreys probes are far reaching. Any person who commits a violent crime can be identified if any of the assailant's

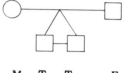

M   T₁   T₂   F

Genetic "fingerprint" showing the hypervariable sequences inherited by identical twins (T₁ and T₂) from their mother (M) and father (F). The arrows point to bands that the twins share with their father but which are absent in the mother and unrelated persons. Such a genetic fingerprint can prove who is a parent and who is not. (Courtesy of A.J. Jeffreys, University of Leicester. Reprinted by permission from *Nature*, vol. 316, pp. 76–79. Copyright © 1985 Macmillan Magazines Ltd.)

cells can be recovered. For example, the tiniest blood stain yields enough DNA to convict a murderer who has left even a few drops of blood behind. The semen recovered from a rape victim yields enough DNA to "fingerprint" a suspect rapist. DNA fingerprinting has already been used to obtain convictions in England and the United States. As more laboratories are equipped to perform this analysis, criminal investigations are going to be helped enormously.

The possibility exists that, in the future, everyone will be genetically fingerprinted and each person's unique pattern of hypervariable regions will be recorded in a computer. Any DNA that is recovered at the scene of a crime could be analyzed and matched up with the computer file of genetic fingerprints. Some people view this possibility with dread because of what it may mean for civil liberties; others see it as the ultimate weapon in combatting crime.

READINGS: P. Gill, A.J. Jeffreys, and D.J. Werrett, "Forensic Applications of DNA Fingerprints," *Nature*, 318, 577 (1985). R. Lewis, "DNA Fingerprints: Witness for the Prosecution," *Discover*. June, 1988. H. Logtenberg and E. Bakker, "The DNA Fingerprint," *Endeavour*, New Series, 12, 28 (1988).

The ability to determine who carries the Huntington gene has raised serious ethical and legal problems. What are the consequences of telling a young child that he or she will come down with a fatal disease sometime in mid-life? What happens to relationships in a family when some members refuse to be tested for the mutant gene? What if insurance companies have access to the information and refuse to issue health or life insurance policies? What if employers refuse to hire persons with the Huntington gene, knowing that they eventually will become sick and die?

Probes for many hereditary and infectious diseases are now available or are being developed (Table 16-1). Although these tests have significantly increased our ability to detect and diagnose diseases, they have also created serious problems for individual persons and for society. Biotechnological advances give us powerful new medical tools, but they do not tell us how or when to use them (see Box 16-1).

Genetic Load

Most of us are born healthy and do not have to worry about hereditary diseases in ourselves or in our children. However, even the healthiest people carry from three to five lethal alleles in their chromosomes and many more alleles that are harmful but sublethal. In a population, the decrease in overall reproductive capacity that results from the presence of mutant genes, whether expressed or not, is called the **genetic load**.

A lethal gene is one that causes the death of the fetus before or after birth if it is expressed; a sublethal gene permits the individual to survive but produces some form of serious disease or disability if expressed. The reason that people survive with lethal genes and can pass them on to progeny is that the expression of recessive mutant alleles is masked by the presence of dominant normal alleles. For most traits, the expression of a dominant normal allele is sufficient to maintain normal function and phenotype in the person.

TABLE 16-1 Inherited and infectious diseases that can be detected by the use of DNA probes as diagnostic tools at present and those for which probes are in some stage of preparation.

Disease	Estimated population	Status of probe
Inherited diseases		
Cystic fibrosis	33,000 affected 132,000 at risk	Test available
Duchenne muscular dystrophy	10,000 affected 50,000 at risk	In development
Huntington disease	25,000 affected 50,000 at risk	Test available
Sickle-cell anemia	50,000 affected 2 million at risk	Test available
Infectious diseases: sexual transmission		
AIDS	70–80 million tests per year	Being tested
Chlamydia	3 million tests per year	Being tested
Herpes I and II	2–4 million cases per year	Being tested
Infectious diseases: bacterial and viral		
Salmonella	5 million tests per year	Used by food companies
Legionnaire	1 million tests per year	First FDA-approved probe
Campylobacter	5 million cases per year	Being tested
Cytomegalovirus	0.5 million cases per year	Being tested
Hepatitis B	300,000 cases per year	Waiting for FDA approval
Rubella (measles)	Risk in pregnancy	In development

The number of people who are carriers for a particular mutant gene can be calculated by knowing the number of persons that are actually affected (the calculation is explained in Chapter 18). Even for hereditary diseases in which affected persons are extremely rare (1 per million), the number of people who carry the recessive allele is quite large—one of five hundred (Table 16-2). For hereditary diseases that occur more frequently in the population, the number of carriers of the mutant alleles becomes very large. Altogether, these lethal and sublethal alleles constitute the genetic load of the human population. As the genetic load increases, fewer people survive. If the number of mutant genes becomes too great, the survival of the population as a whole may be threatened.

Proposals for reducing the genetic load of human populations have generated heated controversy and much resistance. Although a strong case can be made for attempting to reduce the number of babies born with Down syndrome, the arguments become less convincing for diseases such as PKU, which can be treated, or for Huntington disease, in which no symptoms appear until mid-life or later. And the arguments become progressively weaker when defects such as cleft palate and club foot are included in estimates of the genetic load.

It is virtually impossible to draw the line between severe human hereditary defects that almost everyone agrees should be eliminated and those defects that are less severe. What some people would consider an unacceptable defect others would consider quite tolerable. Moreover, trying to change the frequency of genes in the population of a modern society by selective breeding is impossible. As Table 16-2 shows, the number of people who carry recessive genes is enormous, even for traits that are quite rare. No counseling or screening programs can change these numbers appreciably. Furthermore, mutant genes are continually being introduced into the human population by new mutations that cannot be prevented. Despite the new technologies and medical advances, hereditary diseases will continue to occur and to cause suffering in all human populations.

TABLE 16-2   Frequency of disease-causing recessive alleles in a human population.

If the number of persons affected is	The number of carriers is
1 in 10	1 in 2.3
1 in 100	1 in 5.6
1 in 1,000	1 in 16
1 in 10,000	1 in 51
1 in 100,000	1 in 159
1 in 1,000,000	1 in 501

Summary

By convention, chromosomes analyzed in a karyotype are arranged and numbered according to their size, shape, and banding patterns. A chromosome's centromere joins its short arm (p) and its long arm (q), and each arm is divided into numbered regions, which are subdivided into numbered bands. This system is used to map human genes — that is, to identify their locations in the genome.

The fusion of two different lines of somatic cells from different species to form hybrids is a technique for mapping genes by genetic complementation. In the first stage of formation of a somatic-cell hybrid, the chemically treated cells fuse so that their cytoplasms unite within a common membrane. At this stage, the fused cell is called a heterokaryon because it contains two nuclei and two different sets of chromosomes. As the heterokaryon grows, its nuclei fuse and some chromosomes are lost; the result is a somatic-cell hybrid.

To map genes by genetic complementation, somatic-cell hybrids are selected from a mixture of fused and unfused cells from two different cell lines. Each of the two parental cell lines lacks the capacity to synthesize some product (say, product A for one cell line and product B for the other) essential for growth *in vitro*. Growth of each separate cell line is possible only if the medium contains the missing product. However, fused cells in the mixture grow because one cell line contains a gene or genes coding for the product that the other cell line cannot synthesize; the converse is also true. As the hybrid cells grow and divide, some of their chromosomes are lost. If the hybrid is of mouse and human cells, human chromosomes are the ones that are lost. The ability of the hybrid cells to grow depends on the function of a particular protein encoded by a gene on one of the human chromosomes remaining in the hybrid. This chromosome can be identified by a karyotype analysis and the gene assigned to it.

DNA probes are also used to detect and map genes. In a procedure called a Southern blot, restriction enzymes are used to cut large molecules of DNA into fragments of various lengths. The fragments are separated according to size by gel electrophoresis. The DNA fragments are transferred to a sheet of nitrocellulose to which the fragments bind very tightly. The specific fragment containing the gene of interest can then be identified by the hybridization of a radioactive fragment of DNA—the DNA probe—to the nitrocellulose filter. If the DNA probe is known to contain base sequences from a particular chromosome, the gene can be assigned to that chromosome.

A restriction fragment length polymorphism (RFLP) can be used to detect the presence of genes causing hereditary diseases in carriers and affected persons. RFLPs in a population refer to differences among people in a particular restriction-enzyme site in their DNA. For example, Huntington disease can be diagnosed on the basis of a difference in RFLP between normal persons and those carrying the mutant gene.

Genetic load is the decrease in the overall reproductive capacity of a population due to the presence of mutant genes, whether expressed or not. The mutant genes may be either lethal or harmful (sublethal). Lethal genes survive when the expression of recessive mutant alleles is masked by the presence of dominant normal alleles. The mutant alleles may be passed on to progeny.

Key Words

DNA probe A radioactive fragment of DNA carrying a particular cloned gene or sequence used to detect DNA fragments in a Southern blot.

genetic complementation In somatic-cell hybrids, the provision of essential functions by genes on different chromosomes from different cell lines. More generally, a test for whether mutations occur in different functional genes or on two different chromosomes.

genetic load The average number of lethal (or deleterious) genes per member of a population.

heterokaryon A cell containing two nuclei that have not fused into one.

polymorphism The presence of several different forms of a trait, gene, or restriction site among individuals in a population.

restriction fragment length polymorphism (RFLP) Variation among individual members of a population in the location or number of restriction-enzyme sites in their DNA. When DNA from individual members is digested and analyzed, different patterns of DNA fragments (RFLPs) are observed.

somatic-cell hybrid A somatic cell with a single nucleus containing chromosomes from two genetically different cells that fused to form a single one.

Southern blot The digestion of DNA by several different restriction enzymes followed by separation of the DNA fragments according to size by use of agarose gel electrophoresis. The DNA fragments are transferred from the gel to a cellulose filter. Specific DNA fragments can be detected on the filter by using radioactive DNA probes carrying a known gene or DNA sequence.

Additional Reading

Caskey, C.T. "Disease Diagnosis by Recombinant DNA Methods." *Science*, 236, 1223 (1987).

Dworkin, R.B., and G.S. Omenn. "Legal Aspects of Human Genetics." *Annual Review of Public Health*, 6, 107 (1985).

Lewin, R. "DNA Fingerprints in Health and Disease." *Science*, 233, 521 (1986).

Lewis, R. "DNA Fingerprints: Witness for the Prosecution." *Discover*, June, 1988.

Nichols, E.K. *Human Gene Therapy*. Harvard University Press, 1988.

Palmiter, R.D., and R. L. Brinster. "Gene Transplants into Germ Cells." *Hospital Practice*, September 15, 1987.

Pines, M. *The New Human Genetics: How Gene Splicing Helps Researchers Fight Inherited Disease*. NIH Publication 84-662 (1984).

Ruddle, F.H., and R.S. Kucherlapati. "Hybrid Cells and Human Genes." *Scientific American*, July, 1974.

Saltus, R. "Biotech Firms Compete in Genetic Diagnosis." *Science*, 234, 1318 (1986).

White, R., and J.-M. Lalouel. "Chromosomal Mapping with DNA Markers." *Scientific American*, February, 1988.

Study Questions

1 Name three different kinds of legal or criminal matters that can be resolved through the use of RFLPs.

2 What is the location of a human gene with the designation 15q21? 2p22?

3 Why are dyes used to stain chromosomes?

4 What is the difference between a heterokaryon and a somatic-cell hybrid?

5 Name three hereditary human diseases that can be detected by the use of a DNA probe.

6 What is meant by the expression "genetic load"?

7 What enzymes are essential for performing a Southern blot analysis?

Essay Topics

1 Discuss how murder and rape cases can be solved by the use of particular DNA probes.

2 Discuss the pros and cons of using DNA probes to detect carriers of mutant genes that can cause hereditary diseases.

3 Explain how a human genetic map is constructed.

17

Immunogenetics

Antibody Diversity and Clonal Selection

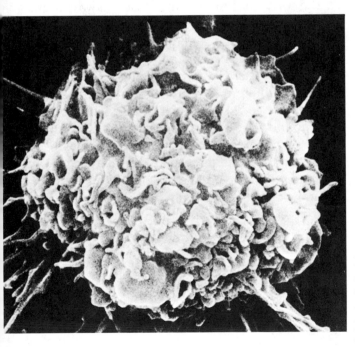

Dishwashers washed tedious dishes for you, thus saving you the bother of washing them yourself, video recorders watched tedious television for you, thus saving you the bother of looking at it yourself; Electric Monks believed things for you, thus saving you what was becoming an increasingly onerous task, that of believing all the things the world expected you to believe.

DOUGLAS ADAMS, *author*

IF GENETICISTS were asked to name the most important and interesting organ system in the human body, many would choose the immune system. Other scientists might select the nervous system, which includes the brain, as being more interesting. But from a genetic, evolutionary, and medical point of view, the immune system has been the source of the most insights. The human immune system is capable of distinguishing the body's own cells from any that are foreign to it either because the foreign cells have invaded the body from outside or because some of the body's own cells have changed, as when cancer cells arise. Our health—even survival—depends on our immune system's ability to recognize and destroy foreign microorganisms and substances before they cause disease. But the immune system is not infallible.

Like any complex biological system, the immune system may go awry. When it mistakenly attacks "self" (normal body cells), **autoimmune diseases**, such as certain kinds of rheumatoid arthritis or lupus erythematosus, may result. In arthritis, the immune system mistakenly attacks cells in joints; in lupus, the immune system may attack cells almost anywhere in the body. We do not yet know why these disease-causing changes occur in the immune systems of certain people.

Allergies are undesirable reactions of the immune system that are poorly understood. Allergic reactions of the nose, skin, lungs, eyes, and other parts of the body are caused by inappropriate responses of certain cells of the immune system to harmless foreign substances such as dust, pollen, or food. Allergies result when environmental substances produce chemical changes in cells that cause the discomforting rashes, wheezes, and sneezes characteristic of allergic reactions.

The most-remarkable aspect of the immune system from a geneticist's point of view is that the millions of specialized cells of this system are genetically programmed during embryonic development to recognize any of the innumerable kinds of viruses, bacteria, and other harmful substances that might be encountered in the course of a human lifetime. Because a person's entire genome probably does not contain more than a hundred thousand genes, how the enormous amount of genetic information needed to construct the immune system can be encoded in the forty-six human chromosomes was a great mystery. In fact, in the past few years it has become apparent that only about three hundred different genes supply all the genetic information necessary for synthesizing as many as 18 *billion* different antibodies—proteins that are synthesized by certain cells of the immune system and that recognize foreign substances.

How the antibody-determining genes of the immune system are organized, rearranged, and expressed during embryonic development is now partly understood and is the subject of this chapter. Also discussed are some of the many research and medical applications that are beginning to emerge from our increased understanding of immunogenetics.

What Is the Immune System?

In vertebrates, the **immune system** consists of numerous cells and proteins that recognize foreign substances and microorganisms. The diverse biological functions of the human immune system are carried out by a variety of

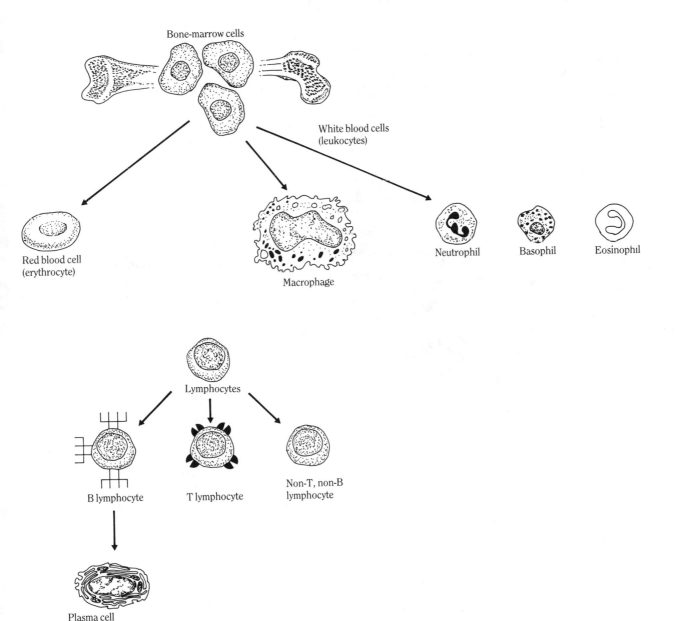

Bone-marrow cells

White blood cells
(leukocytes)

Red blood cell
(erythrocyte)

Macrophage

Neutrophil

Basophil

Eosinophil

Lymphocytes

B lymphocyte

T lymphocyte

Non-T, non-B
lymphocyte

Plasma cell
(synthesizes antibodies)

FIGURE 17-1 The different cells of the immune system originate in bone marrow. Cells are processed in various organs and tissues where they acquire their specialized functions.

white blood cells, called **leukocytes**, that circulate in the blood (Figure 17-1). The expression *white blood cells* is used to distinguish leukocytes from red blood cells (**erythrocytes**), which owe their color to the presence of hemoglobin molecules—transporters of oxygen from the lungs

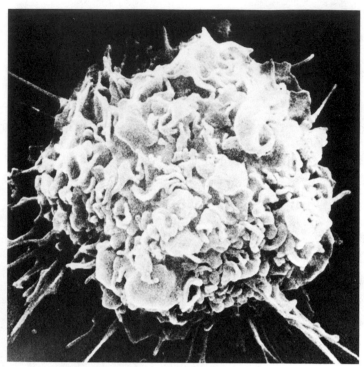

FIGURE 17-2 Electron micrograph of a macrophage. These large cells engulf and destroy foreign substances. Macrophage cells in the stomach, such as this one from a mouse stomach lining, protect animals from foreign particles contained in food. (Courtesy of L. Lin, Cetus Corp.)

to all body cells. Both red and white blood cells are manufactured in bone marrow and are secreted into the circulatory system. Specific kinds of white blood cells are secreted into the fluid of the lymphatic system as well, and for this reason are referred to as **lymphocytes**. These cells are processed by various organs of the lymphatic system, such as the spleen, tonsils, and thymus, in such a way that they become more specialized in their functions. The tonsils are now thought to play a role in the immune system, which is why they are no longer removed unless they become infected.

Three main kinds of immune-system cells circulate throughout the body and protect it in various ways. **Macrophages** are large phagocytic cells that engulf and destroy viruses, bacteria, and other foreign particles (Figure 17-2). The **T lymphocytes**, which are processed in the thymus, are special white blood cells that regulate the responses of other cells of the immune system in ways that are not fully understood. They also assist other cells in the recognition of foreign substances. The functions of the **B lymphocytes** are rather well defined; these white blood cells direct the synthesis of particular antibodies. When B lymphocytes detect the presence of a foreign substance or microorganism — referred to as an **antigen** — a specific B lymphocyte cell that is genetically preprogrammed to respond to that particular antigen becomes activated. The antigen-activated B lymphocyte cell replicates and ultimately produces many **plasma cells**, which in turn synthesize the millions of antibody molecules that eventually inactivate the antigens. Because each B lymphocyte can direct the synthesis of many plasma cells, which together produce one specific kind of antibody, there must be as many different types of B lymphocytes in the body as there are different harmful antigens in nature. If a person is not capable of destroying harmful microorganisms, he or she does not survive long (Figure 17-3).

Before considering how B lymphocytes can direct the synthesis of every conceivable antibody using only a few genes, we need to examine the structure of antibody molecules — a structure that allows them to recognize and inactivate specific antigens.

Antibodies

Antibodies are a large and enormously varied class of proteins found in the blood. Each antibody is a large protein molecule consisting of four polypeptide chains (Figure 17-4). Identical pairs of chains are referred to as heavy and light. Antibody proteins have specific shapes that permit them to recognize and inactive antigens whose shapes they are able to recognize (Figure 17-5).

The four polypeptide chains of an antibody molecule are held together by disulfide bridges. These bridges are formed by chemical bonds between sulfur atoms in certain amino acids in the four polypeptide chains. The two heavy (H) chains are so named because of their larger size; the light (L) chains are smaller polypeptides. The two heavy chains of an antibody molecule are identical in amino acid sequence, as are the

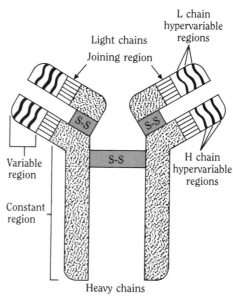

FIGURE 17-4 Details of an antibody molecule. There are two identical large polypeptides called heavy (H) chains and two identical small ones called light (L) chains. The sulfur–sulfur (S–S) bridges hold the four chains together. The constant, joining, and variable regions of each molecule give it its specificity. The constant region consists of an amino acid sequence that is similar in different kinds of antibodies. The variable region contains an amino acid sequence that is different in each different kind of antibody; within this region are segments that are hypervariable. The joining region of each chain links its constant and variable regions together.

FIGURE 17-3 David in his bubble chamber at age eleven. A year later, David died after having undergone a bone-marrow transplant that failed. As a result of a hereditary immune deficiency disease, David's body lacked the ability to synthesize antibodies. He was delivered in a germ-free environment and lived his entire life in a sterile bubble chamber or spacesuit to protect him from infectious organisms. (Courtesy of Baylor College of Medicine, Houston, Texas.)

two light chains. Each light and heavy chain has a *variable* amino acid sequence at one end and a *constant* sequence in the rest of the chain. The constant sequence of amino acids usually does not change within a given class of antibodies even though the various members of that class recognize different antigens. Antigen recognition is principally determined by the *hypervariable regions* located within the variable regions at the ends of both the light and the heavy chains of each antibody. These hypervariable regions consist of amino acid sequences that differ in each different kind of antibody.

The structure of antibodies was figured out by analyzing the most-common class of antibodies found in the blood, but it probably applies to the other classes as well. **Immunoglobulins**, the scientific term for antibody proteins, are divided into five major classes (Table 17-1). These immunoglobulin classes—IgG, IgA, IgM, IgD, and IgE—are distinguishable by their different kinds of heavy chains, as well as by their specific immunological functions and their concentration levels in blood serum. Most immunoglobin molecules found in blood serum are of the IgG type, which recognize a wide range of antigens. The functions of the other immunoglobulins are less well understood; however, it has been shown that the synthesis of small amounts of IgE can cause allergic reactions. Immunoglobulins are found in saliva, tears, stomach secretions, and most other body fluids, as well as in the blood.

If each different heavy and light polypeptide chain in the different IgG immunoglobulins

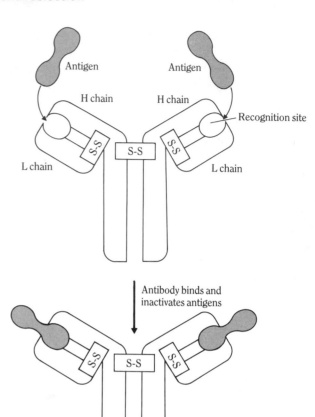

FIGURE 17-5 Antibody structure and function. Each antibody recognizes a specific antigen and can inactivate a pair of them.

TABLE 17-1 The five major classes of immunoglobulin (Ig) molecules and their functions.

Immunoglobulin class	Relative amount in blood serum	Functions
IgG	1.0	Inactivates viruses, bacteria, and toxins
IgA	0.25	Inactivates toxins
IgM	0.1	Inactivates viruses, bacteria, and toxins
IgD	0.03	Normal function is unknown
IgE	0.0001	Activates allergic responses; normal function is unknown

were encoded by a different gene, millions of genes would be needed. But, as noted earlier, it has been determined that only about three hundred human genes participate in the synthesis of antibodies. Thus, the puzzle that confronted immunogeneticists was: How can a virtually unlimited number of different antibodies be encoded by a small number of genes?

Antibody Diversity

By determining the amino acid sequences of many different IgG antibodies, researchers discovered the variable and constant regions in the heavy and light polypeptide chains. They also noted certain recurring patterns in the amino acid sequences of both the heavy and light chains from one antibody to the next. Although antibodies often had the same sequences of amino acids in the constant regions, different antibodies always had different sequences in the variable regions. Eventually, it became apparent that different antibodies could be constructed from a relatively few kinds of heavy and light chains whose amino acid compositions were identical in one part (the constant region) but quite different in another part (the variable region).

Using recombinant DNA techniques, researchers isolated the genes that code for light and heavy chains and determined the sequence of bases in the heavy- and light-chain genes. The amino acid sequences in the chains could then be compared with the base sequences. Each light chain, we now know, is constructed of short polypeptides synthesized from three families of genes: genes that code for the constant region, genes that code for the variable region, and genes that direct the joining of any constant region with any variable region. Several different copies of these genes that differ in base sequence are present in the genome, and so, when they are joined in different combinations, an enormous number of different light chains can be generated.

The heavy chains are constructed in a more-complex manner. The particular class of the antibody molecule and its antigen specificity are determined by four families of genes that direct the synthesis of heavy chains: genes that code for the constant region, genes that code for the variable region, and a fourth group of genes than enhances antibody diversity (D-region). Because genes from the four families may be joined in any combination, an enormous number of different heavy chains can be synthesized (Figure 17-6). Each antibody molecule is assembled from two identical heavy chains and two identical light chains that are constructed from polypeptides synthesized from several different genes.

The puzzle of how the enormous diversity of antibodies arises can now be explained. Individual genes from each of the seven families of genes are brought together by recombination during development of the embryo and end up in the chromosomes of millions of genetically different B lymphocytes. Any gene from one family can be joined to any gene from the other families, and so billions of antibody combinations are possible. Calculations show that even a few hundred different genes can be rearranged by somatic recombination in B lymphocyte cells to generate all conceivable antibodies. For example, by recombining only one hundred different genes from the seven gene families that have already been identified, more than 10 billion gene combinations (and presumably the same number of different antibodies) are possible. Even limiting the mechanisms by which the different genes can be recombined would still produce millions of different antibodies. One important prediction from such a scheme is that each B lymphocyte or plasma cell derived from it should synthesize one, and only one, kind of antibody.

Clonal Selection Explains Diversity

In the early 1950s, Macfarlane Burnet, an Australian immunologist, proposed the **clonal selection model**, which explains how cells of the immune system develop and how the genetic information is rearranged so that B lymphocytes can recognize any conceivable antigen that might be encountered. Burnet's clonal selection model also accounts for the fact (although not

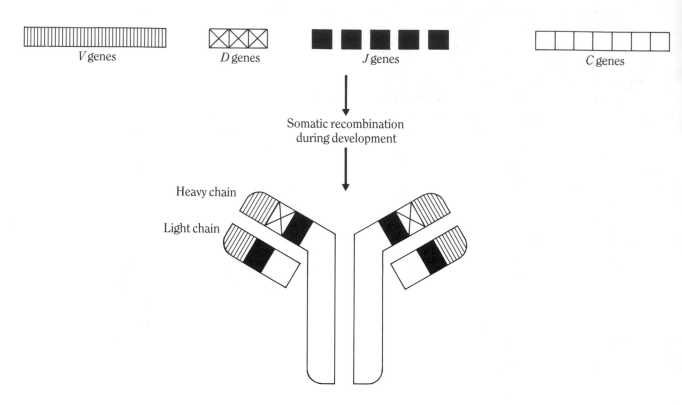

FIGURE 17-6 The families of genes that code for the synthesis of antibodies. Genes from each of the four families recombine in somatic cells during development. Each B-lymphocyte is ultimately capable of synthesizing two specific polypeptide chains and one specific kind of antibody. All cells derived from the original B-lymphocyte (plasma cells) will produce identical antibody molecules.

the mechanism) that the immune system destroys foreign tissues and cells that enter the body yet does not harm the body's own cells or tissues. In other words, the clonal selection model proposes how the immune system distinguishes "self" from "nonself."

According to the clonal selection model, very early in development genes in embryonic

immune-system cells undergo extensive recombination (or mutation). From these embryonic cells of the immune system, a vast number of genetically different bone-marrow cells are generated that are capable of synthesizing all possible antibodies (Figure 17-7). During development of the embryo, any immune-system cell that recognizes a normal antigen of the body is inactivated or destroyed. This ultimately leaves a population of immune-system cells capable of recognizing foreign antigens only.

The clonal selection model was viewed with scepticism at the time that it was proposed. In his original hypothesis, Burnet suggested that the genetically different immune-system cells were the result of **hypermutability**—an extremely high rate of mutation—of these cells. That certain cells could mutate at extremely high rates seemed unreasonable to many scientists. However, Burnet's idea was close to the truth. We now know that antibody diversity results primarily from an unusually high rate of recombination among genes in immune-system cells, but mutations also contribute to the diversity.

If the clonal selection model is correct, a single immunocompetent cell should produce one and only one kind of antibody. This important prediction was tested in an experiment using two immunologically different strains of bacteria, called strain A and strain B in Figure 17-8. A mixture of the two different strains was injected into a rat. After a week, during which the specific immune cells that react with the bacterial antigens proliferated, the rat's spleen was removed and homogenized into individual cells. Some of these cells were the plasma cells able to synthesize antibodies directed against the injected bacterial antigens. From the homogenized mixture, individual spleen plasma cells were placed in tiny dishes, and a mixture of both strains of bacteria was added to each dish. If the spleen cell produced no antibody, both strains of bacteria were unaffected and contin-

ued to swim. If anti-A antibody was produced by the spleen cell, bacteria of type A were inactivated and stopped swimming. If anti-B antibody was produced, bacteria of type B stopped swimming. No dish was observed in which *both* strains of bacteria stopped swimming.

Because no spleen cell was able to stop the movement of both types of bacteria, this experiment showed that only one type of antibody is produced by a single spleen cell. It is now firmly established that each B lymphocyte and plasma cell synthesizes only one kind of antibody and that the diversity of antibodies derives from the

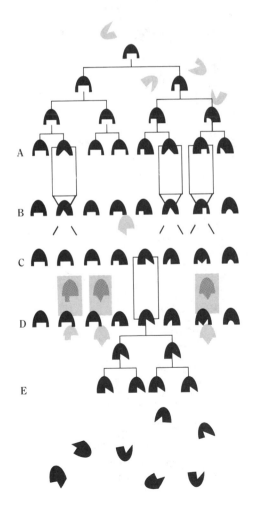

FIGURE 17-7 The clonal selection model: (A) during embryonic development, genes in immune-system cells are rearranged to give every possible combination of light and heavy chains, as well as constant and variable regions of each chain; (B) antigens on normal body cells are recognized by cells of the immune system as the embryo develops and those immune-system cells are destroyed; (C) what remains are millions of immune-system cells that can recognize any foreign substance; (D) if a foreign antigen is detected by cells of the immune system, one specific class of immune cells begins to proliferate; (E) these clonally derived cells synthesize antibodies that inactive the antigen.

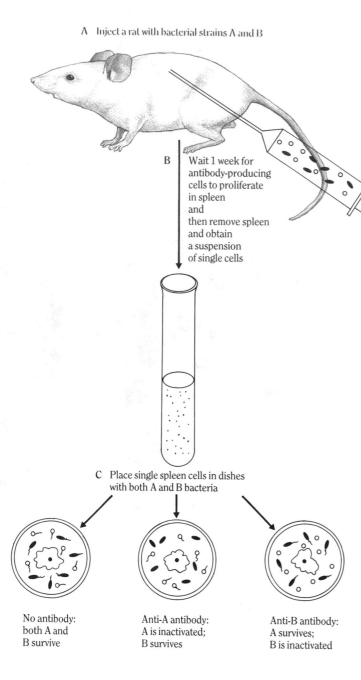

A   Inject a rat with bacterial strains A and B

B   Wait 1 week for antibody-producing cells to proliferate in spleen and then remove spleen and obtain a suspension of single cells

C   Place single spleen cells in dishes with both A and B bacteria

No antibody: both A and B survive

Anti-A antibody: A is inactivated; B survives

Anti-B antibody: A survives; B is inactivated

FIGURE 17-8 Test of the clonal selection model. The model predicts that each plasma cell can produce only one kind of antibody for each kind of antigen.
A. A rat is injected with two different strains of bacteria.
B. After one week, the spleen, which contains the B lymphocytes, is removed and homogenized into single cells.
C. Individual cells are incubated with a mixture of both bacterial strains. Only one of the two strains is inactivated in any dish—never both. Individual spleen cells produce only one kind of antibody.

What prevents antibodies from attacking normal cells in the same way that they recognize and attack foreign cells? According to the clonal selection theory, as normal body cells and their antigens are formed, any immune-system cells that might synthesize antibodies against these normal cells are themselves inactivated or destroyed. The mechanism that selects against these unwanted and potentially destructive immune-system cells is still unknown, but, whatever the mechanism, it is clearly essential to an organism's survival. Every person's immune system is able to distinguish that person's own tissues from those of any other person; if cells from one person are injected or implanted into another, a strong immunological response is generated and the foreign cells are destroyed.

Why Are Tissue Transplants Rejected?

Skin transplantation experiments with inbred strains of mice led to the concept of **histocompatibility**; that is, the condition that determines whether a tissue (histo means "tissue") transplanted from one animal to another will be accepted or rejected by the recipient animal. Strains of mice in which all animals are genetically identical are produced by inbreeding brothers and sisters generation after generation. After many generations, these inbred strains of mice will accept skin grafts from one another

genetic diversity of the cells. Thus, the clonal selection model originally proposed by Burnet, who died in 1985, has been shown to be essentially correct.

but will reject grafts from mice of other in-bred strains.

Genetic crosses can be made between male and female mice of two separate inbred strains to produce F_1 hybrid mice (essentially the same experiment that Mendel performed with his inbred strains of peas). Skin grafts between an F_1 hybrid mouse and its parent mice produce a characteristic pattern of graft acceptance and rejection (Figure 17-9). The F_1 mouse will accept skin grafts from either parent, but neither the black nor white parent will accept a skin graft from its progeny.

These experiments show that acceptance or rejection of tissue in mice is genetically determined by histocompatibility genes. Different strains of mice will reject tissue that is transplanted from a mouse of one strain to a mouse of another strain. However, experiments with **allophenic mice**—mice whose cells are derived from four parents—show that histocompatibility depends not only on particular genes but also on developmental factors.

Construction of allophenic mice begins by mating true-breeding strains of mice, in this case a male and a female having all-black coats and a male and a female having all-white coats. The embryos are removed from the two pregnant females' oviducts shortly after fertilization, generally at approximately the eight-cell stage. The embryonic cells are then separated and the cells from the two embryos are mixed together. New embryos will form that are mixtures of the embryonic somatic cells from the black and the

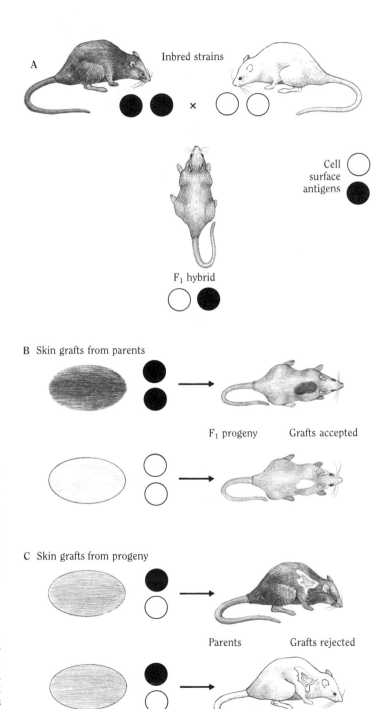

FIGURE 17-9 Pattern of skin-graft acceptance or rejection in inbred strains of mice: (A) mating between a male mouse and a female mouse of different inbred strains produces F_1 hybrid offspring; (B) when skin from parental mice is grafted onto the F_1 progeny, the offspring accept the skin grafts; (C) however, the parents reject skin grafts from their offspring.

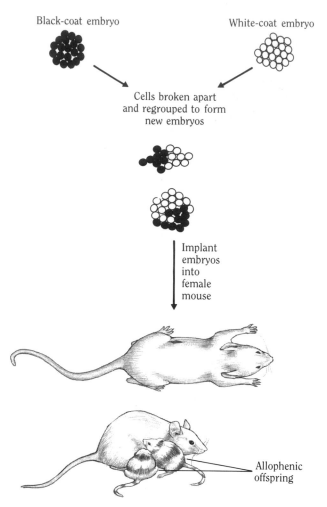

Black-coat embryo   White-coat embryo

Cells broken apart
and regrouped to form
new embryos

Implant
embryos
into
female
mouse

Allophenic
offspring

FIGURE 17-10 Technique for producing allophenic mice, whose cells are derived from four parents. The black-coat embryo was formed by the union of an egg and sperm of female and male parents; likewise for the white-coat embryo.

white embryos. Several of these embryos are then implanted into the uterus of a female mouse that has been hormonally prepared for pregnancy by having been mated with a sterile male. The implanted embryos develop normally and are born as allophenic mice (Figure 17-10).

Allophenic animals are valuable sources of insight into how genes govern the development and differentiation of tissues in animals. For example, by studying the patterns of the black and white fur stripes that appear in the progeny mice, researchers can determine how skin tissue is organized during embryonic development. If an allophenic mouse were born with half of its body black and the other half white or if it were covered all over with black and white spots, either pattern would suggest a different developmental organization than the one that produces black and white stripes (Figure 17-11).

Normally, an animal will reject any foreign tissue that is grafted onto it. However, allophenic mice will accept skin grafts from any one of their four parents (whereas the parents themselves will reject skin grafts from one another). Patches of white fur can be grafted onto the black patches in allophenic mice, even though the black and white cells are genetically different and the graft would normally be rejected. This shows that histocompatibility (immunological tolerance) is a property that is acquired by cells during development and is not just due to the genetic information carried in the cells.

Genetic Basis of Histocompatibility

Particular genes synthesize the proteins that appear on the surface of each cell and that can act as antigens. In the inbred strains of black and white mice (Figure 17-9), these genes produce different cell-surface antigens. The F_1 progeny mice synthesize cell-surface antigens corresponding to both the black and the white parental mice because they have inherited chromosomes and genes from both parents. Thus, the cell-surface antigens characteristic of both the black and the white mice are a normal part of the F_1 hybrid mice and these mice accept skin from either parent. However, skin from the F_1 mouse has cell-surface antigens that are recognized as foreign by the immune systems of both parents; thus, each parent rejects skin grafts from the F$_1$ progeny. These experiments show that histocompatibility is hereditary and that cell-surface antigens are determined by genes.

The genetic rules governing histocompatibility in human beings are quite similar to those discovered in mice. Histocompatibility plays a critical role in the transplantation of organs from one person to another. The more genetically related two people are, the more successful tissue transplants are. Identical twins can accept organ transplants from each other because they are genetically identical. Siblings other than identical twins share 50 percent of their genes and may be more similar in their histocompatibility antigens than unrelated persons.

When tissues are transplanted, the donor is usually unrelated to the recipient. In these cases, histocompatibility antigens of the recipient and of potential donors must be compared to obtain as close a donor-recipient match as possible. Inevitably, however, some antigens will differ between the donor and the recipient. The recipient must therefore be given drugs that suppress the responsiveness of his or her immune system; otherwise the tissue is likely to be rejected.

HLA Genes

Human histocompatibility is determined by four families of genes that are located close to one another on human chromosome 6 (Figure 17-12). These four groups of genes are called **HLA** (**h**uman **l**eukocyte **a**ntigen) **genes** because the antigens were originally identified on the surfaces of leukocytes. However, it is now known that the antigens coded for by three of the HLA genes (A, B, and C) are present on the surfaces of all body cells and that the D antigen appears only on certain cell types.

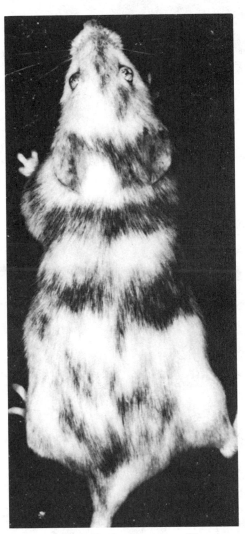

FIGURE 17-11 An allophenic mouse. (Courtesy of Beatrice Mintz, Fox Chase Cancer Center, Philadelphia; from B. Mintz, *PNAS* 58(1967): 344–351.)

There is an enormous diversity of HLA genes and of HLA antigens in human populations. In fact, except for identical twins, the likelihood that any two persons will have the same set of HLA genes or histocompatibility antigens is nil. Many alleles for each gene in the HLA complex exist in the human population. At least eight,

TABLE 17-2 Frequency of particular HLA alleles identified with some human diseases.

Disease	HLA allele	Frequency among patients (percent)	Frequency in control population (percent)	Increased risk factor
Ankylosing spondylitis	B27	71–100	3–12	90.1
Reiter's disease	B27	65–100	4–14	36.0
Myasthenia gravis	B8	38–65	18–31	4.4
	Dw3	23–36	14–19	2.3
Juvenile diabetes mellitus	B8	19–55	2–29	2.4
	Dw3	50	21	3.8
	Dw4	42	19	3.5
Multiple sclerosis	B7	12–46	14–30	1.7
	Dw2	47–70	15–31	4.3
Adult rheumatoid arthritis	Dw4	38–65	18–31	4.4

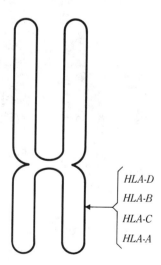

FIGURE 17-12 HLA (human leukocyte antigen) genes are linked together on the short arm of human chromosome 6. The four genetic loci (A, B, C, D) have numerous alleles; that is, the loci are polymorphic.

and in some cases as many as forty, different alleles for each of the four HLA genes have been identified, and the number of HLA antigen combinations that can be constructed is in the billions.

Some years ago, it was noticed that particular HLA alleles are more commonly identified with certain human diseases than would be expected by chance (Table 17-2). These observations raise the possibility that persons at risk for particular diseases might be identified before onset of the disease by determination of their HLA type in infancy or childhood. However, except for ankylosing spondylitis and Reiter's disease, both rare conditions, the correlation between a particular HLA allele and a human disease is not significant. It is important to keep in mind that carrying a particular HLA allele does not *cause* a disease, although it may increase a person's susceptibility to that disease. For example, most people with the HLA B27 allele do not get

TABLE 17-3   ABO blood groups and patterns of transfusion acceptance or rejection.

Blood group	Genotype	Antigens on red blood cells	Transfusions cannot be accepted from	Transfusions are accepted from
O (universal donor)	OO	None	A, B, AB	O
A	AA, AO	A	B, AB	A, O
B	BB, BO	B	A, AB	B, O
AB (universal recipient)	AB	A, B	None	A, B, AB, O

ankylosing spondylitis even though they are at higher risk than the general population.

Pregnancy   Any tissue that carries histocompatibility antigens that are different from those in the animal receiving the tissue will be recognized by the recipient's immune system as foreign and will stimulate an immune response. Yet a common situation exists in which foreign tissue is not rejected: pregnancy. The embryo contains HLA genes contributed by both the mother and father. From what we now understand about the immune system, paternally derived antigens on the surfaces of the embryo's cells should be recognized as foreign by the female's immune system. But somehow the embryo is protected from being destroyed by antibodies synthesized by the mother-to-be. Despite the progress that has been made in understanding how the immune system works, this and many other aspects are still not understood.

Blood Antigens

In the early part of this century, a blood transfusion was often fatal to the recipient because the transfused blood would clot and kill the patient. We now know that such clotting occurs because red blood cells have certain antigens on their surfaces that evoke strong immunological reactions in a recipient. In the early 1900s, Karl Landsteiner established the rules that allow blood transfusions to be made safely between individuals.

Reactions between transfused blood and the recipient's immune system are a special example of tissue rejection. A locus for which there are three different alleles determines the red blood cell surface antigens. The kinds of antigens in turn determine whether the transfused blood will be accepted or will clot (Table 17-3). These three alleles are referred to as the **ABO blood group**, and each person carries two of the three possible alleles: A, B, and O.

Persons with type O blood carry identical alleles that produce neither antigen A nor antigen B on their red blood cells; for this reason, type O people are *universal blood donors*. Because neither antigen A nor antigen B is present, type O blood does not provoke an immune response when transfused into persons carrying any of the four possible ABO blood groups. Persons who are type AB carry both the A and B alleles and have both antigens on their red blood cells. Because these antigens are normal components of their tissues, their immune systems recognize them as "self" and do not react to them. For this reason, type AB persons are *universal recipients*: they can accept blood transfusions from persons carrying any of the four ABO blood groups.

There are many other kinds of proteins on the surfaces of red blood cells that can act as antigens and that differ from one person to the next.

Genes Are Not Passed On in Blood

BOX 17-1

The idea that personal traits and even social status are determined by the kind of blood one has is well established in many societies. Reference is often made to the "royal blood" of kings and queens, to "pure blood lines" of nationalities and races, and to "bad blood" carried by criminals or poor people. The fact is that human red blood cells circulating in the bloodstream carry no hereditary information. Red blood cells are unique in that they are the only cells in the body that have no nuclei and therefore no chromosomes. When human red blood cells are synthesized in bone marrow, they do contain chromosomes like all other cells, but by the time they enter the circulatory system the nuclei have disintegrated.

Many people have suffered because they have been accused of having "inferior" blood. In the early part of this century, some people attributed America's social and labor troubles to "undesirable blood" in workers who struggled for better working conditions. In 1924, Albert Wiggam, a popularizer of eugenic ideas (discussed in Chapter 20), wrote:

> Heredity has cost America a large share of its labor troubles, its political chaos, many of its frightful riots and bombings, the doings and undoings of its undesirable citizens. Investigation proves that an enormous proportion of its undesirable citizens are descended from undesirable blood overseas. America's immigration problem is mainly a problem of blood.

Prejudices over blood were aggravated in World War II, when blood transfusions came into wide use and the American Red Cross established blood banks. During that war and for a period afterward, some American hospitals, particularly in the South, separated blood donated by blacks from blood donated by whites. We now know that what is important in blood transfusions is matching the ABO antigens on the cells; successful transfusions are not influenced by a person's race or ancestry. In some states, blood donors are still asked to give information on their race or ethnic origin. However, this request does not arise from prejudice now but rather from the fact that certain rare blood phenotypes occur only among Afro-Americans, Amerindians, Mexican Americans, and Orientals. Locating the appropriate blood for certain people who need transfusions can be a costly and time-consuming process if such information is not known about donors.

READING: W. J. Miller, "Blood Groups: Why Do They Exist?" *BioScience*. September, 1976.

Generally, however, these other antigens do not stimulate an immune response strong enough to cause the red blood cells to clump together in the recipient's blood when the donor and recipient are genetically mismatched for them. The genetic basis for the ABO blood types, as well as for other less immunologically reactive blood groups, is now well understood, but prejudice and ignorance concerning human blood have had serious social consequences through the years (see Box 17-1).

Another red blood cell antigen, known as the **Rh factor**, is important, particularly during pregnancy. About 15 percent of all women are Rh negative—that is, their red blood cells do not have the Rh antigen. If the woman's mate is Rh positive and she becomes pregnant, the child could inherit an Rh positive gene from the father (depending on whether he is homozygous or heterozygous for the Rh positive allele) and its red blood cells would have the Rh antigen on their surfaces. The Rh positive antigens cause no problems in the first pregnancy. But, during that pregnancy, some of the embryo's blood cells may enter the woman's bloodstream at birth or because of a damaged placenta and cause her immune system to produce antibodies directed against the Rh antigen. If the woman becomes pregnant a second time and if the embryo is again Rh positive, antibodies in her blood can enter the embryo's blood system and destroy its red blood cells. This reaction can produce severe, even fatal, anemia in the fetus.

By determining the Rh blood type of both the mother and the father, physicians can control the potential problems of Rh incompatibility. Nowadays, any woman who is Rh negative and who is carrying an Rh positive child is treated with antibodies immediately after delivery to destroy any Rh antigens in her bloodstream. In this way, her immune system is prevented from responding to the embryo's Rh antigens in a later pregnancy.

Monoclonal Antibodies

When a person is infected by viruses or bacteria, many different antibodies are produced by his or her immune system. Even simple single-celled organisms such as bacteria have dozens of antigens on their cell surfaces that can stimulate the synthesis of many different antibodies. Human blood serum contains thousands of different antibodies that are present in different concentrations and that continually change depending on a person's exposure to particular antigens. This antibody heterogeneity made it impossible to isolate and purify specific antibodies.

In 1975, Georges Köhler and Cesar Milstein (both of whom received Nobel prizes ten years later for their discovery) developed a way of selecting a particular kind of somatic hybrid cell, called a **hybridoma**, that produces only one kind of antibody that is specific for one kind of antigen. Such specially selected somatic hybrid cells produce **monoclonal antibodies** — antibodies of one kind synthesized by a clone of genetically identical cells. The techniques of monoclonal-antibody production have revolutionized immunological research and have already spawned a new industry that manufactures monoclonals that can be used in diagnosing and treating diseases.

The basic procedure for producing monoclonal antibodies begins with the injection of an animal, usually a mouse, with either a specific antigen or a mixture of the antigens of interest (Figure 17-13). After several days, the animal's spleen, which contains particular B lymphocytes and plasma cells that have been synthesized in response to the injected antigens, is removed and homogenized into single cells. These cells are mixed with cancer cells, called myelomas, that can be grown *in vitro* indefinitely. The spleen cells are fused with the myeloma cells in a special nutrient medium to produce hybridoma cells.

After the antibody-producing hybridomas have been grown *in vitro* for several generations, individual hybridoma cells producing the one specific antibody are isolated from all the other antibody-producing hybridomas. This hybridoma is cloned, and each clone is tested for the antibody that reacts only with one antigen. These hybridoma cells can be grown *in vitro*, where they will continue to synthesize significant amounts of the (monoclonal) antibody. Finally, the hybridoma cells can be frozen and revived whenever more of the monoclonal antibody is needed.

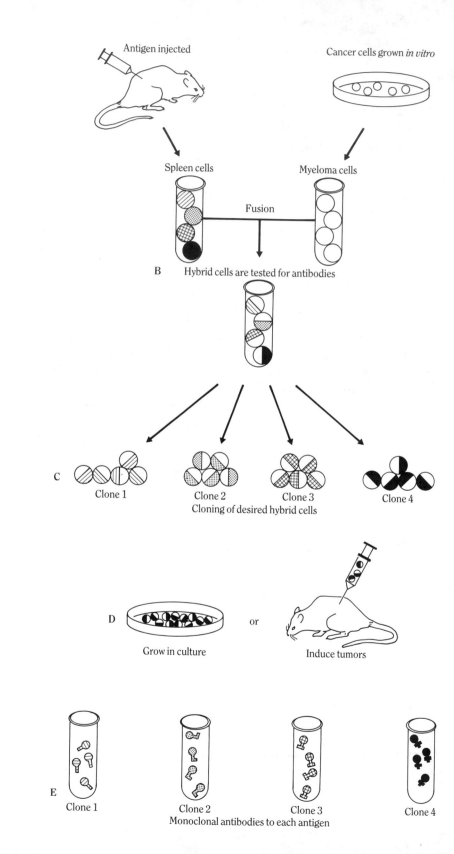

Antigens

Antigen injected

Cancer cells grown *in vitro*

Spleen cells

Myeloma cells

Fusion

B   Hybrid cells are tested for antibodies

Clone 1   Clone 2   Clone 3   Clone 4

C

Cloning of desired hybrid cells

D

Grow in culture   or   Induce tumors

E

Clone 1   Clone 2   Clone 3   Clone 4

Monoclonal antibodies to each antigen

Psychoneuroimmunology
Chemical Links Between the Brain and the Immune System

BOX 17-2

One of the more-remarkable discoveries of modern immunological research is the demonstration that mental states can affect the synthesis and functions of various cells of the immune system.

Experiments with people show that emotions such as grief, despair, hopelessness, depression, and anxiety can impair normal functioning of the immune system (see the accompanying table). Experimental animals that are shocked or stressed in other ways also show diminished immune-system functioning. Thus, modern research seems to be giving new weight to centuries-old observations linking emotional states to disease and healing. Nearly two thousand years ago, the Greek physician Galen observed that cancer appeared more often among women who were depressed. Scientific studies of how emotions

and stress affect the immune system have become possible because new tests are able to measure changes in specific cells of the immune system. Dr. Candace Pert, director of the section on brain biochemistry at the National Institute of Mental Health, believes that the mind and the immune system communicate by means of neuropeptides—small molecules that bind to receptor sites on cells and change how they function. One type of cell receptor that has been identified and studied in great detail is the pain receptor on nerve cells. This receptor interacts with opium and its chemical derivatives such as morphine. When the receptors

are saturated with these drugs, pain perception is blocked.

However, we now know that the brain itself synthesizes natural opiates called *endorphins* during periods of stress or injury. Thus, the brain helps to ease pain by synthesizing morphinelike molecules. Evidence is beginning to show that the outcome of virtually any disease can be affected by what we think and feel. The brain converts our subjective feelings into the synthesis of neuropeptides that can increase or decrease the functions of the immune system. The truth of the phrase "think well, not sick" has found support in the branch of science called psychoneuroimmunology.

READINGS: C. Pert. "The Wisdom of the Receptors: Neuropeptides, the Emotions, and Bodymind," *Advances.* Summer, 1986. B. Dixon, "Dangerous Thoughts," *Science 86.* April, 1986.

Behavior	Immune-system response*
Stress, marital breakup	Reduced number of T cells (T lymphocytes)
Shock, grief, unemployment	Lymphocyte response reduced
Exam anxiety	Increase in helper T cells
Depression	Reduced number of certain receptors on lymphocytes
Meditation	Increase in NK (natural killer) cells of immune system

*T lymphocytes are divided into subsets. Helper T cells stimulate the immune response of other cells; natural killer cells attack and destroy other cells, particularly cancer cells.

◀ **FIGURE 17-13**   The basic technique for producing monoclonal antibodies: (A) a mouse is injected with a specific antigen or antigens and, after a week, the animal's spleen cells are fused with an established line of mouse cancer cells called myeloma cells; (B) somatic hybrid cells are selected and tested for production of the desired antibody; (C) cells producing the antibody are cloned and finally the best clone is selected; (D) quantities of these cells are grown *in vitro* or in an animal; and (E) the desired antibodies are isolated.

Monoclonal antibodies are currently being used to analyze blood types and to diagnose serious bacterial and viral infections much more rapidly than can be done with other techniques. Identification of a pathogenic (disease-causing) bacterium can be accomplished in a few hours through the use of monoclonal antibodies specific for certain pathogens, compared with the several days necessary if other techniques are used.

Monoclonal antibodies are also being used to identify particular receptor sites on cells. Once the receptor sites have been characterized, it may be possible to design more-effective drugs—drugs that interact with specific cellular receptor sites and produce the desired pharmacological effect with a minimum of undesirable side effects. For example, it may be possible to construct effective pain-relief drugs that are nonaddictive (see Box 17-2).

Some scientists believe that, in the more-distant future, monoclonal antibodies may prove useful for delivering drugs or toxins to treat cancers. Because most cancer cells have cell-surface antigens different from those on normal cells, it is hoped that monoclonal antibodies can be produced that will recognize specific antigens on cancer cells. If powerful drugs were chemically attached to monoclonal antibodies, perhaps they could be carried directly to the cancer cells and destroy them without causing harm to other body cells. However, because monoclonal antibodies are isolated from mouse cells, it is likely that these mouse-cell-derived proteins will be recognized as foreign antigens if they are injected into people. The problem now is to develop monoclonal antibodies from human cells, using procedures that have proved successful in mice.

Summary

In human beings and other vertebrates, the immune system consists of cells and proteins that recognize foreign substances (antigens) and microorganisms. The three main kinds of immune-system cells are T lymphocytes, B lymphocytes, and macrophages. B lymphocytes specific to a particular antigen become activated when that antigen is present in the body. An activated B lymphocyte replicates and directs the synthesis of many plasma cells, which in turn produce antibodies specific to the antigen.

Antibodies are large protein molecules. Each antibody consists of four polypeptide chains: an identical pair of heavy chains and an identical pair of light chains. Each light and heavy chain consists of a constant region and a variable region. Antigen recognition is principally determined by the hypervariable regions within the variable regions of both heavy and light chains.

The scientific term for antibody is immunoglobulin, of which there are five major classes: IgG, IgA, IgM, IgD, and IgE. Immunoglobulins are found in saliva, tears, stomach secretions, and most other body fluids, as well as in the blood. The most-prevalent class of immunoglobulins in the blood is IgG; this class recognizes a wide range of antigens.

Relatively few (about three hundred) genes code for a very large number (millions) of different antibodies. The synthesis of all these antibodies is directed by only seven families of genes: for light chains, one family of genes encodes the constant region, another encodes the variable region, and a third directs the joining of the two regions; for heavy chains, another three families of genes encode constant and variable regions, as well as their joining, but, in addition, a fourth family enhances antibody diversity.

The clonal selection model, proposed in the 1950s, explains how cells of the immune system develop and how the genetic information in these cells is rearranged to create antibody diversity. It also explains how the immune system destroys foreign tissues and cells but does not harm the body's own tissues and cells.

Histocompatability determines whether tissue transplanted from one animal to another will be accepted or rejected by the recipient. It depends on both genetic and environmental factors. In human beings, the more genetically related the donor and recipient are, the greater the rate of success of transplants. In genetically unrelated cases, drugs that suppress the responsiveness of the immune system must be administered to the recipient to prevent rejection. Human histocompatability is determined by four groups of human leukocyte antigen (HLA) genes. (It is now known that antigens encoded by three of the four groups of genes are not restricted to leukocytes, or white blood cells, but are found on the surfaces of all body cells.)

Red blood cells have antigens on their surfaces that are encoded by the ABO blood group alleles. Each person has two of the three possible alleles: AA, AO, BB, BO, OO, or AB. If a transfusion of type A blood were given to a person whose blood is type B or if type B blood were given to someone having type A, the recipient's blood would clot, causing death. However, people having type O blood are universal donors because they do not produce antigens A or B; people having type AB blood are universal recipients because they have both A and B antigens and their immune systems recognize such antigens as "self." Rh factor is another type of red blood cell antigen. It is of significance in pregnancies of Rh negative women if their mates are Rh positive.

Monoclonal antibodies specific for a single antigen are produced *in vitro* by somatic hybridoma cells. Such antibodies are used to analyze blood types and to diagnose serious bacterial and viral infections. They are also used to identify receptor sites on cells in an attempt to design drugs that will interact with the receptors to produce the desired pharmacological effects with a minimum of unwanted side effects.

Key Words

ABO blood group   Three alleles that occur in pairs in each person. These determine antigens on red blood cells and whether a donor's blood will be rejected by a recipient.

allophenic mice   Mice whose cells are derived from four different parents.

antibody   An immunoglobulin molecule capable of combining with specific antigens and inactivating pathogenic microorganisms.

antigen   Any foreign substance that stimulates an antibody response in an animal.

autoimmune disease   A disease such as some forms of rheumatoid arthritis or lupus erythematosus, in which the immune system reacts to normal body cells and damages or destroys normal tissues.

B lymphocyte   A white blood cell that carries the genetic information for synthesis of particular antibodies.

clonal selection   Explains how antibody diversity is generated during embryonic development and how the immune system distinguishes "self" from "nonself."

endorphins   Morphinelike substances synthesized in the human brain in response to stress or injury.

erythrocyte   A red blood cell.

histocompatibility   The antigens that determine whether tissue transplanted from one animal to another will be accepted or rejected.

HLA genes   Determine antigens on white blood cells (leukocytes) and other tissues. These antigens determine whether foreign tissue will be accepted or rejected.

hybridoma   A somatic-cell hybrid between a B lymphocyte and a cancer cell; it produces monoclonal antibodies.

hypermutability   An unusually high rate of mutation.

immune system   The numerous mechanisms in vertebrates that recognize and eliminate foreign substances and alien cells.

immunoglobulin   The scientific term for antibody.

leukocyte   A white blood cell.

lymphocyte   A specialized white blood cell present in lymph nodes, spleen, thymus, bone marrow, and blood; lymphocytes mature into B lymphocytes and T lymphocytes (also called B cells and T cells.)

macrophage   A large leukocyte that is able to destroy foreign cells by phagocytosis.

monoclonal antibody   An antibody of a single type that is produced by genetically identical somatic hybrid cells (clones).

plasma cell   A specialized cell derived from B lymphocytes that synthesizes antibodies.

Rh factor   A red blood cell antigen that can cause anemia in a fetus.

T lymphocyte   A white blood cell that regulates responses by other cells of the immune system.

Additional Reading

Ada, G.L., and G. Nossal. "The Clonal Selection Theory." *Scientific American*, August, 1987.

Bach, F.H., and D.H. Sachs. "Current Concepts: Transplantation Immunology." *New England Journal of Medicine*, August 20, 1987.

Buisseret, P.D. "Allergy." *Scientific American*, August, 1982.

Cohen, I.R. "The Self, the World and Autoimmunity." *Scientific American*, April, 1988.

Marrack, P., and J. Kappler. "The T-Cell and Its Receptor." *Scientific American*, February, 1986.

Miller, J.A. "Mouth Immunity." *Science News*, October 5, 1985.

Silberner, J. "Survival of the Fetus." *Science News*, October 11, 1986.

Yelton, D.E., and M.D. Scharff. "Monoclonal Antibodies." *American Scientist*, September-October, 1980.

Young, J.D.-E., and Z.A. Cohn. "How Killer Cells Kill." *Scientific American*, January, 1988.

Study Questions

1   What class of immunoglobulins causes allergic responses?

2   How many different antigens can be recognized accurately by one antibody molecule?

3   How many different alleles are there in the ABO blood group? List the ABO phenotypes in this blood group.

4   What class of genes is responsible for tissue rejection when organs are transplanted?

5   What functional property of DNA is primarily responsible for generating antibody diversity.

6   How many different polypeptide chains are present in each antibody molecule?

7   What kind of cells produce monoclonal antibodies?

Essay Topics

1   Explain how monoclonal antibodies are manufactured and what they are used for.

2   Discuss how stress or emotional upset may influence disease susceptability.

3   Discuss the importance of HLA genes in the transplantation of a heart from a donor to a recipient.

18

Evolution
Populations and Natural Selection

The most shocking fact about evolution is not that we descend from something we probably wouldn't like to meet alone in a forest at night, but that something descends from us which we certainly wouldn't like to meet even at noon in a crowded street.

SRI AUROBINDO, *philosopher*

Probably no area of biology arouses more interest or creates more controversy than evolution, which deals with the origin of living organisms, the genetic diversity of populations, the mechanisms of speciation, and the biological history of our planet. Evolution is concerned with the relatedness of existing organisms, as well as with those that have become extinct but whose earlier existence can be documented from fossil records. Evolution is the branch of biology that attempts to explain how the tens or hundreds of millions of different species of plants, animals, and microorganisms arose in the course of several billion years of earth's history. Evolutionary science analyzes the various forces that cause species to adapt, to change, and to eventually become extinct. Evolution and genetics are inextricably interconnected because the biological changes that occur in organisms through time are due to changes in their hereditary information—changes in their genes.

Although various evolutionary ideas have been proposed through the centuries, the first generally accepted scientific explanation for the origin, adaptation, diversification, selection, and extinction of species came from the detailed observations and brilliant insights of the English naturalist Charles Darwin. He provided a wealth of evidence for evolutionary change by the process that he called *natural selection*. Darwin argued persuasively that individuals who are best suited genetically to survive and reproduce in particular environments are the ones that pass their genes on to succeeding generations (an idea often referred to as "survival of the fittest"). He believed that natural selection is the primary force of evolution—that it is responsible for the origin of new species by

gradual change over long periods of time and for their extinction when new environments arise in which they cannot survive.

Another of Darwin's important insights was that in each generation a certain amount of heritable variability is introduced into individual organisms and that natural selection operates on this genetic variability to continually select new types. Darwin was unaware of Mendel's experiments; so he had no way of incorporating Mendel's ideas on the pattern of inheritance of genes into his evolutionary theories. Nor could Darwin (or anyone else at that time) know that all genetic diversity is ultimately due to mutations that arise in DNA. Darwin did not jump to his controversial conclusions about evolution; quite the contrary. He spent twenty-eight years mulling over and adding to his data from the time he sailed on the *H.M.S. Beagle* in 1831 until he finally published his pioneering work on evolution in 1859.

The unification of Darwin's ideas with those of Mendel and subsequent geneticists eventually resulted in what is now called the **modern synthesis of evolutionary theory** (also referred to as *neo-Darwinism*). It unites the ideas of natural and sexual selection with the patterns by which genes and chromosomes are inherited and with the chemical properties of DNA. This modern evolutionary synthesis, which began in the 1930s and has continued to the present day, is one of the truly great accomplishments of the biological sciences. Indeed, it is the major unifying principle of all biology.

On the one hand, molecular genetics explains the continuity of hereditary information by showing how each DNA molecule is an exact copy of the one from which it was replicated. On

FIGURE 18-1 Berry's World. (Reprinted by permission of NEA, Inc.)

the other hand, evolutionary and population genetics explain how ever-so-slight genetic changes (mutations) can provide the variability that is essential for the selection of new types of organisms, new species, and new populations. Modern evolutionary theories scientifically explain the extraordinary variety of plants and animals that we see all around us in nature (Figure 18-1).

Evolutionists attempt to reconstruct biological history, and it is in this area that much of the controversy arises, especially in the recurrent confrontations between creationists and evolutionary scientists about what the schools should teach regarding biological history and human origins. Because it is an indisputable fact that human beings, along with millions of other species, inhabit the earth at present, it is important to understand how and when they arose. Estimates from fossil records suggest that more than 99 percent of all species that have arisen on earth have become extinct. As the emminent scientist Stephen J. Gould puts it, "The ultimate fate of all species is extinction."

To understand the facts, theories, speculations, and controversies of evolution, it is necessary to first know how evolutionary data are obtained and analyzed and how evolutionary hypotheses are tested. Only then can one decide

which ideas make sense and which do not. This chapter and the next present the basic scientific facts of evolutionary biology and discuss some of the controversies.

What Is Evolution?

The word *evolution* means an unfolding, a process of development and change. Evolution applies to the formation and development of the entire physical universe—atoms, molecules, mountains, planets, stars, galaxies—as well as to living organisms. Geneticists, however, limit their studies to biological evolution (termed evolution for short) and generally define it in a stricter sense.

To a geneticist, **evolution** refers to changes in gene frequencies that arise and accumulate through time in populations of organisms. (It is populations, not individual organisms, that evolve). These cumulative changes in gene frequencies are subject to natural selection, as that process was originally proposed by Charles Darwin. According to Darwin, natural selection determines the reproductive success of one individual compared with another and thereby determines which individual's genes will be

FIGURE 18-2 The fourteen species of finches on the Galapagos and Cocos islands. These finches are closely related yet have evolved into separate species. Some of the species of finches live in trees and feed on insects; others live on the ground and feed on seeds. Darwin's observations on the finches were important in the development of his evolutionary ideas. (From "Darwin's Finches" by David Lack. Copyright © 1953 by Scientific American, Inc. All rights reserved.)

well known of Darwin's many examples was the fourteen species of finches that he studied while visiting the Galapagos Islands (Figure 18-2). He concluded that the slight differences in the physical appearances of the finches on the different islands were the consequences of adaptations that had accumulated over long periods of time in response to the different island environments. The adaptations, Darwin concluded, were due to heritable changes that ultimately resulted in the development of the different species of finches. The physical differences among the species evolved because the finches were reproductively isolated from one another by the water that restricted each population to a particular island.

The term **species** refers to reproductive groups that interbreed among themselves but that do not mate or exchange genes with other reproductive groups in nature, even when they share a common environment. Chickens, geese, and ducks may share the same barnyard, but they mate only with other birds of their own species. **Race** is an arbitrary subclassification of a species based on physical or genetic differences discussed further in Chapter 19. However, individuals of different races can and do interbreed, since they all belong to the same species. In some species such as dogs, the term *breed* is used instead of *race* to denote a subspecies. All breeds of dog belong to the same species, and in principle, any dog can mate successfully with any other dog. However, because of the artificial selection of specific characteristics by dog breeders through the centuries, physical and reproductive problems can arise in matings between certain breeds of dogs. The problems of breeding a male Great Dane with a female dachshund, for example, are formidable (Figure 18-3).

The assignment of organisms to species and races is a method of classification that helps

represented more frequently in subsequent generations. In Darwin's own words:

> As many more individuals of each species are born than can possibly survive; and as, consequently, there is a frequently recurring struggle for existence, it follows that any being, if it vary however slightly in any manner profitable to itself, under the complex and sometimes varying conditions of life, will have a better chance of surviving, and thus, be naturally selected.

Darwin reached his conclusions by observing the kinds of slight phenotypic variations among individuals of closely related species. The most

FIGURE 18-3 Breeds (races) of dogs vary greatly in size and other characteristics as a result of artificial selection through many generations. However, all dogs belong to the same species regardless of how different they appear; all have the same number and kind of chromosomes in their cells.

scientists organize the otherwise bewildering diversity of organisms into groups with some shared characteristics. Races of plants or breeds of animals can be organized by size, color, hair or skin texture, presence or absence of certain genes, or any other physical or genetic characteristics that can be measured. Assignment of persons to human races has often carried with it implications of social or intellectual inferiority or superiority. Historically, persons of different human races have been prevented from interbreeding, but the obstacle has been social and cultural, not biological. In nature, biological mechanisms ensure that any member of a species is fertile with any other member of the same species, regardless of how physically similar or dissimilar the individual organisms may appear. By the same token, other biological mechanisms ensure that matings between individual organisms of different species are infertile.

As the example of a Great Dane and a dachshund shows, dramatic differences in appearance do not necessarily mean that individual organisms belong to different species. It also happens that organisms may appear similar or even identical to the nonexpert yet belong to different species. In the Hawaiian Islands, several hundred species of the fruit fly *Drosophila* have been identified, each adapted to a particular island environment. Flies of one *Drosophila* species will mate only with flies of the same species, although to most people one fruit fly looks pretty much like any other.

Reproductive Isolation

Nature has numerous ways, called **reproductive isolation mechanisms**, that prevent matings or reproduction between individual organisms of different species. Formation of species is important for evolution because it is a means of stabilizing genetic information in a way that optimizes the survival and reproduction of organisms in particular environments. Advantageous genetic information is maintained by matings between members of the same species that carry the same (or similar) genetic information.

Reproduction between animals of different species is usually prevented by the absence of matings. Most animals exhibit sexual behaviors that influence male-female attraction and mating. Many species of birds, for example, display elaborate mating behaviors. Male birds generally have a more-striking physical appearance: they tend to be larger and more extravagantly colored than are females. *Sexual selection* influences which genes are passed on to subsequent generations; hence, sexual selection is also a mechanism for evolutionary change, as is natural selection. Bigger, more-colorful male birds are more successful in attracting females than their smaller, drabber male competitors. Thus, when peacocks fan their gorgeous tail feathers, they are announcing their sexual intentions to females (Figure 18-4).

Darwin regarded sexual selection as only slightly less important than natural selection as a force in evolution. In *The Descent of Man and Selection in Relation to Sex*, which deals specifically with the significance of sexual selection in human beings and other animals, he wrote:

> It cannot be supposed that male Birds of Paradise or Peacocks, for instance, should take so much pains in erecting, spreading, and vibrating their beautiful plumes before the females for no purpose.

However, just because male animals are usually more colorful and aggressive in mating rituals does not mean that male behaviors necessarily determine which genes are passed on to

FIGURE 18-4 A male peacock displays his tail feathers to attract females. (Courtesy of Janis Hansen, Kaaawa, Hawaii.)

progeny. For many species of birds and other animals, it is the female that chooses the male — decides which one is to be her mate — and thereby determines which male genes will be passed on to her progeny.

Reproductive isolation can result from mating behaviors; it can also result from geographical separation. If species of animals, plants, or microorganisms inhabit territories that are physically separated from one another, individual organisms are forced to mate within their particular territory. As mentioned earlier, animals isolated on islands mate with other animals of that species on the same island. As Darwin observed with the finches, geographical isolation can eventually give rise to new species that are adapted to the particular isolated environments.

Biological mechanisms that affect fertilization and reproduction also contribute to the reproductive isolation of different species. Infertility between individual organisms of different species is often due to differences in chromosome numbers. If individual organisms of different species are able to mate, the union of dissimilar gametes generally produces abnormal numbers of chromosomes or abnormal arrangements of genes on chromosomes in the hybrid. Occasionally, viable animal hybrids are formed between different species — the mule, for example, is produced by the mating of a horse with a donkey, which belong to different species. However, mules are almost always sterile (see Box 18-1).

Many species of plants also have biological mechanisms to ensure that eggs will be fertilized only by pollen from the same species. However, plant breeders can often circumvent normal plant reproductive barriers, and they have been able to construct many kinds of hybrid plants that do not occur in nature. An example of a useful plant hybrid is *Triticale*, a new type of grain that was constructed by crossing wheat (*Triticum*) with rye (*Secale*). Wheat has a haploid chromosome number of forty-two and rye has a haploid number of fourteen. The reason that hybrid plants can be propagated is that some hybrid plants are fertile, whereas hybrid animals such as mules are not.

Unusual Birth in a Georgia Zoo

BOX 18-1

A few years ago, quite by accident, a mating took place between two apes of different species in the Grant Park Zoo in Atlanta, Georgia: a female classified by the zoo as a siamang was impregnated by a male classified as a gibbon. To everyone's surprise, a healthy hybrid ape was born some months later. This unique primate was named a *siabon*, because it was a cross between a *sia*mang and a gib*bon* (see the accompanying photograph).

Even though siamangs and gibbons occupy overlapping territories in the jungles of Southeast Asia, the two species are physically very different and presumable never mate in nature. Not only do the two species have different numbers of chromosomes—siamangs have fifty, whereas gibbons have forty-four—the organization of their genes on the chromosomes would be expected to be quite different. And, in fact, staining the chromosomes in cells from a siamang and a gibbon showed that the chromosome banding patterns are very different, an indication that the two ape species are only distantly related.

Despite the physical and genetic differences between these apes, the birth of a viable, normal infant hybrid ape as a result of their mating means that all the genetic information required for normal development and differentiation was present in the hybrid cells of the embryonic siabon. This is all the more surprising because the cells of the siabon have an odd chromosome number (2N = 47), twenty-two chromosomes from the gibbon father and twenty-five from the siamang mother. Because of the odd number of chromosomes and the impossibility for all of them to pair with homologous partners in meiosis, it is likely that the siabon will be sterile.

The fact that such an unusual hybrid ape is viable and is growing normally raises the possibility

that unusual matings might have played a role in primate evolution in the past. In 1982, a detailed comparison of the bands in orangutan, gorilla, chimpanzee, and human chromosomes showed that eighteen of the twenty-three human chromosomes are virtually identical with those found in all three ape species. Human and chimpanzee cells differ in chromosome number by only one, and the banding patterns of both species' chromosomes are very similar.

More and more evidence suggests that slight genetic rearrangements or subtle changes in genetic regulation may have caused the evolution of the different species of primates. These genetic studies certainly raise the question—disturbing to some people—of whether human beings are indeed the culmination of primate evolution. Is it possible that other primates may yet evolve and replace the present human species, even as we replaced other, now-extinct hominids?

READING: Richard H. Myers, "The Chromosome Connection," *The Sciences.* May–June, 1980.

(Photograph by Sister Moore, Atlanta, Georgia, courtesy of Richard H. Myers, Boston University Medical Center.)

Population Genetics

Because the underlying basis for evolution is the accumulation of genetic changes, one of the main ways of studying evolution is to measure the frequency of various genes in populations and to determine how those frequencies change from generation to generation. This is the subject of **population genetics**.

As noted earlier, geneticists define evolution as the cumulative changes in gene (allele) frequencies in populations through time. To calculate how a particular pair of alleles changes in frequency from one generation to the next, we need to make some simplifying assumptions. We will assume that initially our two alleles, A and a, are distributed equally among the members of the population ($\frac{1}{2}\,A$, $\frac{1}{2}\,a$). We will also assume that any individual can mate with any other individual in the population, a process called **random mating**. If no other forces are operating to select for or against these alleles, we can easily calculate the frequencies of the alleles in succeeding generations. (As pointed out earlier, sexual selection means that real matings are not random; so we are discussing an idealized population here.)

Because eggs and sperm combine at random in this idealized population, individual progeny of three distinct genotypes will be created in the next generation in predictable ratios (Figure 18-5A). One-fourth of the progeny will be homozygous for the dominant alleles (AA), one-half will be heterozygous (Aa), and one-fourth will be homozygous for the recessive alleles (aa). Similarly, if the genotypic frequencies of individual members of a population are known, the allelic frequencies can be calculated by counting alleles in the gametes (Figure 18-5B). It is apparent that in this simple example the frequency of alleles (A and a) in the population does not change from generation to generation;

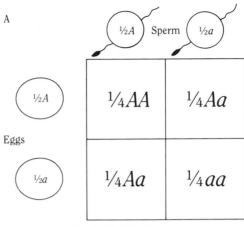

B Next generation ¼AA ½Aa ¼aa

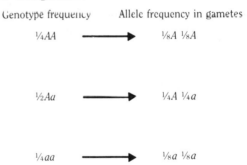

Total allele frequency in gametes:
$\frac{1}{8}A + \frac{1}{8}A + \frac{1}{4}A = \frac{1}{2}A$
$\frac{1}{8}a + \frac{1}{8}a + \frac{1}{4}a = \frac{1}{2}a$

FIGURE 18-5 Genotype frequencies and allele frequencies. A. Random mating in a population means that the frequency of genotypes does not change from one generation to the next. In this example, the allele frequencies are assumed to be equal ($\frac{1}{2}\,A$ and $\frac{1}{2}\,a$), but the conclusion that the genotypic frequencies do not change is valid for any ratio of alleles.
B. The allele frequencies can be calculated from the genotype frequencies. Note that the allele frequency turns out to be the same as that in part A.

the frequency of allele A remains one-half, as does the frequency of allele a. (If we had started with allelic frequencies different from one-half, the frequencies would still remain unchanged in succeeding generations).

TABLE 18-1   Allelic frequencies for M-N blood antigens in various human populations.

Population	Number of individuals	Frequencies of genotypes			Allelic frequencies	
		MM	MN	NN	M	N
U.S. Caucasians	6,129	.29	.50	.21	.54	.46
Navajo Indians	361	.85	.14	.01	.92	.08
Australian Aborigines	730	.03	.30	.67	.18	.82

In 1908, an English mathematician, Godfrey H. Hardy, and a German physician, Wilhelm Weinberg, independently discovered that the relation between allele frequencies and genotypes of individual members of a population can be expressed by the simple formula

$$p^2 + 2pq + q^2 = 1$$

which is known as the **Hardy-Weinberg law**. The letters p and q represent the frequencies of the two alleles in the population; p^2, pq, and q^2 represent the fraction of individuals of different genotypes. Together, the alleles for any genetic locus must equal 100 percent; so $p + q = 1$. (In reality, many genetic loci have numerous alleles. The HLA genes, for example, may have as many as forty different alleles in a human population, which makes the equation more complicated.)

The Hardy-Weinberg formula expresses an ideal situation in which matings occur at random among individuals. It does not take into account other biological mechanisms that cause allele frequencies to change. In populations that are in Hardy-Weinberg equilibrium, allele frequencies remain unchanged from one generation to the next and, consequently, evolution *cannot* occur in such populations. Because the Hardy-Weinberg law shows that allele frequencies would remain unchanged in a randomly mating population, it provides the means for

discovering and analyzing what forces *do* contribute to evolution and how they cause allele frequencies to change in real populations in nature.

In reality, no population in nature is in Hardy-Weinberg equilibrium. In different human populations, for example, allele frequencies are often quite different. The human M-N blood antigens (another blood group, distinct from the ABO group) are synthesized from a single genetic locus with two alleles. Homozygous (*MM* or *NN*) or heterozygous (*MN*) persons can be distinguished by biochemical blood tests, and the frequencies of the *M* and *N* alleles in any population can be calculated from the Hardy-Weinberg formula. As shown in Table 18-1, the *M* and *N* alleles occur with equal frequency among U.S. Caucasians, whereas the *M* allele is found almost exclusively among Navajo Indians and the *N* allele predominates among Australian Aborigines. Many other examples from natural populations show that certain forces or mechanisms must change allelic frequencies to produce the genetic variation that is responsible for evolutionary change.

The Forces of Evolutionary Change

As pointed out in the preceding section, the Hardy-Weinberg law ignores all of the forces

A

B

FIGURE 18-6 Comparison of (A) a wild-type fruit fly and (B) a mutant bithorax fruit fly. Three mutations can be combined in *Drosophila* to produce a bithorax fly with two pairs of wings. It is thought that today's single-winged flies evolved from flies with two pairs of wings. This laboratory-generated mutant demonstrates how muta- tions create new structures, new organisms, and eventu- ally new species. (Curiously enough, airplanes were first designed with two wings and "evolved" into single-winged aircraft.) (Courtesy of Edward Lewis, California Institute of Technology.)

that cause nonrandom matings in populations, such as mating preferences, differential mortal- ity of individuals with certain genotypes, and deviations from randomness in matings. In ad- dition to the forces that influence matings, four other important forces have been identified that change allelic frequencies in populations—*mu- tation, migration, genetic drift,* and *natural selection.*

Mutation is the ultimate source of all ge- netic variation and, consequently, the force that drives all evolutionary change (Figure 18-6). Whereas recombination between chromosomes in meiosis can generate new gene combinations, recombination does not produce new alleles. A mutation is analogous to introducing a joker into a deck of cards. Shuffling a normal deck of fifty two cards rearranges the cards but does not create any new information; inserting a joker into the deck does.

If a mutation arises in sperm or eggs, and if that mutation makes an individual organism

better able to survive and reproduce than other members of the population, the frequencies of the mutant type of organism and of the new allele in the population will increase. This is what Darwin meant by "survival of the fittest" and what geneticists and evolutionists refer to as **fitness**: the reproductive contribution of in- dividual organisms to succeeding generations. A person's fitness may or may not have anything to do with his or her intelligence, health, strength, or looks. However, from the biologist's perspective, fitness refers only to the number of progeny produced. The more progeny, the greater the number of that person's genes in the next generation.

Mutations can have either beneficial or harmful effects in a given environment. If a mutation results in an individual organism that is less fit for its particular environment, the mutant gene carried by such an organism will tend to decrease in frequency, because those

Founder Effects and Bottlenecks

BOX 18-2

Not all genetic diseases are serious enough to interfere with an organism's ability to reproduce, and some disease-causing genes can rapidly increase in frequency in a population. Certain hereditary diseases can be traced to a *founder effect*, in which an allele has a markedly higher frequency in new populations than it had in the population in which it originated. For example, the Afrikaner population in South Africa shows a very high frequency of two different inherited diseases that can be traced to one or two Dutch settlers who emigrated to South Africa in the seventeenth century.

One of these diseases is *variegate porphyria*, a slightly different form of porphyria from that which afflicted King George III. This disease is extremely rare worldwide, but in South Africa more than ten thousand persons suffer from it. Medical geneticists have traced the introduction of the mutant allele into South Africa to a Dutch couple who were married in Cape Town in 1688. All the present-day carriers and affected people in South Africa are their descendants and carry the same mutant alleles.

Another inherited disease that occurs with high frequency among both white and colored South Africans (a "colored" is a person whose ancestry includes one or more blacks and one or more whites) is Huntington disease. All of the cases of this disease that have been identified in South Africa—almost five hundred to date—have been traced to Jan van Riebeeck, who led the first group of Dutch settlers to South Africa in 1658. More than half of all South Africans with Huntington disease are direct descendants of van Riebeeck, and the others are believed to be distantly related to him. Founder effects also explain why certain genetic diseases are much more common in religious communities such as the Amish in Pennsylvania, who tend to intermarry among themselves.

Bottleneck effects also change allelic frequencies in populations and affect the evolution of species. A *bottleneck* occurs when the size of a population shrinks dramatically and then expands again. This results in loss of most of the genetic diversity in the population. An example of a recent bottleneck is the northern elephant seal population that was reduced by excessive hunting to about twenty animals in the 1890s. The population has since recovered and now numbers thousands of animals. However, analysis of genetic variation in present-day animals shows no allelic differences at twenty-four different genetic loci. The California condor (if it does not become extinct) also has experienced a severe bottleneck because only a few birds remain. Some scientists believe that bottlenecks of human populations have contributed significantly to hominid evolution in the past.

READING: M.R. Hayden, J.M. MacGregor, and P.H. Beighton, "The Origin of Huntington's Chorea in the Afrikaner Population in South Africa," *South African Medical Journal.* August 2, 1980.

organisms will not survive and reproduce as well as others. Ultimately, it is the interaction of the individual genotype with the environment that determines fitness. (Box 8-2 describes how a disease-causing allele increases fitness in malarial environments.)

Migration refers to the movement of individual organisms (and their genes) into or out of a population. A dramatic example of a change in gene frequency due to migration is the **founder effect**. If an individual organism introduces a new allele into a population and leaves many progeny, and, if that allele does not adversely affect survival and reproduction, its frequency will increase in the next generation and possibly in subsequent generations (see Box 18-2). If a group of animals leave a population, travel a long distance, and eventually establish a new, isolated population, survival in the new environment may depend on the presence of certain genes that were not subject to strong selection in the original population. As a result of natural selection in the new environment, the frequency of some alleles will increase in the population and the frequency of other alleles will decrease.

Genetic drift refers to the mechanism by which alleles may change in frequency in a population by chance, which plays a particularly important role in allele selection in small populations, just as it does in a limited number of coin tosses. If a coin is tossed many times, chance produces, on the average, equal numbers of heads and tails. However, if a coin is tossed just a few times, a run of all heads or all tails is often observed. Allelic frequencies will fluctuate from generation to generation in small populations because of random genetic drift.

If the population is very small, it is possible for certain alleles to be lost completely. For example, suppose that, in a population consisting of just two females and two males, only three progeny are produced. Suppose further that by chance only the A allele is passed on. In this case the a allele is lost forever, even though the three offspring themselves may leave large numbers of progeny and the population might increase in size (Figure 18-7).

A wealth of evidence, much of which was uncovered by Darwin, supports the idea that of the four evolutionary forces natural selection has the greatest influence. The other three forces—mutation, migration, and genetic drift—do cause allelic frequencies to change,

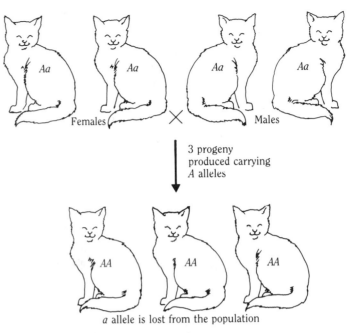

a allele is lost from the population

FIGURE 18-7 Effects of genetic drift. In small populations, an allele may disappear because of random genetic drift. Once the allele has been lost, the chance of it reappearing in the population is negligible, regardless of the size of the population in the future.

but the genetic changes that accumulate through time are principally the result of natural selection. Most biologists believe that Darwin was justified in assigning such great importance to natural selection in the processes of adaptation and speciation and in the overall course of evolution.

The Rate of Evolutionary Change

Natural selection is a slow process. To appreciate how slowly gene frequencies change, consider our example in which the alleles A and a occur with equal frequencies in a population ($A = 0.5$, $a = 0.5$). Let us assume that all homozygous recessive organisms are selected against (die) and so are unable to reproduce or pass on any genes to the next generation. In this extreme case, 25 percent of the population is unable to pass genes on to the next generation. A simple reformulation of the Hardy-Weinberg law can be derived that gives the change in the frequency of the allele being selected against in any future generation:

$$q_n = \frac{q_0}{1 + nq_0}$$

In this formula, q_0 equals the frequency of the a allele (0.5 in the first generation) and n equals the number of generations. What will the frequency of allele a be in 100 generations — about 2,500 years in terms of human lifetimes? Solving the equation:

$$q_{100} = \frac{0.5}{1 + (100 \times 0.5)} = \frac{0.5}{1 + 50}$$
$$= 0.0098 = 1\%$$

Thus, even with a complete selection against homozygous (aa) members, the a allele will still be present in the population at a frequency of 1 percent after 100 generations.

Deleterious genes persist in populations for exceptionally long times because they are carried in the heterozygous state. The frequency of people affected by phenylketonuria (aa) is about 1 per 10,000 in the white population in the United States. Using the Hardy-Weinberg law, we find that the frequency of the recessive allele (q) in the population is 0.01 ($q^2 = 1/10,000$, or 0.0001; $q = 0.01$, the square root of 0.0001), or about 1 percent. Because $p = 1 - q$, p equals 0.99. Therefore, heterozygous carriers of the PKU allele ($2pq = 2 \times 0.99 \times 0.01 = 0.0198$) constitute about 2 percent of the entire U.S. population, or more than 4 million persons.

Further calculations show that matings between heterozygous carriers of PKU produce 98 percent of all the PKU babies in the next generation. The conclusion is inescapable. Preventing those affected with PKU from reproducing would have virtually no effect on the number of PKU babies born in subsequent generations. This conclusion applies to *all* recessive hereditary diseases. Elimination of affected persons scarcely reduces the frequency of the mutant alleles in the population; nor does it reduce the incidence of affected people to any significant degree.

This simple application of population genetics has profound social implications. Uninformed people still advocate sterilization or euthanasia (mercy killing) for people affected by serious hereditary diseases as a means of improving the human gene pool. People who espouse these ideas simply fail or refuse to comprehend the facts of population genetics. Almost all persons who are genetically handicapped as a result of carrying homozygous recessive alleles are the progeny of normal, unaffected heterozygous parents. Eliminating affected people who are judged to be genetically undesirable would reduce the number of affected individuals in the next generation only to an insignificant degree. The population will still contain millions of persons who are phenotypically normal yet carry the recessive alleles. (Eugenics is discussed in Chapter 20.)

Human Inbreeding Is Undesirable

Most human societies have established taboos (prohibitions) against **inbreeding**, or matings between closely related people. Even before the principles of heredity were established, people realized that inbreeding produces more biological problems among progeny than do matings between unrelated people. Homozygosity is increased by inbreeding, which increases the chance that deleterious recessive alleles will come together in offspring.

It is estimated that each one of us carries as many as eight lethal recessive alleles in the heterozygous condition and many more sublethal alleles. Using the Hardy-Weinberg formula, we can calculate the increase in the frequency of homozygous individuals in a population that is self-fertilizing—the most-extreme form of inbreeding, which commonly occurs in plants. The number of heterozygous individuals in self-fertilizing populations decreases 50 percent in each generation, and the number of homozygous individuals carrying recessive alleles increases proportionately (Table 18-2). The extent to which heterozygosity is reduced in a population can be calculated if the amount of inbreeding is known. Inbreeding in human populations results in more people who are homozygous for recessive alleles, which increases the chance for genetic diseases and defects in offspring.

Punctuated Equilibrium, Saltation, and Neutral Mutations

According to the classical Darwinian view, evolution proceeds by a gradual series of heritable changes that accumulate until a new species finally evolves. The mechanism underlying the gradual development of new species is assumed to be natural selection. However, some evolutionary scientists now argue that major evolutionary changes, especially those leading to new species, occur abruptly after long intervals of relatively little change. This hypothesis has come to be called **punctuated equilibrium**. It says that species persist virtually unchanged for long periods (millions of years) and then suddenly evolve into new species in relatively short periods (possibly thousands of years). Propo-

TABLE 18-2 Decrease in frequency of heterozygous individuals as a consequence of self-fertilization.

Generation	Frequency of genotypes		
	AA	Aa	aa
1	¼	½	¼
2	³⁄₈	¼	³⁄₈
3	⁷⁄₁₆	⅛	⁷⁄₁₆

nents of the punctuated-equilibrium model of evolution say that the fossil record shows little evidence of gradual changes but does yield numerous examples of abrupt changes that lead to the sudden appearance of many new species. The new model also explains why transitional forms are rarely found in the fossil record: there simply are not very many.

For years, the fossil record of the horse was cited as a classic example of gradual evolution: the gradual emergence of a single hoof from a three-toed ancestor, and the gradual increase in size. Punctualists reinterpret the fossil record and argue that the modern single-hoofed horse appeared quite abruptly about 3 million years ago and evolved from an animal that had toes flanking its hoofs.

Note that punctuated equilibrium does *not* say that biological evolution has not occurred or that the facts of the fossil record are in dispute. Rather, it suggests an alternative mechanism to Darwin's idea of gradual progressive change. Yet another hypothesis to explain the demise of some species and the origin of others was proposed many years ago. This is the idea of **saltation** (literally, a leap or jump), which proposes that one or more mutations occur that cause a large phenotypic change in organisms that may give them an enormous reproductive advantage. Thus, these new organisms would rapidly replace others in their shared environment and abruptly lead to the appearance of new species.

Another alternative mechanism to natural selection is the **neutral-gene theory** of evolution. According to this theory, bases in the amino acid composition of proteins can change without changing the proteins' functions. Therefore, mutations can occur that would not, in principle at least, be subject to natural selection because the phenotype of the organism has not changed. A variety of mathematical models have shown how accumulation of neutral mutations for many generations can also create the genetic diversity from which new species might evolve.

Evolutionists split themselves into opposing (and often hostile) positions concerning the relative importance of these alternative mechanisms in explaining evolutionary history. Most scientists would agree that natural selection has played a vital role in evolutionary history but acknowledge that neutral mutations, saltation, or punctuated equilibrium may also have contributed. In the public mind, these disagreements are often interpreted to mean that Darwinism or the facts of evolution are in dispute or doubt. Nothing could be further from the truth.

For example, the "neutralists" and the "selectionists" both agree that evolution occurred; their disagreement is over which mechanism has been more significant in producing evolutionary change during the history of life on earth. As is the case with most arguments, both sides are probably partly correct; both natural selection and the accumulation of neutral mutations probably contribute to changes in gene frequencies and, thereby, to evolution.

The Evidence for Natural Selection

In the *Origin of Species*, Darwin established two principles that he believed fundamental to evolution: (1) a certain amount of genetic variation is produced in a population each generation, and (2) this genetic variation is subject to natural selection. Darwin defined natural selection as the "preservation of favorable variations and the rejection of injurious variations." Thus, he perceived evolution in the context of a struggle by individual organisms for survival.

Natural selection has no purpose. Nor can it see into the future to select for traits that might someday be adaptive or useful. The environment of the moment determines which organisms survive and which genes are passed on to progeny. François Jacob, of the Pasteur Institute, has likened natural selection to a tinkerer who "during eons upon eons, would slowly modify his work, unceasingly retouching it, cutting here, lengthening there, seizing the opportunities to adapt it progressively to its new use."

Changes occurring in plants and animals today, as well as the facts of the fossil records, are satisfactorily explained by natural selection. However, like other scientific hypotheses, natural selection must be confirmed by experiments. Among the most-convincing demonstrations of natural selection are the **industrial-melanism** experiments carried out with the peppered moth in England by the late Henry B.D. Kettlewell of Oxford University and his successors.

Peppered moths exist in two distinctive forms: dark-winged (the melanic form) and light-winged (wild-type). On normal, lichen-covered oak trees found in rural areas of Wales, the light-colored moths are virtually invisible against the trees, whereas the dark-colored moths are clearly visible against the bark's light background (Figure 18-8). In heavily industrialized areas of England, the bark of many oak trees has become covered with soot, a form of pollution that was especially common earlier in the century. Light-colored moths are clearly visible on the soot-covered trees, whereas dark-colored moths remain almost indistinguishable from the background.

The principal predators of these peppered moths are birds. Presumably, the moths' coloration helps protect them from being seen and eaten by the birds. The light-colored moths are found where they are harder to spot against the lighter, soot-free trees. If natural selection accounts for this prevelance, then by the same token the dark-colored moths should be found in industrial areas where the trees are covered with soot. Numerous measurements taken by Kettlewell and others have indeed shown that the dark-colored moths prevail in sooty environments, just as the light-colored moths prevail in rural, nonindustrial regions.

Additional experiments carried out by Kettlewell and his successors have shown that survival of the moths depends on how many of each type fall prey to the birds that eat them. To demonstrate that wild-type moths survive better on lichen-covered trees and melanic moths survive better on soot-covered trees, the researchers pinned many moths of both types to trees in rural and industrial areas and then counted the survivors of each kind after several days. As might be expected, the birds more readily found and ate the melanic moths pinned to lichen-covered trees and the light-colored moths pinned to soot-covered trees, results consistent with natural selection.

These ongoing studies of industrial melanism in moths lend strong support to Darwin's hypothesis of natural selection and survival of the fittest. They also show quite dramatically how the frequencies of genes for wing coloration change in the moth population in response to changes in the environment. In the "struggle for

FIGURE 18-8 Industrial melanism. Wild-type (light-winged) moths are nearly invisible on lichen-covered oak trees in rural Wales, whereas the mutant (melanic) form of the moth is quite visible on the same bark, as shown at the bottom. Near English industrial cities such as Liverpool, the oak trees are soot-covered; so the melanic form is better camouflaged than is the wild-type, as shown at the top. Natural selection results in a higher proportion of melanic moths in industrial areas, whereas the wild-type predominates in rural areas. (Courtesy of L.M. Cook, University of Manchester, England.)

FIGURE 18-9   Mimicry. When a bird captures and begins eating a monarch butterfly, it has a severe reaction and learns from the experience not to eat the noxious butterflies. Other butterflies and moths that resemble (mimic) monarchs are also avoided; so their mimicry helps protect them from predators. (Courtesy of Lincoln P. Brower, University of Florida, Gainesville.)

survival," the fitter moths—those that avoid being eaten—are able to reproduce and thereby increase their numbers at the expense of those that do not survive to reproduce.

Natural selection can take many interesting twists, one of which is **mimicry**; moths and butterflies are good examples. Not all moths or butterflies make good meals for birds; some have evolved so as to be able to synthesize substances that birds find unpalatable (Figure 18-9). A bird quickly learns which butterflies are tasty and which are not, and the butterfly population changes accordingly. Natural selection will favor moths and butterflies that contain substances distasteful to the birds. As a result, many species of moths and butterflies closely resemble other, indigestible species but, in fact, are themselves quite palatable. Mimicry (which can entail behavior, as well as physical appearance) protects many species of insects and plants from predators.

Another fascinating example of mimicry is the light flashes emitted as sexual signals by both male and female fireflies. Male fireflies flash and females flash back in signal patterns that are specific to each firefly species. Females of one firefly species are able to attract—and eat!—males of another species because they are able to mimic the light flashes of females of the other species. So natural selection promotes either the survival of individual organisms or, as in firefly mimicry, their destruction.

Certain characteristics have adaptive value in certain environments. The "survival of the fittest" that Darwin described is not a struggle between good and evil. It is simply a fact of nature—the interaction of the organism's genotype with its environment. Biological fitness is simply the genetic contribution of an individual organism to succeeding generations. And natural selection largely determines what the frequencies of particular genes will be in the next generation.

Summary

Biological evolution deals with the origin of living organisms, the genetic diversity of populations, the mechanisms of speciation, and the biological history of the earth. The modern synthesis of evolutionary theory is a unification of Darwin's ideas with those of Mendel and subsequent geneticists — that is, the ideas of natural and sexual selection with the rules governing the inheritance of genes and chromosomes and with the chemical properties of DNA. To geneticists, evolution refers to changes in gene frequencies that arise and accumulate through time in populations of organisms — gene frequencies that change in response to natural selection and other forces.

The term species is defined as a reproductive group whose members interbreed among themselves but do not mate or exchange genes with members of other reproductive groups in nature. Mechanisms of reproductive isolation include mating behaviors, geographical separation, and biological mechanisms that affect fertilization and reproduction.

Population genetics is a method for studying evolution by measuring the frequency of various genes in populations and determining how those frequencies change from generation to generation. The Hardy-Weinberg law, $p^2 + 2pq + q^2 = 1$, describes the relation between allele frequencies and genotypes of individual members of a population in which mating is random. This law does not take into account other biological mechanisms that change allele frequencies. In idealized populations in Hardy-Weinberg equilibrium, allele frequencies are unchanged from one generation to the next; thus, evolution does not occur. However, the Hardy-Weinberg law is a means of discovering and analyzing those forces that do contribute to evolution and cause allele frequencies to change.

The important forces of evolutionary change in populations are mutation, migration, genetic drift, and natural selection. Mutation is the ultimate source of all genetic variation. It can be harmful or beneficial. If a mutation makes an organism better able to survive and reproduce than other members of the population, the frequency of the mutant organism (and the mutant allele) will increase. This reproductive contribution of individual organisms to succeeding generations is called fitness. Migration is the movement of individual organisms (and their genes) into or out of a population. Genetic drift is the mechanism by which the frequency of alleles changes in a population by chance.

Natural selection is a very slow process in producing genetic changes. Even though they are being selected against, deleterious recessive alleles persist for exceptionally long times because they are carried in heterozygous individuals. Almost all genetically handicapped persons who carry homozygous recessive alleles are the children of normal, unaffected heterozygous parents. Inbreeding (mating between closely related people) increases homozygosity, which increases the chance that deleterious recessive alleles will come together in an offspring.

Darwin proposed that evolution proceeds by a gradual series of heritable changes that accumulate until a new species evolves. The principal evolutionary mechanism for this gradual development, according to Darwin, is natural selection. Other hypotheses for mechanisms of evolutionary change are punctuated equilibrium, saltation, and neutral mutations. The punctuated-equilibrium hypothesis states that species remain virtually unchanged for long periods and then suddenly evolve into new species in relatively short periods. Saltation refers to the demise of a species that is replaced by a new species. In this case, mutation causes a great change in phenotype, giving the new organisms an enormous reproductive advantage. The advocates of the neutral-mutation hypothesis hold that bases in DNA can change without causing changes in the functions of proteins or cells. Such accumulated base changes would not be subject to natural selection, yet might eventually result in new genotypes and species.

Key Words

bottleneck A dramatic reduction in the size of a population followed by a large increase.

fitness The reproductive contribution of an individual organism to subsequent generations.

founder effect Demonstration of a gene whose frequency is much higher in a particular population than it is in other populations.

genetic drift Variation in gene frequencies in a population from one generation to the next as a result of chance.

Hardy-Weinberg law A mathematical expression of the relation between allele frequencies and genotype frequencies in an idealized population.

inbreeding Mating between closely related individuals.

industrial melanism The natural selection of dark-colored (melanic) moths in industrial areas of England.

migration Flow of genes from one population to another due to movement of individual organisms or gametes into or out of a population.

mimicry A resemblance of one organism to another, which often protects it from predators.

modern synthesis of evolutionary theory Unites Darwin's ideas of natural and sexual selection with what is now known about the inheritance and chemical properties of chromosomes, DNA, and genes.

neutral-gene theory Proposes that mutations arise that change the amino acid composition of proteins without any change in the proteins' function or in the survival and reproduction of the organism; these "neutral mutations" can spread through a population purely by chance because a small number of all gametes produced contribute to individuals in the next generation.

population genetics Branch of biology dealing with the measurement and calculation of allele frequencies in populations and with how gene frequencies change in successive generations over time.

punctuated-equilibrium theory A model of evolution suggesting that populations remain stable for long periods of time and change abruptly rather than gradually and continuously.

race An arbitrary subclassification of a species based on physical or genetic differences.

random mating Mating in which any organism can mate with any other in a population.

reproductive isolation The inability of organisms to interbreed because of biological differences.

saltation A sudden, large phenotypic change (jump) in a group of organisms.

species A population of organisms that interbreed with one another in nature but do not interbreed with members of other populations, from which they are reproductively isolated.

Additional Reading

Barrett, S.C.H. "Mimicry in Plants." *Scientific American*, September, 1987.

Carson, H.L. "The Process Whereby Species Originate." *BioScience*, November 1987.

Dawkins, R. *The Blind Watchmaker*. Norton, 1987.

Diamond, J. "Survival of the Sexiest." *Discover*, May, 1988.

Kimura, M. "The Neutral Theory of Molecular Evolution." *New Scientist*, July 11, 1985.

O'Brien, S.J. "The Ancestry of the Giant Panda." *Scientific American*, November, 1987.

Sibley, C.G., and J.E. Ahlquist. "Reconstructing Bird Phylogeny by Comparing DNAs." *Scientific American*, February, 1986.

Stebbins, G.L., and F.J. Ayala. "The Evolution of Darwinism." *Scientific American*, July, 1985.

Study Questions

1   What are four biological forces that contribute to evolution?

2   What is the difference between natural selection and sexual selection?

3   What simplifying assumption is made in deriving the Hardy-Weinberg law?

4   Is it possible to eliminate the sickle-cell gene from the U.S. population in a few generations by preventing affected persons from having children? Explain.

5   Why is human inbreeding undesirable?

6   Is it ever possible for animals of different species to mate and produce progeny?

7   About what percentage of all species that ever existed on earth have become extinct?

Essay Topics

1   Discuss some of the mechanisms of reproductive isolation in plants and animals.

2   Discuss why Darwin's idea of natural selection is so important and how it has been tested scientifically.

3   Explain how the various forces of evolution change the frequencies of genes in populations over time.

19

Evolution

Biological History in Molecules and Fossils

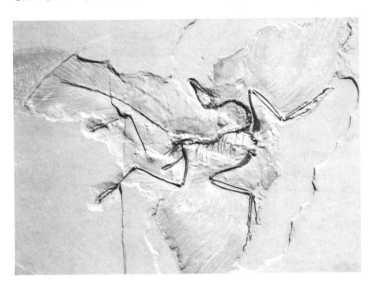

Extinction is the ultimate fate of all lineages, yet it would be absurd to argue that all species are therefore badly designed or poorly adapted. Extinction is no shame.

STEPHEN JAY GOULD, *biologist*

OR LIVING ORGANISMS, evolution is the continual unfolding of new forms of life through time. Scientists attempt to decipher the earth's biological history from observations, measurements, and experiments. All studies of history, whether biological or social, try to reconstruct past events so that we may come to understand more about the earth's past, including human origins, evolution, and behaviors. Even when the history of human societies is well documented—for example, the causes and events leading to World War II or the American Civil War—that history is always subject to conflicting interpretations. By their very nature, biological and social histories are imprecise in comparison with mathematics, physics, or chemistry—sciences that generally yield exact solutions to specific problems. The events and causes of social and biological change are continually being challenged and reinterpreted as new information becomes available.

Scientists reconstruct biological events of the past from existing data but, as new facts are discovered, their hypotheses and interpretations change. The fact that organisms lived and became extinct in the past is as well established as the fact that our present society evolved from past societies and is still changing. The fossil records supply undeniable factual evidence of evolution and of biological change that has occurred for billions of years. For example, we know without a doubt that many species of

dinosaurs flourished on earth in the past; we even know that all the species of dinosaurs became extinct within a relatively short period of time about 65 million years ago (see Box 19-1). What is less certain are the specific evolutionary mechanisms that produced the dinosaurs in the first place and the specific environmental events that caused their extinction (Figure 19-1).

As a result of recent **hominid** (humanlike) fossil discoveries, mainly in Africa, it is now possible to trace hominid evolution over a period of 3 to 4 million years. However, modern human beings (*Homo sapiens*) appeared quite recently and occupy but a tiny fraction of the overall period of earth's biological history, which spans from about 4 billion to 5 billion years. As recently as 35,000 years ago, hominids were living in Europe who were physically different from us yet similar enough to be classified as a subspecies of modern human beings. *Homo sapiens neanderthalensis*—or Neanderthals, as they are

FIGURE 19-1 All species of dinosaurs became extinct abruptly about 65 million years ago. Until their disappearance, they had inhabited earth for about 150 milion years. (Courtesy of Kent and Donna Dannen.)

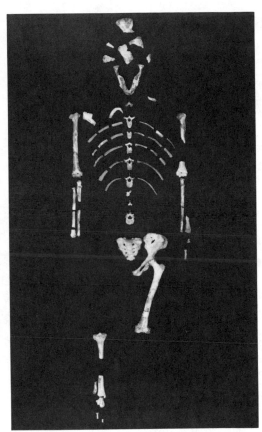

FIGURE 19-2 Lucy, the most-complete, early hominid fossil unearthed to date. Lucy was uncovered in Ethiopia and is dated at about 3.5 million years. (Courtesy of Donald C. Johanson, Institute of Human Origins.)

commonly called—lived in Europe for more than 40,000 years before us and, for reasons unknown, became extinct. Donald C. Johanson, an anthropologist who was the codiscoverer of Lucy (Figure 19-2), a remarkably complete female skeleton estimated to be 3.5 million years old and classified as *Australopithecus afarensis*, describes Neanderthal man as someone who "could make change at the subway booth and recognize a token."

The biological history of human beings and other organisms can be followed not only from fossil remains but also from differences that are observed in the macromolecules of organisms that are alive today. Researchers can determine the similarities between the genes and proteins of different organisms, which serve as sources of insight into evolution. For example, the amino acid sequences of hemoglobin molecules in the blood cells of dogs, pigs, rats, monkeys, apes, and human beings have been determined. These different amino acid sequences can then be used to reconstruct the evolutionary relatedness of these species, which is referred to as their **phylogeny**, or family tree. And newly developed techniques for determining the sequence of bases in DNA are another accurate means for determining the genetic relatedness of different species.

This chapter discusses the various ways in which researchers study evolution and reconstruct earth's biological history. To properly evaluate the scientific facts of evolution, we must understand how the ages of fossils and rocks are determined, as well as what the differences between biological molecules mean. Only then can we appreciate our unique position in evolutionary history. Darwin summed up his own view of evolution in the sentence that concludes his *Origin of Species*:

> There is grandeur in this view of life, with its several powers having been originally breathed into a few forms, or into one; and that, whilst this has planet gone cycling on according to the fixed laws of gravity, from so simple a beginning endless forms most beautiful and wonderful have been, and are being, evolved.

The Age of the Earth

Almost all astronomers and physicists agree that the universe came into existence as a result of an explosion of incomprehensible magnitude

What Happened to the Dinosaurs?

BOX 19-1

About 65 million years ago—give or take a few million—something truly catastrophic happened on earth. According to the fossil record, in this period of the earth's history more than half of all the existing genera became extinct. Animals and plants, particularly species of marine life, were annihilated over a period that may have encompassed millions of years, although many species, including all of the dinosaurs, vanished more rapidly.

This worldwide biological extinction is evident in the sediments that separate the Cretaceous and Tertiary periods. Fossils of dinosaurs are abundant in the Cretaceous period but are missing from the more-recent Tertiary rocks. Many other kinds of fossils found in Cretaceous rocks also do not appear in more-recent rock formations, suggesting that some catastrophic event took place at the end of the Cretaceous.

Various ideas have been proposed to explain this relatively sudden global extinction of so many life forms: a sharp rise in the earth's temperature, dust from numerous volcanic eruptions, a drop in the levels of the oceans, a severe disruption in the food chain. Now it is thought that the earth was struck about 65 million years ago by an enormous extraterrestrial object,

probably a meteorite about 10 miles in diameter.

In 1980, scientists reported an unusual geological finding. They detected large amounts of iridium, a rare platinumlike metal, in the narrow band of clay sediments that characterizes the boundary between Cretaceous and Tertiary geological formations. Most rocks contain virtually undetectable amounts of iridium because this metal is a rare element in the earth's crust. However, iridium is abundant in meteorites.

An analysis of the chemical composition of the clay layer separating the Cretaceous and Tertiary formations in Italy revealed that the iridium concentration was thirty times as great as that in the rock formations on either side. Since the initial discovery, unusually high iridium concentrations have been found in geological samples from Denmark, Spain, and New Zealand and in sediment samples from the floors of the Atlantic and Pacific oceans. The high levels of iridium are always found in the layer that separates the Cretaceous and Tertiary rock formations (see the accompanying photograph).

The original discoverers of the iridium anomaly hypothesized

that 65 million years ago the earth was struck by an enormous extraterrestrial object that exploded on impact and resulted in the iridium deposits. Indeed, a few massive craters, created by the impact of huge meteorites in prehistoric times, are well known to geologists. The Meteor Crater near Flagstaff, Arizona is about half a mile in diameter, but it is dwarfed by two other meteor craters. Each of these craters—one in Ontario, Canada, the other in South Africa—is more than 80 miles across.

The scientists who proposed that the iridium came from an exploding meteor also believe that the environment of the earth was so disrupted that most life forms simply could not survive. The atmospheric ozone layer that protects the earth from intense radiation may have been disrupted temporarily. Or the millions (or billions) of tons of dust ejected into the atmosphere may have obscured the sun for years, causing the earth's temperature to change abruptly. Whatever climatic changes occurred, the fossil record shows that a majority of the species, including the dinosaurs, were destroyed.

Small mammals were among the few survivors. What different paths might evolution have followed if this catastrophe had not occurred? What if the dinosaurs

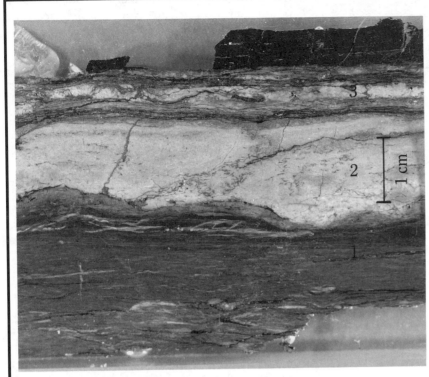

had not become extinct? Would mammals still have evolved into primates and human beings? It is entirely possible that the human species owes its very existence to a catastrophic accident that marked the end of the Cretaceous period 65 million years ago.

READINGS: Norman Myers, "The End of the Lines," *American Scientist*. February, 1985. Ben Patrusky, "Mass Extinctions: The Biological Side," *Mosaic*. Winter, 1986–1987

A rock slab showing the Cretaceous-Tertiary boundary from a site near Trinidad, Colorado. Layers 1 and 2 are shale and clay from the Cretaceous. Layer 3 is the impact layer that contains the highest concentration of iridium. Layer 4 is coal from the Tertiary. (Courtesy of Glen Izett, U.S. Geological Survey, Denver.)

some 10-to-20 billion years ago—an event that has popularly become known as the Big Bang (briefly described in Box 4-1). In the instant that began time itself, this cosmic explosion created everything: matter, energy, space, and all the laws of physics and chemistry that govern the properties of matter and energy from that moment until now. The Big Bang is the farthest back in history that science can probe. Before the Big Bang there was nothing—or, at any rate, whatever existed must remain a total mystery to science. The creation of the universe and of earth is also the subject of myth and religion.

The solar system, which includes our sun and its nine planets, was not created at the beginning of the universe. In fact, compared with the universe, the solar system is quite young; radio-isotope dating of earth and moon rocks indicates their age to be about 4.5 billion years. In the early history of the universe, stars and galaxies were just beginning to form from the vast clouds of hydrogen atoms that were slowly being condensed by the gravitational force of their own enormous mass. Not until an enormous number of stars had been born, aged, and then destroyed in gigantic nova or supernova explosions could a solar system such as our own have come into existence.

It is in these stellar explosions that many of the heavier elements that are found so abundantly in the earth's crust were created. No life forms could have existed during the early history of the universe, for cells contain elements

FIGURE 19-3   A lunar rock collected by astronauts from Apollo 17. The age of this rock as determined by radioisotope dating is about 4.5 billion years. It is among the oldest known rocks and supports the idea that the earth and moon were formed at the same time. (Courtesy of the National Aeronautics and Space Administration.)

that are created only by stellar explosions. Fortunately, nature has supplied us with radioactive elements that provide a means for determining the ages of earth and moon rocks and the ages of fossils that were trapped in rocks eons ago (Figure 19-3).

Radioactive Clocks Measure Time

Radioactivity was discovered in 1896 by Antoine Henri Becquerel, a French scientist. Since his discovery, many applications of that knowledge have been developed. Naturally occurring radio-

active isotopes, or radioisotopes (discussed in Chapter 4), serve as remarkably accurate "atomic clocks" by which the ages of rocks and fossils can be measured.

Nuclei of radioactive elements spontaneously decay at a constant rate that never varies even over millions or billions of years. In radioactive decay, atomic nuclei are altered by the emission of energy or particles such as protons and neutrons. The emitted particles and energy are measured as radioactivity. The unchanging rate of nuclear decay serves as an exceptionally accurate atomic clock that has been running in rocks and fossils ever since the particular radioactive elements were trapped in them.

As stated in Chapter 4, the *half-life* of a radioisotope is the time—in seconds, weeks, or millions of years—that it takes for half of the nuclei to decay and be converted into some other element. It is impossible to predict which particular atoms will decay in any radioactive sample, but half of them will always decay in a certain time period, half of the remaining atoms during the next, identical time period, and so forth (Figure 19-4).

FIGURE 19-4   Radioisotope dating. There are half as many radioactive atoms in a sample at the end of each half-life interval as there were at the beginning of the interval. If the radioisotope's rate of decay is known, and if the remaining amount of the radioisotope and its decay product can be measured, the age of the sample can be calculated.

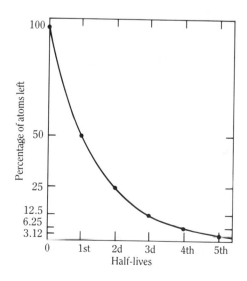

TABLE 19-1 Radioisotopes used to determine the ages of rocks and fossils.

Material to be dated	Original isotope	Half-life (millions of years)	Decay product
Rocks and older fossils	Rubidium-87	47,000	Strontium-87
	Uranium-238	4,510	Lead-206
	Potassium-40	1,300	Argon-40
	Uranium-235	713	Lead-207
Recent fossils	Carbon-14	5,730 years	Nitrogen-14

Certain stable isotopes are formed only as a result of the decaying of radioactive elements; so these radioisotopes make particularly useful atomic clocks. For example, lead-204 (^{204}Pb) is the most-abundant stable isotope of lead, but other stable lead isotopes (^{206}Pb and ^{207}Pb) are formed exclusively as the decay products of radioactive isotopes of uranium (^{238}U and ^{235}U). The amounts of different stable isotopes can be determined with great accuracy by an instrument known as a mass spectrograph, which measures the amount of any isotope according to its weight. Because ^{204}Pb can be distinguished from ^{206}Pb in a mass spectrograph, the particular decay products of uranium radioisotopes can be determined with great accuracy.

When volcanos spewed forth molten rock during the earth's early history, uranium, rubidium, and other radioactive elements became trapped in the rocks as the lava cooled and hardened. The atomic clocks in the newly formed rocks began to run. The radioactive elements had been decaying all along but they did not begin to function as clocks until they became trapped in rock from which their breakdown products could not escape. The clock works by making available measurable amounts of both the radioactive element and its breakdown products, and it is these *relative amounts* that enable researchers to determine when the clock started "ticking."

By carefully measuring the amount of strontium-87, lead-206, and lead-207 in rock samples and determining the remaining number of radioactive atoms, scientists can calculate the number of half-lives that must have been necessary to produce the relative amounts of the various isotopes. Because many rock samples can be dated by several independent atomic clocks, the age of the earth is now known with a great degree of certainty (Table 19-1).

The Earth's History Is Divided into Geologic Intervals

In 1654, James Ussher, Archbishop of Armagh (in Ireland), argued that the earth was created in the year 4004 B.C.; he arrived at this number by counting the successive "begats" recorded in the Bible. By the nineteenth century, geologists favored an earth age of several million years, based on rates of sedimentation and the degree of salinity of the oceans. In 1846, the British mathematician and physicist Lord Kelvin calculated the age of the earth (incorrectly as it turned out) to be 20 million years, based on what he perceived to be its rate of cooling. Today, atomic-clock measurements of the oldest known rocks from both the earth and the moon give an age of 4.6 billion years. To appreciate how long a time this is, consider an imaginary clock in which each hour equals about 190

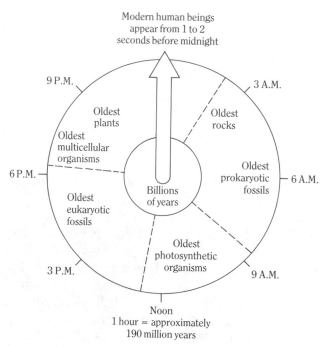

Modern human beings
appear from 1 to 2
seconds before midnight

9 P.M.

3 A.M.

Oldest
plants

Oldest
rocks

Oldest
multicellular
organisms

6 P.M.

Oldest
prokaryotic
fossils

6 A.M.

Billions
of years

Oldest
eukaryotic
fossils

3 P.M.

Oldest
photosynthetic
organisms

9 A.M.

Noon
1 hour = approximately
190 million years

FIGURE 19-5 A hypothetical evolutionary clock. The age of the earth is known to be about 4.6 billion years from the radioisotope dating of rocks. If this total time is set equal to a 24-hour day, the different major life forms can be placed at particular times on the clock. In relation to the total age of the earth, modern human beings have been around for only a few seconds of a 24-hour day.

million years (4.6 billion years divided by 24 hours). On such a time scale, hominids arose only about one minute ago. Modern human beings appeared only about one second ago (Figure 19-5).

The geological history of the earth is divided into eras, periods, and epochs. Geologists base these classifications both on the various kinds of rock formations that can be characterized and on the fossils in these rock formations (Table 19-2). Because of the continual evolution of new life forms and the extinction of others during the several billion years of earth history, geological formations can often be identified by the presence of certain microfossils that have become trapped in the rocks. Geological formations and their fossil remains are a means of cross-checking the relative ages and events of both geological and biological history. Atomic clocks determine absolute age.

The lovely Pacific coral islands, or *atolls*, such as Bikini, Eniwetok, and Kwajalein, were formed by volcanos erupting and afterward sinking beneath the ocean floor. Corals grow on the lava rocks beneath the ocean, and some Pacific coral reefs contain the skeletal remains of the oldest animals on earth. Corals that are still growing today contain fossil remains that are more than 100 million years old; thus, these corals began to grow in the Jurassic period.

The coral skeletons on Eniwetok extend almost a mile below the ocean's surface; from changes in the patterns of reef formation, ocean scientists have been able to figure out when the ocean's level rose and fell during millions of years of geological history. At some periods in the past, the ocean level dropped by as much as 180 meters from what it is now. Samples of the coral reef skeletons taken from below the surface range in age from 160,000 to 700,000 years.

Fossils: Evidence of Biological History

Historians rely on written documents to reconstruct human history and to evaluate the significance of past events. **Paleontologists**, the historians of earth's biological history, study the fossils that have been preserved in rocks and sediments. From the fossil record, they attempt to reconstruct biological history (Figure 19-6).

Fossil Remains

Most people are familiar with the spectacular large-fossil finds: the bones and tracks of such animals as dinosaurs, saber-toothed tigers,

TABLE 19-2   The geological time scale and the appearance of various life forms.

Era	Period	Epoch	Millions of years before present	Life forms and major events
Cenozoic (Age of Mammals)	Quaternary	Recent	0.01	Modern human beings
		Pleistocene	2	Early hominids
	Tertiary	Pliocene	10	Apes
		Miocene	25	Grazing mammals
		Oligocene	35	Flowering plants (angiosperms)
		Eocene	55	
		Paleocene	65	First placental mammals appear
Mesozoic (Age of Reptiles)	Cretaceous		135	First flowering plants
	Jurassic		185	
	Triassic		225	First dinosaurs
Paleozoic	Permian		275	Extinction of much marine life
	Carboniferous (Pennsylvanian, Mississippian)		350	Amphibians: giant swamp trees
	Devonian		400	Age of fishes
	Silurian		430	First land plants
	Ordovician		480	Earliest known fishes
	Cambrian		600	Marine invertebrates; algae
Precambrian			4,600–600	Single and multicellular soft-bodied organisms

A

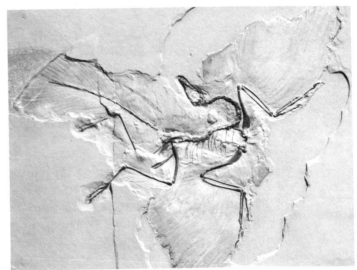

B

FIGURE 19-6   Fossil skeletons.
A. A fossil fish skeleton from Fossil Butte National Monument in Wyoming. This fish lived from 40 million to 65 million years ago. (Courtesy of Kent and Donna Dannen.)
B. *Archaeopteryx*, one of the transition forms between reptiles and birds. This fossil was found in rocks that are about 150 million years ago. (Courtesy of the American Museum of Natural History.)

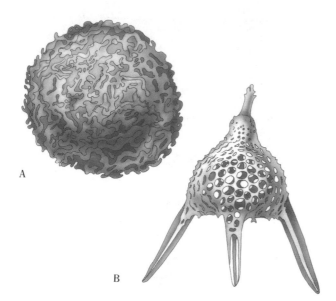

FIGURE 19-7 Microscopic fossils: (A) a plant spore from the Mississippian period; (B) a marine protozoan from the late Tertiary. Certain fossils are present only in rocks of particular geological periods. Paleontologists reconstruct biological and geological history from the microfossil record.

billion years old suggests that simple cells arose on earth within a billion years or so after its formation. One of the most fortunate aspects of biological history is that so many fossils, both large and microscopic, have been preserved. It is the enormous diversity and age of the fossil record that substantiates the overriding fact of evolution—namely, that all life forms on earth have been in a continual state of evolution for billions of years.

Carbon-14 Dating of Fossils

The biological and geological history presented so far goes back hundreds of millions, even billions, of years. What kind of information has been preserved regarding more-recent biological events—say, within the past 100,000 years, as human civilizations have developed? The carbon-14 (^{14}C) clock is the most-valuable technique for dating organic materials, such as

mammoths, and mastodons; and the impressions left by the leaves, stems, and other parts of ancient plants. However, important as these large fossils are in understanding biological history, even more valuable information is obtained from microscopic fossil remains such as pollen grains, protozoa, and bacteria (Figure 19-7).

Microfossils discovered in rocks from northern Minnesota are believed to be about 2 billion years old. Even older rock formations have been found in western Australia that appear to be fossilized mounds of bacteria. These moundlike structures, called **stromatolites**, are formed by layer-by-layer deposits of primitive microorganisms and minerals, which preserve the bacterial remains (Figure 19-8). The existence of stromatolites and other forms of fossilized microorganisms in rocks that are between 2 billion and 4

FIGURE 19-8 Stromatolites in western Australia. These are recent stromatolites, aggregates of microbes cemented together by minerals from the sea. Similar stromatolites have been found that are between 2 billion and 3 billion years old, indicating the presence of primitive cells that long ago.

bones, mummies, plants, and seeds of more-recent vintage. While they are alive and growing, all organisms must assimilate carbon atoms from the environment. The most-abundant isotope of carbon is ^{12}C, but the environment also contains a small amount of the radioactive isotope ^{14}C. Plants assimilate both carbon isotopes as they grow, because they cannot distinguish between them in most cellular chemical reactions—nor can the animals that eat the plants.

The carbon gases in the air contain a fixed ratio of $^{12}C/^{14}C$ atoms, and it can be assumed that this ratio was much the same in the recent past as it is today. When an organism dies, the carbon atoms in its tissues remain there, and the ^{14}C atomic clock begins to run. Every 5,730 years, half of the ^{14}C atoms decay into stable isotopes of nitrogen; the older the organism, the fewer ^{14}C atoms remain, and so the $^{12}C/^{14}C$ ratio in the sample increases with age. The ^{14}C clock is accurate for dating organisms as old as 40,000 years; beyond that, too few ^{14}C atoms remain to be counted accurately.

The accuracy of the ^{14}C clock has been verified by calibrating it with samples from the bristlecone pines of California, some of which are more than 5,000 years old (Figure 19-9). Most of the wood in the bristlecone pines (like wood in other trees) is dead; only a thread of living tissue at the center of the trunk keeps them alive. Small core samples removed from the trunk do not destroy the tree but do allow its age to be determined from the faint but visible growth rings. The ages of sections of the wood are then used to calibrate the ^{14}C clock.

FIGURE 19-9   A bristlecone pine. These trees are used to calibrate carbon-14 atomic clocks. Some bristlecones are 5,000 years old. The carbon-12/carbon-14 isotopic ratio changes in sections of trees that can be dated by counting rings of yearly growth. (Courtesy of Janis Hansen, Kaaawa, Hawaii.)

Molecular Clocks as Measures of Evolution

In addition to the fossil record, similarities and differences between genes and proteins in various species lend strong support for evolution. If organisms from widely different species are indeed related by descent, no matter how distantly, similarities in their genes and other molecules should be apparent. By the 1960s, techniques for determining the amino acid sequences in proteins had advanced to the point where functionally identical, yet structurally different, proteins from different organisms could be analyzed.

Changes in Amino Acids

Cytochrome c is a protein that participates in the biochemical production of ATP, the main energy-storing compound in every living cell. Cytochrome c proteins have been extracted from dozens of different species and their complete amino acid sequences have been determined. If amino acid differences in cytochrome c between two organisms are few, we know that they are closely related. If amino acid differences are extensive, the two species are separated by a long period of evolutionary time. For example, cytochrome c proteins in human beings differ by only one amino acid from cytochrome c proteins in Rhesus monkeys; eleven amino acids separate human beings from dogs, and forty-five differences separate human beings from wheat.

A phylogenetic (family) tree can be constructed using the amino acid differences in cytochrome proteins to show the relative relatedness among species (Figure 19-10). By estimating the mutation rates in DNA that must have caused the accumulation of the amino acid changes, biologists can calculate the time of divergence between various organisms. The ancestral ages that are determined from amino acid differences are then compared with family trees that have been constructed by other physical and physiological measurements — often with remarkably good agreement.

Although it is unlikely that amino acid substitutions have occurred at a uniform rate in organisms through millions of years, it is nevertheless possible to construct a "molecular clock" that gives an estimate of the rate of evolution. Measurements of amino acid changes in seven different proteins from many different species of organisms indicate that the molecular clock runs at an average rate of about 100 amino acid changes per 240 million years. Not all proteins and not all oragnisms obey this clock,

however. Moreover, many assumptions are made in calculating these rates, and molecular clocks are not thought to be nearly as accurate as atomic clocks. The general rule is, however, that the more similar the amino acid sequences in functionally comparable proteins from different organisms are, the more closely related the organisms.

Certain amino acid changes in a protein may alter or destroy its function; in fact, this is one of the more-demonstrable consequences of mutations. Yet many amino acid changes in proteins do not destroy their function, as shown by the structurally different cytochrome c proteins in various species. In fact, analysis of the sequences of bases in genes that code for functionally identical proteins in different organisms gives a surprising result: as many as half of the bases in a gene can be changed without changing the biochemical properties of some proteins. These experimental observations also contributed to the development of the neutral-gene theory of evolution that was discussed in Chapter 18.

Hemoglobin and Human Evolution

New genes arise not only by changes in bases in DNA but also by duplication of segments of DNA that allow an organism to preserve genes essential for its survival while creating new genes. These new genes may then evolve some

FIGURE 19-10 A phylogenetic tree constructed from amino acid differences in cytochrome c. The table shows the number of amino acid differences in cytochrome c between any two species listed. Note that the number of amino acid differences between human beings and baker's yeast (45) is about the same as that between baker's yeast and *Neurospora crassa* (41), both of which are fungi. In reference to evolutionary divergence, yeast and *Neurospora* are as distantly related to each other as human beings are to either fungal species. (Redrawn from R.E.Dickerson and I. Geis. *Structure and Action of Proteins*, Harper & Row, 1969.)

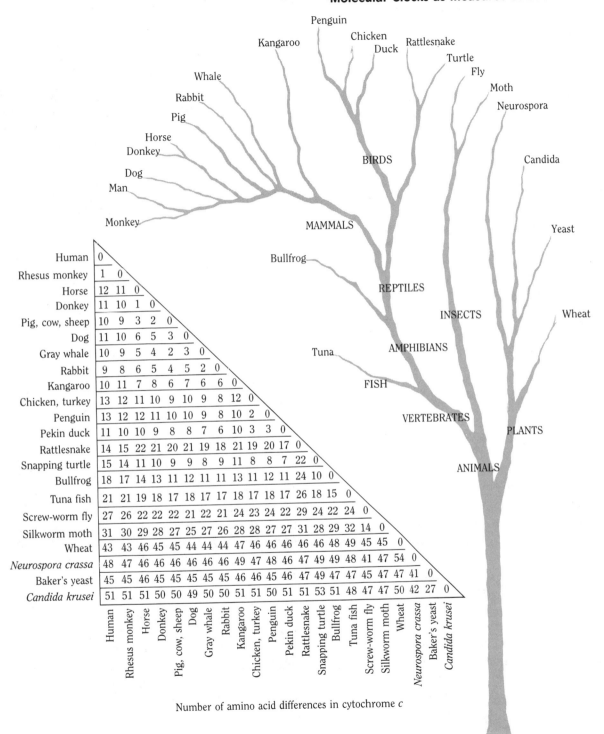

	Human	Rhesus monkey	Horse	Donkey	Pig, cow, sheep	Dog	Gray whale	Rabbit	Kangaroo	Chicken, turkey	Penguin	Pekin duck	Rattlesnake	Snapping turtle	Bullfrog	Tuna fish	Screw-worm fly	Silkworm moth	Wheat	Neurospora crassa	Baker's yeast	Candida krusei
Human	0																					
Rhesus monkey	1	0																				
Horse	12	11	0																			
Donkey	11	10	1	0																		
Pig, cow, sheep	10	9	3	2	0																	
Dog	11	10	6	5	3	0																
Gray whale	10	9	5	4	2	3	0															
Rabbit	9	8	6	5	4	5	2	0														
Kangaroo	10	11	7	8	6	7	6	6	0													
Chicken, turkey	13	12	11	10	9	10	9	8	12	0												
Penguin	13	12	12	11	10	10	9	8	10	2	0											
Pekin duck	11	10	10	9	8	8	7	6	10	3	3	0										
Rattlesnake	14	15	22	21	20	21	19	18	21	19	20	17	0									
Snapping turtle	15	14	11	10	9	9	8	9	11	8	8	7	22	0								
Bullfrog	18	17	14	13	11	12	11	11	13	11	12	11	24	10	0							
Tuna fish	21	21	19	18	17	18	17	17	18	17	18	17	26	18	15	0						
Screw-worm fly	27	26	22	22	22	21	22	21	24	23	24	22	29	24	22	24	0					
Silkworm moth	31	30	29	28	27	25	27	26	28	28	27	27	31	28	29	32	14	0				
Wheat	43	43	46	45	45	44	44	44	47	46	46	46	46	46	48	49	45	45	0			
Neurospora crassa	48	47	46	46	46	46	46	46	49	47	48	46	47	49	49	48	41	47	54	0		
Baker's yeast	45	45	46	45	45	45	45	46	46	45	46	47	49	47	47	45	47	47	41	0		
Candida krusei	51	51	51	50	50	49	50	50	51	51	50	51	51	53	51	48	47	47	50	42	27	0

Number of amino acid differences in cytochrome *c*

Family tree

TABLE 19-3 Number of amino acid differences between the β-hemoglobin chains of various primate species and human hemoglobin.

Human compared with	Number of amino acid differences
Chimpanzee	0
Gorilla	1
Gibbon	3
Rhesus monkey	8
Squirrel monkey	9

new function. The best example of gene duplication and subsequent evolution of new functional genes is the family of genes that direct the synthesis of different hemoglobin proteins. It appears that primitive animals possessed a single ancestral globin gene that was duplicated numerous times during evolution and that eventually gave rise to a modern hemoglobin protein that consists of four globin chains: two α-chains and two β-chains, which can be synthesized from any of about a dozen different hemoglobin genes.

Human hemoglobin genes have been studied extensively by analyzing both the amino acid sequences of the hemoglobin chains and the base sequences of the genes that produce the α- and β-chains. The α-genes in human genes are on chromosome 16, and the β-genes are on chromosome 11. Different α- and β-genes are expressed at various stages of human development, producing embryonic, fetal, and adult hemoglobins, all of which differ in their amino acid sequences but perform the same function of transporting oxygen. In the past few years, hemoglobin pseudogenes also have been found—that is, genes that have the capacity to synthesize β-chains but that are never expressed in cells. It is possible that these "silent"

genes served a cellular function in the past but were permanently switched off at some stage of evolution.

Differences in the amino acids of the five different β-chains found in the various classes of human hemoglobins have been determined and, from the number of those differences, the approximate time that the gene duplications occurred in the past can be calculated by use of the molecular clock. The original β-gene arose in vertebrates about 70 million years ago and was duplicated between 30 million and 40 million years ago; the most-ancestral globin gene appeared from 400 million to 500 million years ago.

Hemoglobin proteins and their genes provide a molecular record of human evolution. Furthermore, a comparison of the amino acid sequences in human hemoglobin with those of other animals gives evidence for their relatedness. For example, a comparison of adult human hemoglobin with the hemoglobin of other primates shows that the β-chains of chimpanzee and human hemoglobin are identical in amino acid sequence; gorilla hemoglobin differs from human hemoglobin by a single amino acid (Table 19-3). These extremely small differences between the hemoglobins of human beings, chimpanzees, and gorillas can be explained by assuming that all three species evolved from a common ancestral primate, possibly from 15 million to 20 million years ago.

This conclusion concerning our relatedness to apes is strengthened by examining the chromosomes in these three primates. Eighteen of the twenty-three human chromosomes are virtually identical in structure and banding patterns with chromosomes in chimpanzees and gorillas. The other five human chromosomes are also very similar to those in the apes. As William Winwood Reade expressed it a century ago, "It is a shabby sentiment . . . which makes men prefer to believe that they are degenerated angels rather than elevated apes."

As a matter of fact, much of the controversy over evolution has to do with *human* evolution: Did human beings evolve from apes? This notion was extremely distasteful to the Victorians of Darwin's time. One eminent English lady expressed her view of the idea as follows: "Let us hope that it is not true but if it is let us pray that it will not become generally known." And when, in 1860, Thomas Henry Huxley read Darwin's scientific report before the Royal Society, he was asked sarcastically by Bishop Wilberforce whether he claimed descent from monkeys on his mother's or his father's side of the family. Huxley, a distinguished biologist and a formidable advocate of Darwinism who came to be known as "Darwin's bulldog," turned the tables on his interrogator:

> I asserted—and I repeat—that a man has no reason to be ashamed of having an ape for his grandfather. If there were an ancestor whom I should feel shame in recalling, it would rather be a *man*, a man of restless and versatile intellect, who, not content with an equivocal success in his own sphere of activity, plunges into scientific questions with which he has no real acquaintance, only to obscure them by an aimless rhetoric, and distracts the attention of his hearers from the real point at issue by eloquent digressions and skillful appeals to religious prejudice.

Today we know that human beings are not descended from monkeys or apes but that all have a common ancestor.

DNA Similarities Among Primates

Another molecular technique used to measure the degree of similarity of genetic information in different organisms is **DNA-DNA hybridization**. This technique consists of extracting DNA molecules from the cells of different organisms, separating the individual strands of DNA by heating, and measuring the extent to which the separated DNA strands from different organisms come together to form a stable double helix (Figure 19-11).

The degree to which the strands of DNA join depends on how similar their base sequences are. If the DNA strands were from the same organism, the base sequences would be virtually identical and the correspondence of the bases would be perfect. Thus, the degree of nonbonding between strands is a measure of the genetic dissimilarity between organisms. (The degree of nonbonding is measured by the specific melting temperature of the hybrid DNA—that is, the temperature at which enough of the hydrogen bonds are broken so that half of the DNA strands are unable to stay joined stably.) If this hybridization technique is applied to DNAs extracted from the cells of various primates, human DNA strands hybridize perfectly with other human DNA strands and almost as well with chimpanzee DNA strands. Human-chimpanzee DNA is the most closely related in base sequences; other primate DNA hybridizes less well (Figure 19-12).

Why, if human beings and chimpanzees are so nearly identical in their proteins, chromosomes, and genes, are the two species so phenotypically different? No one can mistake a chimpanzee for a person. The answer is that differences must derive from only a few genes or from different patterns of gene expression in the two species. But as yet we know very little about how genes affect development in either human beings or primates.

Polymorphisms and Genetic Diversity

Individual members of a population have different phenotypes because of the different

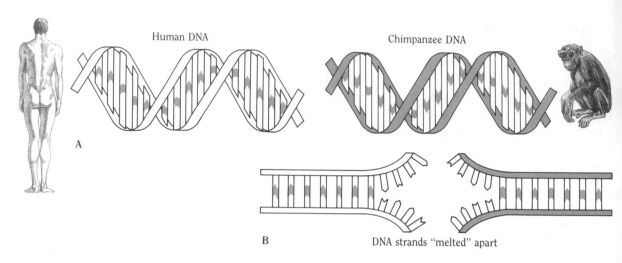

Human DNA

Chimpanzee DNA

A

B   DNA strands "melted" apart

FIGURE 19-11 DNA-DNA hybridization, a technique for measuring the degree of homology in the DNAs of different organisms: (A) DNA molecules are extracted from the cells of human beings and chimpanzees; (B) the DNA molecules are heated *in vitro* to separate (melt apart) the two strands by breaking the hydrogen bonds between base pairs; (C) the separated strands of DNA from the two species are mixed together *in vitro*, and the ability of the strands from each species to hybridize (join to form a double helix) is measured.

C

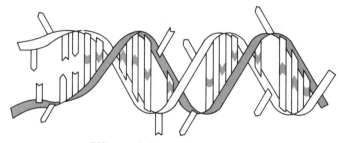

DNA strands from each species join

alleles that can be carried at any genetic locus. *Polymorphism* (literally, "many forms") refers to the presence of several forms of a trait (or of an allele) in a population. For example, among human beings, the various ABO, M-N, and Rh blood groups are polymorphic. Phenotypic differences are usually the result of slight variations in the proteins that are synthesized in different individual members of a population. Because alleles determine the kinds of proteins that are synthesized, protein polymorphisms are a measure of the population's genetic diversity.

Polymorphisms are revealed by examining the proteins synthesized in different members of a population (Figure 19-13). Because amino acid substitutions usually affect the electrical charge on proteins, they can be separated by the process of electrophoresis. Although electrophoresis does not detect all amino acid substitutions (some amino acid changes are electrically neutral), it does reveal some of the genetic variation (polymorphism) in a population.

The extent of polymorphisms varies widely among species. In the many species of the fruit fly *Drosophila*, approximately half of all the genes are polymorphic—that is, more than one allele exists at a particular genetic locus. About 30 percent of all genetic loci in amphibians and fishes are polymorphic. Mammals exhibit the fewest polymorphisms; only about 20 percent of the genes of mammals have multiple alleles.

One of the more-unexpected results of studying polymorphisms is that the amount of genetic diversity *within* a population sometimes

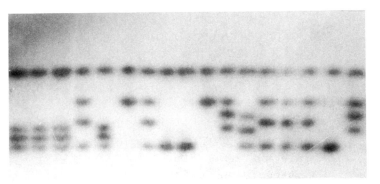

FIGURE 19-13 Electrophoresis of proteins extracted from plant cells. This photograph of a gel that has been stained for a particular enzyme shows the different structural forms of the enzyme synthesized from different alleles in a population. Each column shows the forms of the protein synthesized in an individual plant. There are more than two forms because the subunit polypeptides from pairs of different alleles associate in various combinations. (Courtesy of Leslie Gottlieb, University of California, Davis.)

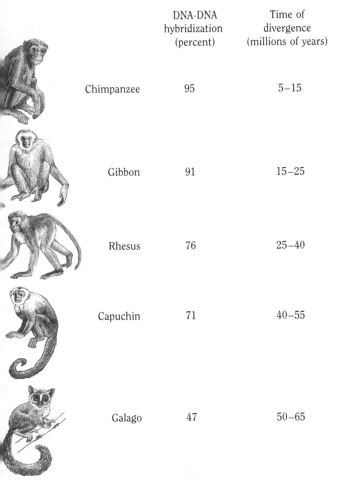

ıman being compared with

	DNA-DNA hybridization (percent)	Time of divergence (millions of years)
Chimpanzee	95	5–15
Gibbon	91	15–25
Rhesus	76	25–40
Capuchin	71	40–55
Galago	47	50–65

FIGURE 19-12 Degree of genetic and evolutionary relatedness among primates as indicated by DNA hybridization. The primate most closely related to human beings is the chimpanzee; other apes and monkeys are more distantly related.

turns out to be greater than the amount of diversity *between* populations—even between species. For example, the genetic diversity in a population of fruit flies of the same species living in a tropical swamp may be greater than the genetic diversity (as measured by polymorphic proteins) between that population of flies and a different species of flies that lives in an arid desert.

The implication of this finding is that most protein changes may not be particularly important in the evolution of new species. It is more likely that changes in the regulation and interaction of genes are what cause new species to evolve. As pointed out in the preceding section, the proteins and genes of human beings and chimpanzees are quite similar, but the overall phenotypic differences are enormous.

Human Polymorphisms and Race

The overall genetic differences between various human populations can be studied by measuring the genetic polymorphisms among individuals. The assignment of people to *races*, a term used to divide a species into subgroups, can be based on morphological differences, on the number of different alleles, or on any other characteristics one chooses to use to distinguish subgroups.

Use of the term *race* by politicians, sociologists, biologists, or other persons who seek, for whatever reason, to divide the human species into subgroups has contributed to racial prejudice. Harold J. Morowitz, professor of biochemistry at Yale, points out what racial prejudice really signifies: "When we hate an individual for his race, we are hating genes over which he has no control. . . . Prejudice means disliking the genes of a . . . person without paying that much attention to the individual that goes with the genes. . . . When you get down to it, what we are hating are DNA molecules."

Noticeable traits such as skin color, hair texture, and facial features traditionally have been used to assign people to one race or another. At various times in this century, anthropologists have divided the human species into as few as two races (straight-haired folks or wooly-haired folks) and as many as thirty races based on an arbitrary selection of traits. (The expression *human race* has come into common use, but nevertheless, human beings are a species, not a race.)

Analysis of polymorphisms in human populations shows that most of the genetic variability in the human species is found among persons who look alike—not among people of different races. To begin with, all people, regardless of how they look, are genetically identical (monomorphic) at 75 percent of all loci. This means that differences (polymorphisms) are found in only about 25 percent of all loci. By studying polymorphisms in different human populations, geneticists calculate that about 85 percent of all the different alleles in the entire species are found within small populations.

As Richard C. Lewontin, professor of population genetics and natural history at Harvard University, explains:

> Genetic variation between one Spaniard and another, or between one Masai and another, is 85 percent of all human genetic variation. . . . If everyone on earth became extinct except for the Kikuyu of East Africa, about 85 percent of all human variability would still be present in the reconstituted species.

Thus, the term *race* as applied to human populations is a cultural and political concept that, when examined scientifically, is not justified by the overwhelming genetic similarities among people.

The frequencies of most alleles in persons with either black or white skin are quite similar. For the five polymorphic loci listed in Table 19-4 (keeping in mind that 75 percent of *all* enzymes are identical), differences are slight. They certainly could not be used to predict whether a person came from a black or a white population. However, there are a few alleles among the hundreds of loci that have been examined that do show marked differences in frequencies between races. For example, one allele of the Duffy blood group is never found in black populations, whereas another Duffy allele is virtually absent in all white populations. Although one allelic difference is not sufficient to unambiguously assign a person to a black or white population, selection of several polymorphisms that are present in high frequency in one population but absent or present in low frequency in another does permit the assignment of persons to a particular population.

The significance of human differences, whether they pertain to genes, race, sex, or intelligence has been controversial in the past

FIGURE 19-14 Human faces show great diversity despite the fact that all people belong to a single species. (Courtesy of the United Nations.)

and is likely to be so in the future. In 1964, biologists working for the United Nations (UNESCO) issued a report on race that is still scientifically accurate.

1 All humans living today belong to a single species, *homo sapiens*, and are derived from a common stock.

2 Pure races do *not* exist in the human species.

3 Differences between individuals within a race are often greater than the average differences between races.

TABLE 19-4 Differences in allelic frequencies at five polymorphic loci in European and African populations. In the five enzymes coded by these genes, the differences between blacks and whites are negligible.

Locus (enzyme)	Europeans			Africans		
	Allele 1	Allele 2	Allele 3	Allele 1	Allele 2	Allele 3
Adenylate kinase	0.95	0.05	0.00	1.00	0.00	0.00
Peptidase A	0.76	0.00	0.24	0.90	0.10	0.00
Peptidase D	0.99	0.01	0.00	0.95	0.03	0.02
Adenosine deaminase	0.94	0.06	0.00	0.97	0.03	0.00
Phosphoglucomutase-I	0.77	0.23	0.00	0.79	0.21	0.00

4   From the biological point of view, it is not possible to speak of a general inferiority or superiority of this or that race.

The study of human genetic differences is in itself a valid scientific pursuit and yields useful data on the genetic variability of existing human populations. It is sad indeed that the division of people into racial, ethnic, religious, and national groups has resulted in much human suffering. Once again, it is the human use of information that can be dangerous, not the information itself.

Human Origins

All the molecular evidence accumulated since Darwin's time—the close relatedness of chromosomes, proteins, and genes between human beings and apes—supports Darwin's basic hypothesis that both have a common ancestor and are closely related. The fundamental question raised by human-ape biological similarities is how there can be such an enormous difference in phenotype when there is so slight a difference in the genotypes of human beings and apes. In seeking answers to this question, anthropolo-

gists have searched for fossil evidence that might enable them to unravel the course of hominid evolution. In Darwin's time, hominid fossils were virtually unknown; in recent years, however, intensive efforts in searching for fossils have established the existence of several hominid species in Africa dating from 3 million to 4 million years.

All primates constitute a mammalian order comprising human beings and extinct hominids, apes, monkeys, and smaller, more-primitive animals such as the lemur. Hominids refer specifically to members of the human family Hominidae. What concerns anthropologists today is not whether hominids and apes evolved from a more-primitive species, but rather reconstruction from the fossil record of the particular path of evolution that culminated in *Homo sapiens*. As mentioned earlier, evolutionists debate the relative importance of evolutionary mechanisms such as natural selection or punctuated equilibrium in the formation of new species. However, the facts of evolution are not being questioned in these debates. Similarly, anthropologists argue and haggle over the particular evolutionary tree that they favor. All anthropologists agree, however, that human beings and apes evolved from a common ancestor within the past 10 million to 15 million years.

Chimpanzees (and gorillas) are genetically very closely related to modern human beings, as evidenced by the dramatic similarities in their chromosome banding patterns and number of chromosomes (Figure 19-15). The close genetic relatedness of human beings and chimpanzees is revealed by the similar appearance and behaviors of their young, although subsequent growth and development in the two species are totally different (Figure 19-16). The specific genetic changes that led to the difference in the evolution of apes and modern human beings are unknown. However, differences in gene regulation that affect development of the embryo, particularly brain development, must be largely responsible for the great differences between a person and a chimpanzee.

All human beings belong to a single human species and human populations interbreed freely with one another in nature. Evolution *could* have given rise to more than one human species, just as it produced more than one species of closely related apes (chimpanzees and gorillas). The fact that one human species has come to dominate all other species on earth has profound implications for further human evolution and survival.

In thinking about the evolution of human beings and other species, it is worth noting that there are no higher or lower forms of life in nature. The designations "higher" and "lower" have as their bases human value systems and human perceptions of nature. All organisms in nature evolve to become as perfectly adapted to a particular environment as their genes allow. As populations evolve genetically and as environments change, some species become extinct and others emerge to replace them. As mentioned earlier, the fossil record shows that more than 99 percent of all species that ever existed

on earth are now extinct, and there is no reason to suppose that the fate of the human species will be different. Even the dinosaurs, who lasted for approximately 150 million years, eventually became extinct.

That the human species may eventually become extinct is a conclusion that many people find hard to believe or accept for a variety of reasons. Many people believe that human beings are separate from the biological forces that contributed to evolution and to the extinction of species. However, understanding human existence and evolution in a biological sense can also provide a perspective that may contribute to our survival. Human beings are part of nature and depend on the preservation of the earth's environment, including other species, for continued survival.

The controversies over evolutionary facts and theories will no doubt continue in the future, particularly as they relate to human origins and human evolution. The relative importance of natural selection, sexual selection, neutral mutations, genetic drift, and founder effects in biological history are still being studied and argued. Anthropologists argue about hominid lineages, evolutionists debate the merits of gradualism compared with punctuated equilibrium, and molecular geneticists are concerned about the role of introns and transposons in evolution. To the bewildered nonscientist, these concerns and disputes may seem to be emanating from a Tower of Babel where one scientist cannot understand what another is talking about. However, despite the acrimony and controversy, today more is understood about the evolutionary history of human beings and other organisms than ever before.

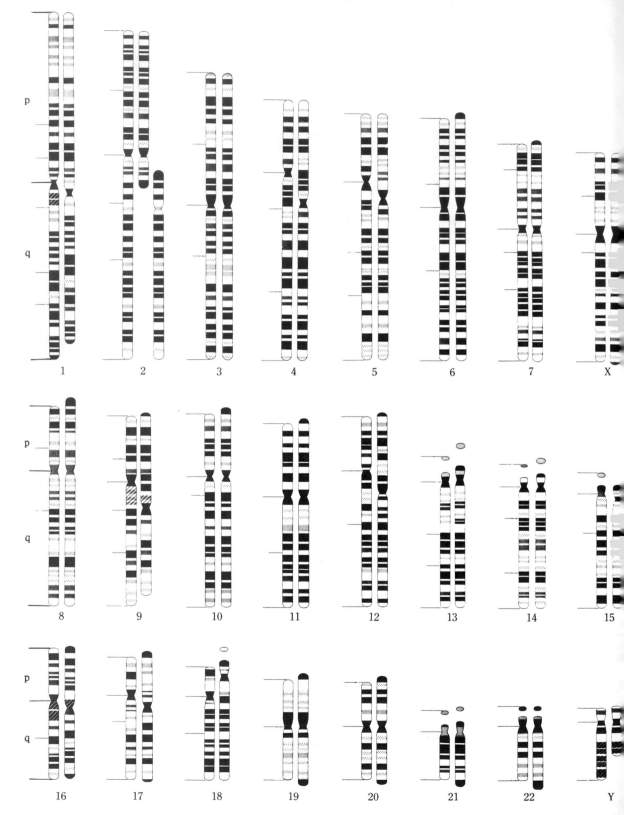

A

B

FIGURE 19-16 Young apes exhibit many behaviors that are also observed in human infants and children: (A) a young chimpanzee clings to its mother's neck; (B) a young gorilla learns to beat his chest with his fists. (Photographs by Phillip Coffee/Jersey Wildlife Preservation Trust.)

◀ **FIGURE 19-15** A schematic representation of human chromosomes (at the left) and chimpanzee chromosomes (at the right). The haploid number of chromosomes in chimpanzee is 24; two chimpanzee chromosomes fused in the course of evolution to produce chromosome number 2 in human beings. Karyotype analysis of the chromosomes of the two species shows that approximately a thousand bands are identical in both species. [Redrawn from J.J. Yunis, J.R. Sawyer, and K. Dunham, *Science*, 208 (1980): 1147.

Summary

The biological history of the earth spans approximately 4 billion years, but modern human beings constitute only a small fraction of that history, having existed for about 35,000 years. The biological history of human beings and other organisms can be reconstructed not only from the fossil record, but also from comparisons of the macromolecules of living organisms.

Radioactive elements (radioisotopes) present in rocks and fossils serve as "atomic clocks" for measuring age. The nuclei of radioisotopes decay at a constant rate, emitting energy or particles such as protons and neutrons. The emissions are measured as radioactivity. As radioisotopes decay, certain stable isotopes form as breakdown products. The amounts of these stable products formed relative to the amount of the radioisotope element lost in a rock or fossil sample enable researchers to determine its age.

The diversity and age of the fossil record, including both large fossils and microscopic ones, tells us that all life forms on earth have been in a continual state of evolution for billions of years. Carbon-14 is the radioisotope used for dating relatively recent organic materials — that is, those of the past 40,000 years. All living organisms assimilate carbon from the environment, including carbon-14 and carbon-12, and the carbon atoms remain in bones and tissues when an organism dies. The ratio of carbon-14 to carbon-12 present in a fossil give its age.

Similarities and differences between genes and proteins in various species lend strong support for evolution. A comparison of the amino acid sequences of a protein common to many species reveals how closely or how distantly related two species are: if amino acid differences are extensive, the two species are separated by a long period of evolutionary time, whereas closely related species have few differences in amino acids. Estimates of mutation rates that caused the accumulated differences in amino acid sequences are used to calculate the time of divergence between species. Thus, a "molecular clock" can be used to measure rates of evolution.

Human hemoglobin proteins and their genes provide a molecular record of human evolution. A comparison of amino acid sequences in human hemoglobin with those of other primates reveals the degree of relatedness.

DNA-DNA hybridization is another technique for measuring similarities and differences in genes. DNA molecules from two different species are melted apart into single strands and mixed together. The extent to which single strands from the different organisms come together to form stable double helices is a measure of their relatedness.

Polymorphism refers to the existence of several forms of a given trait (or of an allele) in a population; thus, polymorphisms can be used to measure the population's genetic diversity. Contrary to popular belief, human polymorphisms are more numerous among individual members of a race than between races.

Although the specific genetic changes that led to the evolution of human beings are unknown, we do know that higher primates are closely related. Human beings, chimpanzees, and gorillas evolved from a common ancestor in the past 10 million to 15 million years.

Key Words

DNA-DNA hybridization   A technique used to measure the similarity in base sequences between DNA molecules from different organisms.

hominid   A primate whose characteristics are more humanlike than apelike.

paleontology The study of the life forms of past geological periods by means of the fossil record.

phylogeny The relationships between groups of organisms that are revealed by their evolutionary history; a
family tree.

primates A mammalian order comprising human beings, apes, monkeys, and certain more-primitive animals, such as the lemur.

stromatolite Large, ancient, rocklike structure formed
by bacteria and the minerals trapped by the bacteria.

Additional Reading

Amato, I. "Tics in the Tocks of Molecular Clocks," *Science
News*, January 31, 1987.

Cloud, P. *Oasis in Space: Earth History from the Beginning.*
Norton, 1988.

Godfrey, L.R., and J.R. Cole, "Blunder in Their Footsteps."
Natural History, August, 1986.

Gould, S.J. *The Panda's Thumb.* Norton, 1981.

Hawaiian Evolutionary Biology. *Trends in Ecology and Evolution*, July, 1987.

Jolly, A. "The Evolution of Primate Behavior." *American
Scientist*, May-June, 1985.

Lambrect, F.L. "Trypanosomes and Hominid Evolution."
BioScience, November, 1985.

Lewin, R. *Bones of Contention.* Simon and Schuster, 1987.

Lewontin, R.C. "Are the Races Different?" *Science for the
People*, March-April, 1982.

Martin, L. "Which Ape Is Man's Closest Kin?" *The Sciences*,
March-April, 1988.

Morell, V. "Announcing the Birth of a Heresy." *Discover*,
March, 1987.

Patrusky, B. "Mass Extinctions: The Biological Side." *Mosaic*, Winter, 1986-87.

Vigue, C.L. "Murphy's Law and the Human Beta-Globin
Gene." *American Biology Teacher*, February, 1987.

Study Questions

1 What is the most-accurate way to measure the age of a
fossilized rock or skeleton?

2 Rank the following fossils from oldest to most recent:
dinosaurs, stromatolites, wooly mammoths, Lucy.

3 What protein has the same sequence of amino acids in
human beings and chimpanzees?

4 What do geneticists mean by the term race? Polymorphism?

5 Is it true that species no longer become extinct?

6 About what fraction of genes are likely to be identical
in two persons picked at random from anywhere in
the world?

7 Why are all persons alive today classified as belonging
to the same species?

Essay Topics

1 Discuss your own views of hominid evolution based on
your personal beliefs and the facts presented in
this chapter.

2 Explain how the fossil record allows scientists to reconstruct biological history and evolution.

3 Describe one or more molecular techniques that are
used to establish the degree of relatedness among
different species.

20

Heredity and Environment

Interactions and Issues

Just as we swallow food because we like it and not because of its nutritional content, so do we swallow ideas because we like them and not because of their rational content.

RICHARD ASHER, *psychologist*

EVERY INDIVIDUAL organism is different from all others in genes, traits, and behaviors. Even identical twins are not phenotypically the same in all respects; a casual observer will detect physical and behavioral differences that distinguish one twin from the other (Figure 20-1). Similarly, genetically identical plants—those that have been cloned—will not grow at the same rate and will not have the same amount of leaves, flowers, stems, or roots. Nor are ants or even bacteria phenotypically identical, although the differences between individuals may not be readily apparent. Taken all together, these observations indicate quite clearly that genes alone do not determine the physical traits and behaviors of organisms; nongenetic factors also play a role.

The nongenetic factors that help to determine the phenotypes of organisms are collectively called the **environment**. For plants, the environment includes such factors as sunlight, water, nutrients, diseases, and pests. For human beings, the environment includes all these factors and many more—not only the physical, chemical, and biological components of the environment, but also the many social and psychological factors that shape our lives. Love, sorrow, pity, loneliness, and anger can and do strongly affect our development and traits.

All of the genetic instructions essential for human development, growth, and reproduction are contained in each person's chromosomes, but the expression of genes is affected by environmental factors to a greater or lesser degree. For example, the presence of even small amounts of certain chemicals in the body of a pregnant woman can drastically alter normal gene expression and development in the embryo; small amounts of alcohol at crucial stages of embryonic development, for instance, adversely affect physical and mental development.

Having "normal" genes guarantees neither normal development nor a normal phenotype. A genetic constitution that contains "flawless" information is like a perfectly recorded tape of music: even if the most-advanced techniques are used to record the tape, the quality of the sound will still depend on the quality of the playback equipment. A cheap cassette player will not produce high-fidelity music; weak batteries or a torn speaker cone will distort the sound even though the recorded tape is perfect.

The opposite effect may also be obtained. The sound on an imperfectly recorded tape or record can be greatly improved by electronic techniques that suppress noise and other unwanted sounds while enhancing the overall musical quality. In an analogous fashion, expression of an "imperfect" genetic constitution may be markedly improved in the appropriate environment. The question, then, is: What are the *relative* contributions of genome and environment to a particular human trait? This question is at the center of all controversies about *nature* (heredity) versus *nurture* (environment).

Breeders have known for centuries that the characteristics of plants and animals can be altered by selectively breeding for desired traits and by controlling environmental factors. In the past century, plant and animal breeders have developed methods for quantifying the relative contributions of genes and the environment to economically important traits in their crops and flocks. For example, the yield of grain from wheat, the protein content in corn, and the crispness of lettuce are traits that can be increased by selective breeding. However, a trait

FIGURE 20-1 Identical twins have identical genes but their appearances may be different. Note that in this photograph one twin is looking at the camera while the other twin looks down. Is this behavior determined by genes (nature) or environment (nurture)? (Photograph courtesy of Steve Dunn.)

can be improved by breeding only if it is largely genetically determined. If, for example, 95 percent of a trait (such as sickness due to a viral infection) is due to environmental factors, it would be futile to attempt to alter it by breeding.

This chapter describes how geneticists measure the relative contributions of both genes and the environment to plant and animal traits. It also examines some of the scientific problems and social consequences of human sociobiology and eugenics.

Biological Determinism

All human traits, including weight, strength, height, sex, skin color, hair texture, fingerprint pattern, blood type, intelligence, and aspects of personality (for example, temperament), are ultimately determined by the information encoded in DNA. However, environmental factors modify most physical and behavioral traits to some degree. Although the environment plays almost no role in modifying blood type or fingerprint pattern, it strongly influences intelligence and personality. If a person is genetically programmed for type AB blood, environmental factors cannot cause that person to have type O blood. Similarly, the environment cannot alter a person's fingerprint pattern after birth. However, we know that factors in the uterine environment affect fingerprints because even those of identical twins are not identical. And, although the expression of many genes helps to determine a person's height, weight, and strength, all of these traits are greatly affected by environmental factors.

Pellagra
Tragic Consequences of the Idea of Biological Determinism

BOX 20-1

In 1914, the U.S. Public Health Service gave epidemiologist Joseph Goldberger the task of discovering what causes pellagra, a debilitating and often fatal disease that was commonly found among among poor and institutionalized people, especially in the South, during the first half of this century. Pellagra, which means "angry skin," is characterized by rough red skin eruptions, and its more serious symptoms include diarrhea, dizziness, lassitude, and, in advanced cases, "feeblemindedness" and death.

Goldberger, an expert on infectious diseases, quickly disproved the idea that pellagra was caused by some infectious agent. He inoculated monkeys with blood and tissues from human victims of pellagra, yet none of the monkeys showed any symptoms of the disease. A few years later, after Goldberger became completely convinced that pellagra was a "disease of poverty" and was simply the result of poor nutrition, he injected blood and skin from pellagra patients into himself, his wife, and other human volunteers to emphasize his claim that pellagra is a vitamin B deficiency disease.

Goldberger's discoveries regarding the dietary cause of pellagra were reported in the leading scientific journals between 1914 and 1917, and he was acclaimed by his medical colleagues for his discoveries. But despite the irrefutable evidence that Goldberger had amassed, nothing was done to eradicate pellagra in the United States for more than twenty-five years. Millions of economically disadvantaged Americans continued to suffer, and hundreds of thousands died from pellagra itself or from their increased susceptibility to infections.

Why did the truth about the causes of pellagra remain hidden from general public view for at least a generation? The explanation involves complex social and economic policies, but a primary reason can be traced to the efforts of a group of American geneticists who believed in *eugenics*, a branch of genetics that advocates the improvement of human traits and behaviors by selective breeding. These eugenicists, whose leader and spokesman was the influential geneticist Charles Benedict Davenport, believed that most human traits and behaviors are genetically determined.

Davenport headed the Pellagra Commission, an influential body set up by the New York Medical School in 1912 to study the causes of pellagra. In the Commission's report, published in 1917, Davenport and other members all but ignored Goldberger's widely acclaimed discoveries, and as late as 1920 Davenport published an article in the *Journal of the American Psychological Association* claiming that feeblemindedness, criminality, and pellagra were all genetic conditions. Davenport's views, along with those of other eugenicists, greatly influenced public health policies in the United States in the first half of this century and—tragically—prevented the dietary changes that would have alleviated great human suffering and saved the lives of many Americans who were too poor to afford adequate diets.

READING: Allan Chase, "The Great Pellagra Cover-Up," *Psychology Today*. February, 1975.

Biological determinism is the idea that the attitudes and behaviors of a human society are due more to its biology than its cultural history—that is, most variation in human traits and behaviors can be attributed to differences in the genes that people carry (see Box 20-1). Proponents of biological determinism believe that social status, intellectual ability, and even sexual preferences are principally due to hereditary differences. It follows from the notion of biological determinism that "genes are destiny"; for each person, his or her genes determine the biological characteristics, the behaviors, and the place of that person in society. Although even the strictest biological determinists do not deny the effects of environment (nurture), they assert that *most* of the variation in human traits and behaviors (such as intelligence, laziness, homosexuality, selfishness, aggression, and so forth) is due to the alleles that people carry. The nondeterminist does not deny the role of heredity but asserts that environmental factors are exceedingly important and override the particular alleles in determining human traits and behaviors.

The crucial problem for geneticists has been to devise experiments that quantify the separate contributions of heredity and environment for a specific trait. For plants and domestic animals, the appropriate genetic experiments are straightforward and have been carried out on many different species. However, the same experiments cannot be performed on people because it would be immoral (and illegal) to cross human beings and manipulate their environments, as is done with plants and domesticated animals. To evalute the pros and cons of biological determinism, we must first know something about the different kinds of genetic experiments that measure the hereditary and environmental contributions to a trait. After examining how these experiments apply to

plants, it will be possible to critically examine the use of alternative genetic techniques in human populations.

Definition of Heritability

The behavioral and physical traits of each person have a strong genetic basis. Without the appropriate genes, a person would not develop nerve cells, a brain, or a mind. Therefore, intelligence, personality, and even emotions are hereditary in the sense that genes are required for these human traits to exist. These traits vary from one person to the next but, before we can say that the variation has a genetic basis, we need a method for measuring *how much* of the variation is determined by different genes and *how much* by different environments.

Heritability (which is *not* the same as heredity) is a measure of the amount of phenotypic variation in a population that can be attributed to genetic differences. Heritability is not a direct measure of the genes that a person carries; rather, it is a measure of the amount of the phenotypic differences that can be accounted for by genetic differences. Heritability is a quantity that can vary between zero and one; a value of *zero* means that environmental factors alone cause all the phenotypic variation among individual people, whereas a value of *one* means that all the variation is due to genetic differences between them.

Most normal polygenic-multifactorial traits such as stature, skin color, personality, intelligence, blood pressure, cholesterol levels, and so forth, show a continuous variation of values in a population within certain limits. The number of genes and the environmental factors that determine these traits are unknown. However, heritability values are a source of insight into the relative genetic contribution to the overall variation to the extent that the methods for measuring the heritability of human traits are valid. The oft-stated claims that intelligence, high blood pressure, and schizophrenia are genetic diseases stem from heritability measurements whose validity is controversial.

Mathematically, the heritability of a trait is defined as the amount of phenotypic variation resulting from genetic differences divided by the total amount of variation (environmental plus genetic) for that trait observed in the population. Heritability is expressed by the formula

$$H = \frac{V_G}{V_T} = \frac{V_G}{V_G + V_E}$$

in which the symbols are defined as follows:

H = heritability
V_T = total phenotypic variance in the trait
V_G = phenotypic variance due to genetic differences
V_E = phenotypic variance due to environmental differences

(The measure of variation is called the *variance* in statistics.)

Although heritability is a well-defined concept, in practice it is often difficult to measure. In plants, heritability can generally be measured accurately because inbred lines can be constructed in which all the genes governing the trait are all identical (homozygous). In a cross between inbred plants, all of the F_1 progeny are genetically identical, and so all of the variance in the trait must be environmental (V_E). However, in the F_2 generation, the plants are genetically heterogeneous because of segregation of the genes, and the variance is caused by both environmental and genetic differences (V_T). Because V_T and V_E can be measured, V_G can be calculated ($V_G = V_T - V_E$).

It is also important to realize that heritability cannot be determined for traits that do *not* vary among individuals, even though such traits are clearly determined by genes. An obvious example of a genetically determined trait whose heritability cannot be measured is the number of organs in a body. Every person has one nose. However, because there is absolutely no variation in the number of noses in a population, regardless of differences in the environments, heritability cannot be calculated for this trait. On the other hand, the shape of the nose does vary in human populations; so the phenotypic variation can be measured (at least in principle) and the heritability can be calculated (Figure 20-2).

Measuring Heritability in Plants

The concept of heritability is best illustrated with plants, for which the genotype and the environment can be manipulated experimentally. Early in this century, the Danish plant geneticist Wilhelm Johannesen showed that variation in the weights of beans could be partitioned into genetic and environmental components. Johannesen measured the variance in bean-seed weights from bean plants grown in a uniform environment and plotted the weight distribution (Figure 20-3). He realized that, if the variation in bean weights among plants was due partly to particular alleles, he should be able to increase the frequency of those genes by selecting and crossing particular plants. Johannesen was successful in selectively breeding bean plants that would yield a greater weight of beans per plant. By crossing plants that produced heavier beans, he produced strains whose average bean weight was greater than the average weight in the original parent population.

If the heritability of bean weight were high—that is, if the variation were due to different genes in the population—the new statistical

FIGURE 20-2 In 1910, Francis Galton developed a method for quantifying important features of peoples' profiles. For each person, a profile, like that shown at the left, is traced from a portrait, and the profile on the right is derived from a formula developed by Galton. Galton concluded, "Peculiarities of profile, as a racial or family characteristic, can be expressed numerically by an extension of this system in a way that promises to be serviceable for eugenic records." (Reprinted by permission from *Nature*, vol. 83, pp. 127–130. Copyright © 1910 Macmillan Magazines Ltd.)

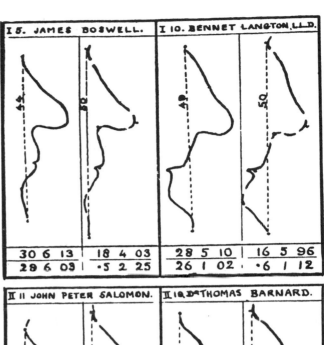
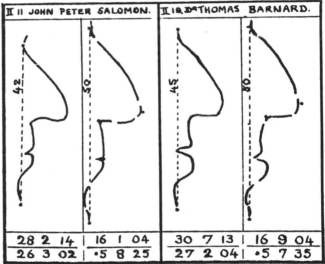

mean would be close to the mean of the selected parent plants. If the heritability were low, the new mean would not be appreciably different from the mean of the original parent population. In fact, the heritability of bean weight can be calculated by comparing the statistical means of the parent and offspring populations, as shown in Figure 20-4. The heritability for plants is expressed by a modified formula, $H = G/D$, in which

H = heritability
G = the selection gain
D = the selection differential

(See legend for Figure 20-4 for definitions.)

Johannesen showed that many traits in plants have a high heritability and the frequency of particular traits can be increased in a population

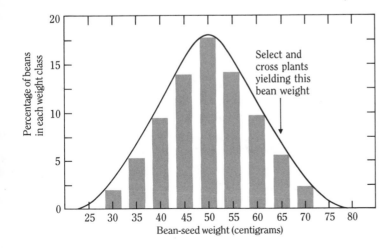

FIGURE 20-3 Seed-weight variance in a population of bean plants. The plants are grown in a uniform environment. If the heritability of seed weight has a large genetic component — that is, if V_G is large — crossing plants from the heavy end of the distribution should increase the average weight of seeds in the next generation.

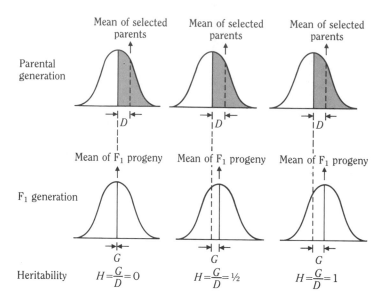

Heritability

$$H = \frac{G}{D} = 0 \qquad H = \frac{G}{D} = \frac{1}{2} \qquad H = \frac{G}{D} = 1$$

FIGURE 20-4 Measuring heritability in plants. Plant breeders calculate a value (between 0 and 1) for the heritability of a trait by measuring the differential selection (D) in the parent plants and the selection gain (G) in the F_1 generation. The value for D is determined by measuring the difference between the mean in the parent plants selected for the cross and the mean in the population. For example, if the average height of the plants is 12 feet and the plants selected for the cross are 16 feet tall, then $D = 4$ feet. Similarly, the value for G is obtained from the difference in the average height of the F_1 generation plants and the average height of the parental plants. If the mean of the F_1 generation progeny is the same as that of the parents selected for the cross, the heritability is 1. This shows that all of the variation in height is due to genetic differences among plants in the population. If the mean of the F_1 generation plants is unchanged from the mean of the parent population, the heritability is 0. This shows that all of the variation in height is due to environmental factors.

by crossing individual plants with the desired characteristics.

Heritability is a useful measure for plant breeders, who use it in their efforts to improve crops by selective breeding. Heritability is more difficult to measure in animal populations because appropriate crosses are more difficult and environments cannot be rigidly controlled (Table 20-1). These difficulties are even greater in human populations. Thus, human heritability measurements are unreliable and sometimes result in unwarranted conclusions that appear to support the idea of the biological determinism of human behavior. More will be said about this later in the chapter.

Environments Affect Heritability

In each experiment that measured the variance of bean weights, heritability was determined for a particular population of plants grown in a uniform environment. Because heritability is a function of both genes and the environment,

TABLE 20-1 Approximate heritability values for various traits in animal populations. High heritability values mean selection is possible; low values indicate that the trait is largely determined by environmental factors. Errors in heritability values range from 2 to 10 percent.

Trait	Heritability
Cattle	
Adult body weight	0.65
Milk yield	0.35
Pigs	
Weight gain per day	0.40
Litter size	0.05
Chickens	
Egg weight (at 32 weeks)	0.50
Egg production (to 72 weeks)	0.10
Mice	
Tail length (at 6 weeks)	0.40
Body weight (at 6 weeks)	0.35

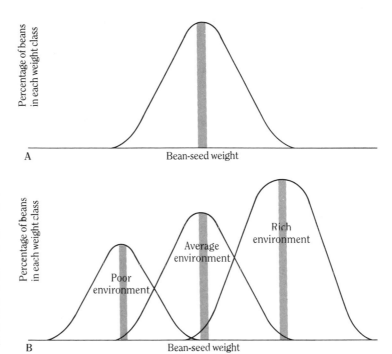

FIGURE 20-5 Variation in the average distribution of bean weights when plants are grown in different environments.
A. A large, random population of bean plants grown under average field conditions gives the distribution shown.
B. If seeds from the same batch are cultivated under three different environmental conditions, the average distributions of bean weights are quite different. Three different heritability values are obtained, even though all the original seeds are from the same batch and, on the average, have the same genes.

changing the environment in which the plants are grown would also change the heritability value. Heritability comparisons are valid only *within* the populations in which the measurements are made; it is a fundamental (and frequent) error to apply a heritability value obtained in one population to a different population.

To understand why a particular heritability value cannot be assigned to different populations, consider a hypothetical experiment. If a large, random sample of bean seeds is collected from a genetically heterogeneous population of bean plants grown in a field with uniform conditions of soil, water, and sunlight and these beans are then planted in the same field, the distribution of bean weights will be about the same in the parent and progeny populations. However, if the same sample of beans is planted in a rich environment—for example, in a field that is heavily fertilized and optimally watered—the average bean weight and the varia-

tion in bean weight of the progeny plants will be greater than that of the parents (Figure 20-5). Similarly, if the beans are grown in a poorer environment than the parental one, the average weight and the variation in bean weight will be lower. Thus, if the environment changes, the heritability also changes, even though the overall genetic constitution of the three populations is identical. Thus, we see that heritability comparisons are valid only *within* populations that are exposed to similar, and if possible identical, environments. This situation is rarely possible to achieve in human heritability studies; so invalid comparisons are often made between different human populations (as in comparing the heritability of intelligence between black and white human populations).

This example shows that heritability values must be interpreted cautiously. In many instances—particularly those in which the heritability of human traits is being measured—the

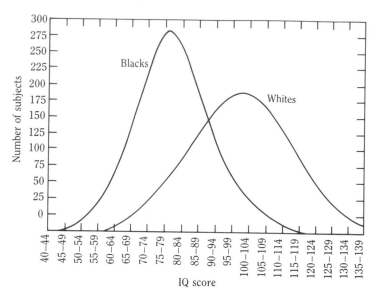

FIGURE 20-6 A typical distribution of IQ scores obtained from tests administered to black and white school children. The curves are based on scores from a number of studies. The average difference in IQ score distribution is from about 15 to 20 points between the two groups. Note, however, the considerable overlap in the two distributions: many blacks scored higher than many whites. Moreover, these data do not indicate how much of the difference in the distributions is due to genetic factors and how much to environmental factors.

genetic and environmental components of phenotypic variation simply cannot be separated from each other. An attempt is made to do so in human studies of identical twins, but, as will be discussed later, these studies have serious flaws. The genetic manipulation of plants can be controlled and plants can be grown in well-defined, uniform environments so that, generally speaking, heritability values are accurate and are useful for improving traits in crops. However, when applied to human populations, heritability measurements are not nearly so reliable. This unreliability makes heritability measurements of human traits a dangerous business, particularly because the values obtained (which may be incorrect) can have serious consequences for individual people and for society if such data are used to influence social attitudes and governmental policies.

Heritability and Intelligence

There is no doubt that genes affect intelligence to some degree. Mutations in any one of hundreds of human genes can result in severe mental retardation and lowering of intelligence. However, environmental factors also play an important part in determining intelligence. Some environmental factors that affect intelligence are nutritional deficiencies, elevated levels of heavy metals such as lead or mercury in the body, and educational deficiencies. Thus, the quesion again is, To what degree is the *variation* in intelligence in a human population the result of different genes or different environments?

The discussion that follows assumes that intelligence is a trait that shows phenotypic variation and that it can be measured by an intelligence (IQ) test. The question "Is intelligence inherited?" then becomes "Can the genetic component of intelligence be separated from other positive or negative environmental factors that also affect IQ?"

Virtually all of the IQ controversy in the United States derives from studies in which the average IQ of black populations has been found to be consistently lower than the average IQ of white populations (Figure 20-6). Despite claims that the IQ tests were given to persons who were raised in similar environments, it is impossible to define what would constitute a "similar" environment for the development of a human trait as complex as intelligence. Moreover, we have seen why comparisons of heritability values between populations exposed to different environments are not regarded as valid by geneticists.

The heritability of human intelligence has been estimated by some psychologists to be as high as 0.8, on the basis of studies of the IQs of identical twins who were raised separately from each other after birth. Acceptance of such a high estimate for IQ heritability leads to the conclusion that differences in human intelligence are mainly genetically determined. Such a conclusion would have devastating personal, social, and political implications for American blacks because no amount of educational help could be expected to significantly improve their overall performance on IQ tests. It could thus be argued (erroneously) that educational-enrichment programs are a waste of time and money. It is important to place the IQ controversy in historical perspective. Measurements of intelligence have been shown to be mired in fraudulent science for the past two centuries. This constitutes one of the most-bizarre chapters in the history of science (see Box 20-2).

The issue of race and intelligence was of such importance in the United States in the 1970s that members of the Genetics Society of America took a position on the issue and endorsed a statement on race and intelligence, a part of which stated:

> It is particularly important to note that a genetic component for IQ score differences *within* a racial group does not necessarily imply the existence of a significant genetic component in IQ differences *between* racial groups; an average

difference can be generated solely by differences in their environments. The distributions of IQ scores for populations of whites and of blacks show a great deal of overlap between the races, even in those studies showing differences in average values. Similar although less severe complexities arise in consideration of differences in IQ between social classes. It is quite clear that in our society environments of the rich and the poor and of the whites and the blacks, even where socioeconomic status appears to be similar, are considerably different. In our views, there is no convincing evidence as to whether there is or is not an appreciable genetic difference in intelligence between races.

Twin Studies and Human Heritability

As already pointed out, the heritability of plant traits can be determined by appropriate genetic crosses in controlled environments. Although we do not cross or raise people like plants, nature does provide "experimental" human material in the form of twins. **Monozygotic (MZ) twins** develop from a single fertilized egg that divides very early in development to produce two genetically identical persons. **Dizygotic (DZ) twins** develop from two separate eggs that have been fertilized by different sperm, and DZ (fraternal) twins are genetically different from each other. If traits such as intelligence, personality, height, or susceptibility to certain diseases are primarily determined by genes, then a comparison of MZ and DZ twin pairs can be used to obtain a measure of heritability for these traits.

As stated in Chapter 11, the degree to which a trait is similar in pairs of MZ or DZ twins is called *concordance*. For most traits, MZ twins have a higher concordance than do DZ twins for the same trait. These findings are interpreted to

Fraudulent Science and IQ

BOX 20-2

In the nineteenth century, before the discoveries of Darwin and Mendel, an American physician named Samuel George Morton spent much of his life attempting to prove that people differ in intelligence according to their race. Morton, like many others, equated human brain size with level of intelligence (an idea that is unsupported by any evidence — Albert Einstein's brain was quite average in size, for example). Morton measured the brain sizes of human skulls that he collected from a variety of racial groups over a period of twenty years. He would carefully fill the skull with lead shot and afterward weigh the shot to determine the brain size.

Morton concluded from his data that the intellectual superiority of Caucasians was well substantiated by their larger brains. American Indians and African blacks had the smallest skulls, which, according to Morton, would account for their lower intelligence. Morton's results and opinions were highly regarded and widely accepted by whites in the nineteenth century. The New York *Tribune* wrote in his obituary that "probably no scientific man in America enjoyed a higher reputation among scholars throughout the world than Dr. Morton."

Morton's summary table of cranial capacity by race showed that the average Caucasian had a brain capacity 9 cubic inches greater than the average brain capacity of an Ethiopian and 5 cubic inches greater than the average American Indian. In 1978, Professor Stephen Jay Gould of Harvard University carefully reexamined all of Morton's calculations. Gould was able to show that Morton had, either inadvertently or deliberately, manipulated his data to conform to his prejudices. When Gould recalculated Morton's original data, the corrected values for cranial capacity showed essentially no differences among races (see the accompanying table).

Was Samuel George Morton prejudiced or simply careless? No one can say for sure, but an indication of Morton's racial views is given by his description of the Shoshonee Indians:

Heads of such small capacity and ill-balanced proportions could only have belonged to savages; and it is interesting to observe such remarkable accordance between the cranial developments, and mental and moral faculties. Perhaps we could nowhere find humanity in a more debased form than among these very Shoshonees, for they possess the vices without the redeeming qualities of the surrounding Indian tribes; and even their cruelty is not combined with courage. . . . A head that is defective in all its proportions must be almost inevitably associated with low and brutal propensities, and corresponding degradation of mind.

The disturbing saga of "scientific racism" continues into the twentieth century. An English psychologist, Cyril Burt, accumulated vast amounts of data that consistently supported the prevailing view of upper-class English society that intelligence is determined by heredity. Burt was the foremost expert on educational psychology in Britain from about 1913 to 1932, and he administered IQ tests to thousands of English children. The data collected by Sir Cyril Burt (he was knighted in 1946 for his "contributions") was cited by scholars for more than three decades as the bulwark of the scientific proof that human intelligence is at least 80 percent determined by heredity.

Burt died in 1971; by 1974, doubts were being raised about the validity and accuracy of his twin-study data. Much of the initial reinvestigation of the Burt data was done by an American psychologist, Leon Kamin. When Kamin discovered systematic flaws and discrepancies in the Burt studies and announced his findings, psychologists Arthur

Jensen and Richard Herrnstein came to Burt's defense. Jensen stated in 1977 that, "The central fact is that absolutely no evidential support for these trumped-up charges of fakery and dishonesty on the part of Burt has been presented by his accusers. The charges, as they presently stand, must be judged as the sheer surmise and conjecture, and perhaps wishful thinking, of a few intensely ideological psychologists." Herrnstein regarded the suggestion of fraud as "so outrageous I find it hard to stay in my chair. Burt was a towering figure of twentieth-century psychology. I think it is a crime to cast such doubt over a man's career."

By 1978, the "towering figure" had been toppled and the only "crimes" were those committed by Burt himself, who fabricated the data that were used to prove that human intelligence is hereditary. In the definitive biography of Burt published in 1979, L.S. Henshaw confirmed that Burt fabricated the data from IQ scores in twins, was deliberately deceptive, and even invented collaborators who never existed to support his own biased ideas about intelligence. All scholars and psychologists, including Jensen, Herrnstein, and other former supporters, now agree that the Burt data are fraudulent.

These two proven examples of scientific fraud have had profound educational and social consequences. Burt, for instance, set up the tier system of education in Britain, which excluded many students from a university education. Although admitting to these past scientific frauds, some scientists and psychologists argue that new studies still support the hypothesis that human intelligence is primarily hereditary.

READINGS: Stephen Jay Gould, *The Mismeasure of Man*. Norton, 1981. Stephen Jay Gould, "Morton's Ranking of Races by Cranial Capacity," *Science*. May 5, 1978. Nigel Hawkes, "Tracing Burt's Descent to Scientific Fraud," *Science*. August 17, 1979.

Cranial capacities of skulls as calculated by Morton and recalculated by Gould.

	Morton's values		Gould's values	
Race	Number of skulls	Cranial capacity (cubic inches)	Population	Cranial capacity (cubic inches)
Caucasian	52	87	Native American	86
Mongolian	10	83	Mongolian	85
American Indian	147	82	Modern Caucasian	85
Malay	8	81	Malay	85
Ethiopian	29	78	Ancient Caucasian	84
			African	83

mean that variation in these traits has a significant genetic basis. For example, the physical traits of MZ twins have a higher concordance than do the same physical traits in DZ twins, as might be expected given that MZ twins are genetically identical. Twin studies have been employed not only to show that variation in height or intelligence is largely biologically determined, but also to prove the genetic basis of other behaviors. Criminality, schizophrenia, obesity, and other traits such as susceptibility to tuberculosis (TB) are often said to be biologically determined on the basis of high heritability values obtained from twin studies. For example,

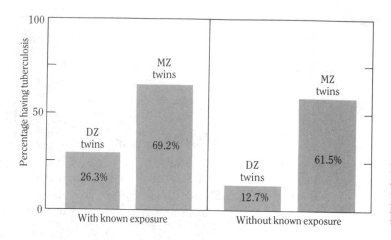

FIGURE 20-7 Concordance in identical (MZ) and fraternal (DZ) twins for tuberculosis. At the time at which these studies were performed, the higher concordance among identical twins was interpreted to mean that TB was a hereditary disease.

in the 1940s, the high concordance of tuberculosis in identical twins was cited by some people as proof that susceptibility to TB was genetically determined (Figure 20-7).

Once again, the issue is not whether particular genes may predispose certain people to contract tuberculosis but rather whether the relative amounts of genetic and environmental contributions can be measured by twin studies. The increased concordance of TB for identical twins compared with fraternal twins might be explained by differences in uterine development or by the more closely shared environment of MZ twins, which exposes each twin to the same infections to a greater extent than each member of a DZ pair is so exposed. Deaths from TB have declined dramatically in the United States not by changing genes but by eliminating or controlling the environmental factors that cause TB.

Overall, genes in the U.S. population have not changed in the past eighty years, nor has the causative agent of TB, *Mycobacterium tuberculosis*, disappeared from the environment. What *has* changed are the diets, sanitary habits, and living conditions of people. The high heritability seemingly demonstrated for TB from the degree of concordance in twin studies could have led to the conclusion that the incidence of TB could not be changed by changing the environment — a conclusion that not only would have been false, but also detrimental to curtailing the spread of the disease.

The most-significant fact to emerge from twin studies through the years is that some people expend enormous effort to determine whether human behaviors and diseases are genetically determined. Examination of all the scientific evidence indicates unambiguously that the environment is responsible for TB and for criminality, yet these traits also have a high concordance in MZ twin studies. Thus, the degree of concordance does not reliably indicate the degree of genetic contribution. The environment clearly is important in the variation in intelligence among individual people and in susceptibility to TB and schizophrenia.

Many twin studies compare MZ twins that are raised together with MZ twins that are separated immediately after birth or when they are quite young. It is assumed that the separated twins experience different environments, whereas the twins raised together are assumed to share the same environmental factors.

However, many uncertainties and problems exist even with these twin studies. For example, not all twins classified as MZ are actually identical; some may be DZ twins that happen to look alike. Without laboratory confirmation, one cannot be absolutely certain of zygosity. There is also confusion over what constitutes a different environment. Often separated twins are raised by close relatives and the environmental factors may be quite similar for each twin. Also important is the age at which the twins were separated, particularly for traits like personality, IQ, or criminal behavior, which may be affected at a very early age. Finally, no matter what the calculated heritability value turns out to be, for any person environmental factors may play the decisive role.

It is important to realize that, although we can change environments, we cannot change our genes. Twin studies continue to produce misleading results (and unfortunate social consequences) because the relative contributions of genes and the environment simply cannot be unambiguously separated and measured for complex human traits.

What Is Sociobiology?

The term *sociobiology* was introduced as early as 1946, but it attracted little notice until a book titled *Sociobiology: the New Synthesis*, by Harvard biologist Edward O. Wilson, appeared in 1975. Wilson, who is generally regarded as the founder and most-influential spokesman of sociobiology, defines it as "the scientific study of the biological basis of all forms of social behavior in all kinds of organisms, including man." Most of the book consists of descriptions of the biology, evolution, and social organization of insects and other animals. It is only in the final chapter that Wilson extrapolates his observa-

tions from animal studies to human behaviors and societies and postulates that such human behaviors as aggression, homosexuality, criminality, and morality are biologically (genetically) determined and are a consequence of evolution and natural selection.

The central thesis of **sociobiology** is that the behaviors of animals and societies are determined to a large degree by genes whose existence in present-day populations has resulted from evolutionary mechanisms, principally natural selection. Sociobiologists are ardent evolutionists and apply the Darwinian ideas of adaptation, fitness, and natural selection to explain animal behaviors whenever possible. One of the goals of sociobiology is to extend the principles of evolutionary and population genetics to human behaviors and cultures and to reorient such disciplines as anthropology, sociology, psychology, philosophy, and economics toward biological and evolutionary explanations.

From its very inception, sociobiology has been kept sharply criticized as yet another attempt to mask society's ills with genetic explanations. When Wilson's book appeared in 1975, critics attacked him for lending support to the biological determinists who sought to explain existing social evils by the presence of "defective" genes in human populations. Wilson's critics not only denounced the scientific flaws in his arguments—which they claimed were serious— but also pointed out the potentially dangerous social consequences of human sociobiology.

Wilson argued in response that he and other sociobiologists were concerned only with discovering the scientific truths underlying animal behaviors. He pointed out that establishing the genetic basis of human behavior was not their primary goal. However, in retrospect, the critics' warnings appear to have been justified, because three years later another book by Wilson appeared—*On Human Nature*—in which he did attempt to establish the biological basis for human social behaviors. The purpose of this book, Wilson asserted, "is simply the extension of population biology and evolutionary theory to social organization."

Too often nature-versus-nurture arguments are based on what appear to be reasonable inferences that may, however, be misleading. For example, some sociobiologists argue that, because many people in different cultures have phobias (intense fears) of snakes or spiders, these phobias must have a genetic basis and must have been selected for in the course of evolution. Yet such arguments ignore the fact that other phobias are caused by very recent inventions and modern social situations. For example, many people are phobic about flying in airplanes—a fear that cannot have been selected for in millions of years of evolution.

The furor over sociobiology that flourished in the 1970s seems to have faded with time. It is possible that in the 1980s it is easier and scientifically more rewarding to clone genes and sequence DNA than to measure human behaviors and heritability.

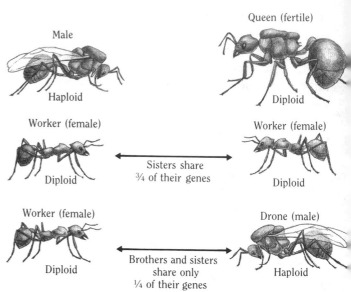

FIGURE 20-8   Insects such as ants exhibit altruistic behaviors. Because males are haploid and females are diploid, sterile female worker ants share, on average, three-fourths of their genes with one another. They share only one-fourth of their genes with their brother drones. By caring for the queen to help her produce more progeny, female worker ants pass on more of their genes to the next generation than if the workers themselves had progeny (see text for further explanation).

Altruism

An important demonstration of a biologically determined behavior is that of altruism found in some species of insects. **Altruism** is a behavior that benefits another individual at some expense to the doer. Altruistic behaviors can reduce the individual organism's chances of survival; hence, the organism's genes will not be passed on if reproduction has not yet occurred. For example, providing food for others while going hungry oneself is altruistic behavior; saving another person from death at the risk (or cost) of one's own life is an extreme form of altruism.

William D. Hamilton, a British scientist, showed that altruism can be explained genetically and, at least in theory, could have been selected for in the course of evolution. Hamilton showed that sister female ants are genetically more related to each other than to their brother ants or even to their parents (Figure 20-8), for the following altruistic reasons. Female worker ants are sterile and expend most of their time and energy feeding the queen (altruistic behavior) to help her produce more progeny ants. Because the female workers are sterile, they produce no offspring themselves. How, then, are their genes transmitted to future generations? The answer is that their altruistic behavior in supplying the queen with food and caring for the nest ensures that their genes are passed on to succeeding generations through the queen. All sterile females produced by the queen share three genes out of four with one another. Thus, the female ants' altruistic behavior is, genetically speaking, advantageous to them. If worker ants were not sterile and produced their own progeny, they would share only half of their genes with the next generation instead of three-fourths.

The evolutionary mechanism that explains altruistic insect behavior is called **kin selection**, a process that is different from sexual or natural selection. Darwin showed that new traits arise by natural selection and give an advantage to individual organisms who are reproductively more fit as a consequence of those traits. In kin selection, however, it is the genetic relatedness of the entire family group that is subject to selection, not the individual organism. Individuals per se are not important in kin selection. In fact, an individual organism's death may ensure that more of that organism's genes are passed on by its surviving relatives. For example, a male lion that is killed protecting his cubs may have ensured that his genes survive and are passed on by his offspring.

Years ago, John B.S. Haldane, a famous British geneticist, quipped that he would give up his life for three brothers or nine cousins. The mathematical logic of his statement derived from the number of genes that he shared with his brothers or cousins. The number of genes shared by various family members is expressed by a **coefficient of relatedness** (Table 20-2). If Haldane's relatives were to produce more progeny than Haldane himself did, then more of "his" genes would be present in the next generation (though they would be distributed among many persons).

To sociobiologists, the evolutionary mechanisms of kin and group selection are thought to

TABLE 20-2   Coefficients of relatedness (*r*) betwen specified relatives.

Relatives	r
Parent–progeny	½
Brother–sister	½
Grandparent–grandchild	¼
First cousins	⅛
Second cousins	¹⁄₃₂
Nephew or niece–uncle or aunt	¼

*Coefficients of relatedness express the mathematical probabilities of shared genes between close relatives. However, the calculations do not take into account the biological fact—derived from measurements of the number of polymorphic gene loci in human populations—that from 70 to 80 percent of all genes in unrelated persons are identical. Although coefficients of relatedness are useful in population genetics and for genetic counseling, the values are misleading. Close relatives are only a few percent more similar in their overall genetic constitution than are unrelated people.

be important in changing the frequency of genes in populations—particularly genes that affect behavioral traits. Individual organisms no longer need to be regarded as the only units on which selection acts, as envisioned by Darwin.

Eugenics: Breeding Better Human Beings

The eugenics movement began in England in 1883 when Francis Galton, a cousin of Charles Darwin, published a book titled *Inquiries into Human Faculty*. This was followed several years later by his book *Hereditary Genius*. In the first book, Galton defined **eugenics** (from the Greek *eu*, "good" or "well," plus *genes*, "born") as "the science which deals with all influences that improve inborn qualities of a race." The American eugenicist Charles B. Davenport gave this definition: "Eugenics is the science of improvement of the human race by better breeding."

The Race Betterment Movement Aims

To Create a New and Superior Race thru EUTHENICS, or Personal and Public Hygiene and EUGENICS, or Race Hygeine.

A thoroughgoing application of PUBLIC AND PERSONAL HYGIENE will save our nation annually:

1,000,000 premature deaths.

2,000,000 lives rendered perpetually useless by sickness.

200,000 infant lives (two-thirds of the baby crop)

The science of EUGENICS intelligently and universally applied would in a few centuries practically

WIPE OUT

Idiocy   Insanity   Imbecility   Epilepsy

and a score of other hereditary disorders, and create a race of HUMAN THOROUGHBREDS such as the world has never seen.

Evidences of Race Degeneracy

Increase of Degenerative Diseases - - - - { Cancer / Insanity / Diseases of Heart and Blood Vessels / Diseases of Kidneys / Most Chronic Diseases / Diabetes

Increase of Defectives { Idiots / Imbeciles / Morons / Criminals / Inebriates / Paupers

Diminishing Individual Longevity

Diminished Birth Rate

Disappearance, Complete or Partial, of Various Bodily Organs - - - - { According to Wiedersheim there are more than two hundred such changes in the structures of the body

FIGURE 20-9 Examples of eugenics proganda in the United States in the early years of this century. (From *Official Proceeding of the Second National Conference on Race Betterment*, August 4–8, 1915, Battle Creek, Michigan. Published by the Race Betterment Foundation, p. 147.)

From about 1870 to 1905, poverty, "feeble-mindedness," and insanity were commonly regarded as resulting from hereditary defects, and, between 1905 and 1930, the idea that *all* human weaknesses are the result of defective genes was widely accepted (Figure 20-9). The eugenics movement was particularly strong and influential in the United States in the early 1900s. Eugenic ideas, promulgated by geneticists and other scientists, were readily embraced by politicians, aristocrats, and successful businessmen who were quite ready to be convinced that their superior socioeconomic status was due largely to their superior genetic endowment. It also seemed quite logical to most people then (and to many even now) that the way to improve the human species was to eliminate the "undesirables" and "defectives"—or at least prevent them from breeding—while at the same time to encourage breeding among "superior" people. Eugenic ideas are not supported by any *scientific* evidence then or now.

Eugenic ideas are based on prejudice and on misinterpretation of genetic principles. Applications of eugenic policies have caused a great deal of individual suffering and have had serious political and social consequences through the years. For example, the passage of the Johnson-Lodge Immigration Restriction Act in 1924 was a major accomplishment of the American eugenics movement. Eminent geneticists and psychologists testified that American racial stock was deteriorating because of the influx of immigrants from southern and eastern Europe. Much of the testimony that helped passage of the Act centered on data collected by a fervent American eugenicist, Henry Goddard. In 1912, Goddard was paid to administer IQ tests to immigrants and he found the following frequencies of feeblemindedness: Italians, Jews, and Hungarians, 83 percent; Russians, 87 percent. It should be noted that the tests were given in English to immigrants who did not speak the language. As a consequence of the test scores, immigration quotas for the four groups were lowered. The Johnson-Lodge Act remained in effect until 1962.

The central beliefs of the eugenics movement are enshrined in a lecture delivered in 1903 by a respected population geneticist, Karl Pearson, who told his audience: "We inherit our parents' tempers, our parents' conscientiousness, shyness, and ability, even as we inherit their stature, forearm, and span." This statement clearly implies that everything you are or become is determined by your genes. This is simply not true. Yet such views are still held by some scientists even today. In 1978, Edward O. Wilson wrote in his book *On Human Nature*: "The question is no longer whether human social behavior is genetically determined, it is to what extent. The accumulated evidence for a large hereditary component is more detailed and compelling than most persons, including even geneticists, realize. I will go further: it is already decisive." When scientists state their beliefs as scientific fact, both science and society are the worse for it.

The eugenics movement was powerful enough in the United States to force the enactment of laws even more discriminatory than the Immigration Act of 1924. To prevent the infiltration of "undesirable genes" into the white population (and possibly of "desirable genes" into black populations), thirty-four states passed **miscegenation** laws, making marriage illegal between blacks and whites and, in some states, between Orientals and whites, too. These miscegenation laws remained in effect in many states until 1967, when they were declared unconstitutional by the U.S. Supreme Court.

About thirty states also passed laws authorizing sterilization for a variety of supposedly inherited conditions and behaviors, including feeblemindedness, tendency to commit rape, alcoholism, and criminality. Even the brilliant and liberal Supreme Court Justice Oliver Wendell Holmes was strongly influenced by the eugenics movement. In his decision upholding the compulsory state sterilization laws, he wrote: "Experience has shown that heredity plays an important part in the transmission of insanity and imbecility." Between 1930 and 1935, about 20,000 eugenic sterilizations were performed in the United States—about half of them in California. Also during the 1930s, an estimated 375,000 people were sterilized in Germany by the Nazis for disorders such as feeblemindedness, blindness, and deafness.

The ideas of the eugenics movement were also used by the Nazis to justify the incarceration and extermination of "undesirables" to prevent what the Nazis referred to as "the menace of race deterioration." This genocidal policy was intended principally for Jews, but other European ethnic groups—such as gypsies—were also victims. Before and during World War II, the Nazis practiced genocide on an incomprehensible scale, exterminating millions of people solely on the basis of their religion, nationality, or supposed inferiority.

No doubt, heredity plays a crucial role in some cases of mental retardation, such as that in persons suffering from Down syndrome or untreated PKU. However, defective genes are not what caused the "feeblemindedness" in the thousands of American pellagra victims suffering from a vitamin deficiency. Nor are defective genes the cause of poverty, starvation, and homelessness in the world today. Human experience is not a particularly good tool for measuring the hereditary contribution to a trait. In the course of human history, experience has told people that the earth is flat, that disease is caused by evil spirits, and that the sun revolves around the earth. Experience may "tell" a banker that he gained his wealth because of his superior genes, or it may "tell" a family of musicians that their talent was inherited, but these beliefs are not supported by genetic or other scientific evidence.

The ideas of the eugenics movement that were so scientifically and politically influential during the first half of this century have now been discredited and, for the most part, discarded. Because history has shown us how dangerous the personal and social consequences of eugenics and biological determinism can be, we need to be exceptionally cautious in accepting assertions by sociobiologists or others that human intelligence, homosexuality, criminality, or aggression are determined primarily by genes. People need to rely on clearly established genetic principles if individual and social injustices are to be reduced and prevented in the future.

Summary

Human chromosomes contain all of the genetic instructions for development, growth, and reproduction, but the environment affects the expression of genes to a greater or lesser degree. Proponents of biological determinism believe that, for the most part, genes largely determine a person's biological characteristics and behaviors, perhaps even social status. Geneticists have devised experiments for quantifying the contributions of hereditary and the environment for specific traits. However, such experimentation is restricted to plants and domestic animals, given the moral and legal constraints regarding genetic manipulations of human beings.

Heritability is a measure of the amount of phenotypic variation in a population that can be attributed to genetic differences. It is a quantity that ranges from zero (variation is due to environment alone) to one (variation is due solely to heredity). Mathematically, the heritability of a trait is defined as the amount of phenotypic variation resulting from genetic differences divided by the total amount of variation (environmental plus genetic) for that trait observed in the population.

Heritability is a useful measure for plant breeders, but it is difficult to obtain for animal populations because the appropriate crosses are difficult to achieve and environments are less easily controlled. For various reasons, heritability measurements of specific traits in human populations are unreliable. Nevertheless, heritability studies of human intelligence have been undertaken, but the conclusions are controversial.

The central thesis of sociobiology is that the behaviors of animals and societies are determined to a large degree by genes whose existence in present-day populations has resulted from evolutionary mechanisms — especially natural selection. Altruism, a behavior that benefits another organism at some expense to the doer, is an example of a biologically determined behavior found in certain species of insects. The evolutonary mechanism that explains altruistic insect behavior is kin selection, in which the genetic relatedness of the entire family group is subject to selection rather than the individual organism.

The goal of the eugenics movement was to breed better human beings. Advocates of eugenics believed that human weaknesses and undesirable traits are caused by defective genes. Thus, they concluded that the way to improve the human species was to eliminate the "defective" people or prevent them from reproducing and to encourage breeding among "superior" people. The ideas held by eugenicists are not supported by scientific evidence. The misconceptions of eugenicists and their misinterpretation of genetic principles have had serious consequences—politically, socially, and individually—in many ethnic groups and nationalities.

Key Words

altruism   Self-endangering behavior that benefits other individuals.

biological determinism   The idea that most of the variation in human traits and behaviors is genetically determined.

coefficient of relatedness   The degree to which family members are genetically related.

dizygotic (DZ) twins   Twins that develop from separate eggs fertilized by different sperm and are genetically different.

environment   All nongenetic factors that affect the phenotypes of organisms.

eugenics   The science of improving people by breeding.

heritability   The amount of phenotypic variation in a population that is caused by genetic differences between individual organisms.

kin selection   Selection for altruistic behavior in a group of individuals that are genetically related.

miscegenation   Marriage or cohabitation between a white person and a person of another race.

monozygotic (MZ) twins   Twins that develop from a single fertilized egg and are genetically identical.

pellagra   A vitamin B deficiency disease in human beings.

sociobiology   The study of the biological (genetic) basis of all forms of social behaviors in all organisms, including human beings.

Additional Reading

Gould, S.J. "Carrie Buck's Daughter." *Natural History*, July, 1984.

Griffin, D.R. "Animal Thinking." *American Scientist*, September-October, 1984.

Hewith, J.K. "Heritability." *Science Progress* 17, 37 (1987).

Kevles, D.J. *In the Name of Eugenics: Genetics and the Uses of Human Heredity.* Knopf, 1985.

Montagu, A. (ed.). *Sociobiology Examined.* Oxford University Press, 1980.

Trivers, R. *Social Evolution.* Benjamin-Cummings, 1985.

Wingfield, J.C., G.F. Ball, A. Dufty, Jr., R.E. Hegner, and M. Ramenofsky. "Testosterone and Aggression in Birds." *American Scientist*, November-December, 1987.

Study Questions

1   How would you define the difference between sociobiology and eugenics?

2   What does it mean for a trait to have a heritability of 1.0? What is the heritability of the number of tails in a population of dogs?

3   Which human chromosome has been said to participate in determining criminal behavior?

4   How could altruistic behavior be selected for in evolution?

5   What is meant by the expression "nature or nurture"?

6 Is it valid to compare the heritability of egg production in flocks of chickens from China with those from the United States? Explain.

7 What are two problems with human twin studies as a means to measure heritability?

Essay Topics

1 Discuss the benefits and dangers of eugenics as you see them.

2 Explain why you think there are differences in IQ test scores between different races or ethnic groups.

3 Discuss some of the goals and results of sociobiology research.

Answers to Study Questions

Chapter 1

1 The trait is determined by a single gene consisting of dominant and recessive alleles. The trait obeys Mendel's law of segregation.

2 Four alleles.

3 Seven traits.

4 No. The expression is affected by other genes and environmental factors.

5 Incomplete dominance, codominance, overdominance, epistasis, variable expressivity, incomplete penetrance.

6 Quantitative genetics, classical genetics, population genetics, evolutionary genetics, molecular genetics, behavioral genetics.

7 Genotype refers to the particular set of genes in a person's chromosomes; phenotype refers to a person's observable traits.

Chapter 2

1 Chromatin includes all of the substances found in chromosomes, such as nucleic acids, histones, and proteins. Chromosomes are structures in the nuclei of cells that carry genetic information. Chromatids are the replicated pairs of identical chromosomes.

2 To determine if there is any observable chromosomal defect in a person's cells.

3 The primary differences are:
   (a) meiosis occurs in reproductive cells, mitosis in somatic cells;
   (b) meiosis reduces the diploid number (2N) of chromosomes to the haploid number (N), and mitosis maintains the diploid number;
   (c) homologous chromosomes pair up (synapse) in meiosis but not in mitosis;
   (d) recombination occurs between homologous chromosomes in meiosis but not in mitosis.

4 G_1, preparation for DNA replication; S, DNA replication; G_2, preparation for mitosis; M, mitosis and cell division.

5 Attachment of spindle fibers, which pull chromosomes to opposite poles of the cell.

6 Meiosis (a) maintains the diploid number of chromosomes from generation to generation; (b) ensures genetic diversity; (c) generates new combinations of alleles through recombination.

7 A sex-linked trait refers to any gene on an X chromosome. A fragile X chromosome is structurally abnormal.

Chapter 3

1 44; 22; 22; 44.

2 The alleles being studied can be present in either the male or the female parent and the results of the cross (the frequency of the various progeny) are the same.

3 Down syndrome, Turner syndrome, Klinefelter syndrome.

4 XX/XY.

5 By carrying out a karyotype analysis.

6 Because of the emotional, psychological, and financial consequences of caring for a mentally retarded person for as long as fifty years.

7 Aneuploidy refers to having one chromosome more or less than the normal set of chromosomes. Polyploidy means having one or more extra sets of chromosomes; for example, 3N, 4N, 5N.

Chapter 4

1 An element cannot be broken down into anything simpler by chemical or physical means. Molecules can be broken down into elements (atoms). Alcohol is an organic molecule; salt is an inorganic molecule.

2 The number of electrons in the outer shell of the atom.

3 Covalent, ionic, and hydrogen bonds. Covalent bonds are the strongest; hydrogen bonds are the weakest.

4 Isotopes are different forms of an element that differ in weight. Radioisotopes are unstable isotopes that emit radioactivity.

5 Photosynthesis.

6 In nova and supernova explosions of stars.

7 Enzymes.

Chapter 5

1 The unifying principles are:
   (a) all organisms are composed of cells;
   (b) the overall chemical composition of all cells, whatever their function, is basically the same;
   (c) all organisms are related to one another by evolutionary descent.

2 About 4,000.

3 See Table 5-1.

4 Mitochondria.

5 Teratogenic substances.

6 Proteins are composed of amino acids; nucleic acids are composed of bases. Proteins contain sulfur atoms; nucleic acids do not. Nucleic acids contain phosphorus atoms; proteins do not. Porteins may have disulfide bridges; nucleic acids do not.

7 Glucose, ATP, glycogen, carbohydrates, sugars, and so forth.

Chapter 6

1 DNA transformation.

2 Because proteins were complex and highly variable like traits.

3 30% A, 30% T, 20% G, 20% C.

4 None.

5 In DNA, the amount of adenine equals the amount of thymine and the amount of cytosine equals the amount of guanine; that is, A = T, G = C.

6 Hydrogen bond.

7 The DNA transformation experiments in pneumococcus by Avery and co-workers; the phage experiments by Hershey and Chase in which protein and DNA were labeled with radioactive sulfur and phosphorus, respectively.

Chapter 7

1 Replication, recombination, mutation.

2 Two genetic loci separated by one map unit will produce 1 percent recombinants in a genetic cross.

3 Mutation and recombination in both bacteria and human beings.

4 Acquired traits and skills can be inherited (passed on to progeny).

5 One thousand.

6 $\dfrac{100 \times 10^6 \times 46}{10^4} = 4.6 \times 10^5$ days = 1,260 years.
If one assumes that both strands will have to be sequenced separately, then it could take twice as long.

7 No. Mitochondria are inherited only from the female egg.

Chapter 8

1 Messenger RNAs carry information from different genes. Ribosomal RNAs are essential in the structure and functions of ribosomes. Transfer RNAs align each amino acid with its correct codon in messenger RNA.

2 RNA polymerase recognizes and attaches to a promoter site in DNA.

3 Ribosomes attach to a messenger RNA and translation begins at the first AUG codon.

4 A segment of DNA in a eukaryotic gene that carries genetic information that is translated into a sequence of amino acids. Bacterial genes do not have introns or exons.

5 Only one.

6 Human mRNAs are processed by having the first base "capped" and a string of adenine nucleotides added to the last base. Also, introns are "spliced out" of human mRNAs. None of these processes occurs in bacterial mRNAs.

7 Hemoglobin. Valine replaces glutamic acid at position 6.

Chapter 9

1 Universal: the same codon means the same amino acid in every organism.
Degenerate: more than one codon specifies each amino acid.

2 Start: AUG. Stop: UAG, UAA, UGA.

3 Six; one.

4 A frameshift mutation causes all amino acids to be inserted incorrectly during translation beyond the point of the mutation. A missense mutation causes

the insertion of only one incorrect amino acid. A nonsense mutation causes translation to terminate at the point of the nonsense mutation.

5 Ionizing radiation, mutagenic chemicals, viruses.

6 Any base change that does not change the amino acid in the protein. For example, both UUU and UUC code for phenylalanine.

7 Not usually, but it could if it occurred in a region of the DNA where both strands code for proteins—so-called overlapping genes.

Chapter 10

1 Many adults, particularly among Asians and blacks, are *lac* negative and cannot metabolize the lactose in milk.

2 Lactose, repressor, CAP, cyclic AMP, RNA polymerase.

3 Regulation of transcription, mRNA processing, mRNA transport, regulation of translation, processing of proteins.

4 Insulin.

5 No.

6 No.

7 Inducers turn on the transcription of genes; enhancers increase the expression of distant genes; hormones regulate the expression of genes and the functions of cells and tissues.

Chapter 11

1 Cells in a benign tumor remain localized; malignant cells can spread throughout the body and cause many tumors to grow.

2 Many mutagens (substances able to cause mutations) are also carcinogens (able to cause tumors in animals). Most carcinogens are also mutagens.

3 Converts the viral RNA into a DNA copy that can be integrated into a chromosome, possibly disrupting an essential gene.

4 Cigarette smoke.

5 Less than 5 percent.

6 Viruses, ionizing radiation, chemicals that damage DNA.

7 Ultraviolet (UV) light.

Chapter 12

1 Restriction enzymes.

2 Isolation of DNA carrying the gene to be cloned; synthesis of the gene from its mRNA; synthesis of the gene from knowledge of the amino acid sequence of the protein encoded by that gene.

3 Lambda.

4 A genetic map constructed from the localization of restriction-enzyme sites in chromosomes.

5 The proper regulatory sites and signals are generally missing. More importantly, human genes consist of introns and exons that cannot be "spliced out" in bacteria.

6 All of the genes of an organism are present in plasmids or viruses carried in microorganisms. The cells can be grown and the genes studied more easily.

7 Bacteria that have lost the protein that causes ice to form on the surfaces of plants.

Chapter 13

1 X and Y chromosomes. Hormones.

2 Turner syndrome (sterile); normal female; abnormal female (sterile); normal male; Klinefelter syndrome (sterile); "normal" male (extra Y chromosome, fertile).

3 The phenotypic differences between males and females of a species.

4 The same as that of the mother.

5 Changes in hormone levels or changes in receptor sites (proteins) on cells that respond to hormones or both kinds of changes.

6 Chromosomal abnormalities.

7 The low success rate and repeated attempts.

Chapter 14

1 Any defect observable at birth is congenital. Not all congenital defects are due to heredity; many are developmental.

2 Down syndrome is caused by a chromosomal abnormality (trisomy 21). PKU is caused by a pair of recessive alleles.

3 Alcohol is a teratogen and causes developmental defects including mental retardation.

4   The three rules are:
    (a)   a Mendelian pattern of inheritance;
    (b)   a chromosomal abnormality;
    (c)   a biochemical defect.

5   Sickle-cell anemia and cystic fibrosis. Hyperlipidemia and hypercholesterolemia.

6   A polygenic multifactorial disorder is caused by many genes and environmental factors. An X-linked disorder is caused by a single gene in one or both X chromosomes.

7   From the mother. Males are most commonly affected and they inherit their X chromosome from their mother.

Chapter 15

1   Anencephaly, hydrocephalus, spina bifida.

2   Ultrasound scanning, amniocentesis, chorionic villus sampling, fetoscopy.

3   No. Most abnormalities have occurred by the time the child is born. Also, an entire extra chromosome would have to be removed from all cells—clearly, an impossible task.

4   Phenylketonuria (PKU).

5   Sickle-cell anemia.

6   Adenosine deaminase deficiency (ADA), purine nucleotide phosphorylase deficiency (PNP), hypoxanthine-guanine phosphoribosyl transferase deficiency (HPRT).

7   Genetic screening detects individual members of a population who carry a defective gene; genetic counseling is communicating to prospective parents the risks of having a defective child.

Chapter 16

1   Paternity or maternity disputes, identification of rape or murder suspects.

2   Region 2, band 1 on the long arm of chromosome 15. Region 2, band 2 on the short arm of chromosome 2.

3   To produce the characteristic dark and light bands that help identify chromosomes and chromosomal defects.

4   A heterokaryon contains two unfused nuclei; a somatic-cell hybrid is a cell with a single nucleus and chromosomes derived from two different kinds of cells.

5   Sickle-cell anemia, Huntington disease, cystic fibrosis (see Table 16-1).

6   The average number of lethal genes carried per individual in a population.

7   Restriction enzymes.

Chapter 17

1   IgE.

2   One.

3   Three alleles: A, B, and O. AB, AA, BB, AO, BO, OO.

4   HLA, or histocompatibility, genes.

5   Somatic recombination in DNA of immune-system cells in the course of development.

6   Four. Two heavy chains and two light chains.

7   Special somatic hybrid cells called hybridomas.

Chapter 18

1   Natural selection, migration, genetic drift, mutation, sexual selection.

2   Natural selection refers to the survival and reproduction of the "fittest" organisms. Sexual selection refers to the individual organisms that choose to mate and reproduce.

3   Random matings among individual members of a large population.

4   Not in any reasonable time period. Most sickle-cell genes are in unaffected carriers of the trait.

5   Increases the likelihood of homozygosity for disease-causing recessive alleles.

6   Yes. Not in nature, generally, but in zoos or between domestic animals.

7   More than 99 percent. Some estimates say 99.999 percent of all organisms have become extinct.

Chapter 19

1   Radioisotope dating.

2   Stromatolites (oldest), dinosaurs, Lucy, Neanderthal man (recent extinction).

3   The beta chain of hemoglobin.

4   A subspecies based on similar gene frequencies or phenotypic characteristics. The presence of two or

more forms of a trait (determined by different alleles) in a population.

5 No. Species become extinct all the time.

6 About 75 percent.

7 All human beings can and do interbreed and produce fertile progeny.

Chapter 20

1 Sociobiology is the study of the genetic basis of behaviors in organisms. Eugenics is the science of improving people by breeding for desirable traits.

2 The trait is entirely determined by genes. Heritability is zero because there is no variation in the population.

3 The Y chromosome.

4 More genes get passed on to progeny as a result of altruistic behaviors.

5 Whether a trait is determined by heredity (nature) or by the environment (nurture).

6 No. Heritability cannot be compared *between* different populations.

7 Accurate determination of zygosity and significant differences in the environments in which twins are raised whether together or separated.

Glossary

ABO blood group Three alleles that are present in pairs in each person. These determine antigens on red blood cells and whether a donor's blood will be rejected by a recipient.

achondroplasia An autosomal dominant disorder causing dwarfism due to retarded growth of long bones.

acrosomal cap Proteins surrounding the head of a sperm that facilitate attachment to an egg.

activating enzymes The different enzymes responsible for attaching tRNAs to their corresponding amino acids.

adapter A transfer RNA molecule.

adenosine triphosphate (ATP) The energy-yielding molecule in cells that is used to drive chemical reactions.

alleles Alternative functional states of the same gene in homologous chromosomes.

allophenic mice Mice whose cells are derived from four different parents.

alpha-fetoprotein (AFP) A protein produced during embryonic development; its level in the mother's blood can be used to detect neural-tube defects.

altruism Self-endangering behavior that benefits other individuals.

amino acids The twenty different small molecules that are found in all proteins.

amniocentesis A procedure in which a sample of amnionic fluid and fetal cells is removed from the uterus and examined for genetic or biochemical defects.

anaphase The phase of mitosis following metaphase; chromosomes move toward opposite poles of the cell.

androgen insensitivity syndrome A syndrome characteristic of women who have an XY chromosome constitution.

androgens A group of male hormones (the major one is testosterone) that are produced early in development and result in the formation of testes.

anencephaly A congenital defect in which a newborn is missing all or a major part of the brain.

aneuploidy The characteristic of having one or more chromosomes too many or too few than the number normally found in an organism.

antibody An immunoglobulin molecule capable of combining with specific antigens and inactivating pathogenic microorganisms.

anticodon Three adjacent bases in transfer RNA that pair with three complementary bases in a codon.

antigen Any foreign substance that stimulates an antibody response in an animal.

ataxia telangiectasia An inherited disorder affecting the ability of human cells to repair damage to DNA caused by x-rays.

autoimmune disease A disease such as some forms of rheumatoid arthritis or lupus erythematosus, in which the immune system reacts to normal body cells and damages or destroys normal tissues.

autosomes All chromosomes in eukaryotes, excluding the sex chromosomes.

bacterial strain Bacteria of the same species that differ from one another in a specific way that can be measured.

Barr body The condensed X chromosome observed in the nucleus of somatic cells of women and other female mammals.

bases The chemical subunits in DNA and RNA whose sequence encodes the genetic information.

benign tumor Unregulated, localized growth of cells that usually does not cause cancer.

beta-galactosidase An enzyme that breaks apart lactose into the sugars glucose and galactose.

biological determinism The idea that most of the variation in human traits and behaviors is genetically determined.

biotechnology The application of scientific knowledge by industries that produce biological products such as food supplements, enzymes, drugs, and so forth. Some companies use recombinant DNA techniques.

bivalent A pair of homologous, synapsed chromosomes.

B lymphocyte A white blood cell that carries the genetic information for synthesis of particular antibodies.

bottleneck A dramatic reduction in the size of a population followed by a large increase.

cancer The unregulated growth and reproduction of cells.

capacitation Physiological changes in a sperm as it moves up the reproductive tract that enable it to penetrate an egg.

carbohydrate A large molecule consisting of chains of sugars; cellulose and starch are carbohydrates.

carcinogen Any agent, such as radiation, a chemical, or a virus, that causes cancer.

carcinoma Cancer of skin and membrane cells.

carrier   An individual who is heterozygous for a recessive allele.

catabolite activator protein (CAP)   The positive controlling element for glucose-sensitive operons in bacteria.

catalyst   Any substance that increases the rate of a chemical reaction without being used up itself in the reaction; enzymes are catalysts.

cells   The fundamental units of living organisms. All cells are capable of reproducing themselves.

cellulose   A long chain of sugar (glucose) molecules found in wood, paper, cotton, and other fibers.

central dogma   The rules that govern the exchange of information between DNA, RNA, and protein molecules.

centromere   The attachment site on chromosomes for spindle fibers, which participate in the segregation of chromosomes in mitosis and meiosis.

chemical reaction   Interactions between atoms and molecules that lead to formation of other substances.

chiasma (pl. chiasmata)   The visible physical crossing-over between sister or nonsister chromatids of homologous chromosomes at meiosis.

chorionic villus sampling   A technique for sampling embryonic cells between the eighth and tenth week of pregnancy to look for genetic defects.

chromatid   One of the two daughter strands of a duplicated chromosome.

chromatin   The substances of which chromosomes are composed; for example, nucleic acids, proteins, histones, and so forth.

chromosomes   Structures in the nuclei of eukaryotic cells that carry genetic information in the form of DNA molecules.

clonal selection   Explains how antibody diversity is generated during embryonic development and how the immune system distinguishes "self" from "nonself."

clone   A group of genetically identical cells or organisms. Monozygous human twins are clones, as are bacteria in a single colony growing on a surface.

cloning vehicle (vector)   Self-replicating molecules such as plasmids or viruses that carry segments of foreign (cloned) DNA.

codominance   Two alleles that are expressed equally.

codon   Three adjacent bases in messenger RNA that specify each of the twenty amino acids.

coefficient of relatedness   The degree to which family members are genetically related.

complementarity   Refers to the base-pairing rules in which A pairs with T (or U) and G pairs with C.

conceptus   A fetus; the product of conception.

concordance   The degree to which identical (monozygotic) twins or fraternal (dizygotic) twins are alike with respect to a trait.

congenital defect   Any abnormality detectable in a newborn at birth.

conjugation   Mating between a donor bacterium and a recipient one such that DNA is transferred from the donor and stably maintained in the recipient.

covalent bond   A chemical bond created by the equal sharing of electrons by two atoms.

crossing-over   The exchange of segments between pairs of homologous chromosomes in meiosis.

cyclic adenosine monophosphate (cAMP)   An important small regulatory molecule in prokaryotic and eukaryotic cells.

cystic fibrosis   An autosomal recessive disorder characterized by abnormal mucus production in the lungs and other organs.

cytoplasm   Everything in a cell exclusive of the nucleus.

degenerate code   Refers to the fact that each amino acid (except methionine and tryptophan) is specified by more than one codon.

deletion   Loss of one or more base pairs from DNA in a chromosome.

diabetes   A disease caused by abnormal insulin production and blood glucose levels.

differentiation   The process by which cells become increasingly (an irreversibly) specialized in their functions in tissues and organisms during development.

diploid   The chromosomal state of a cell or organism in which each different chromosome is present in two copies (2N).

disaccharide   Two sugar molecules joined together.

disulfide bridge   A covalent bond between two sulfur atoms; a disulfide bridge binds two cysteine amino acids in a protein together.

dizygotic (DZ) twins   Twins that develop from separate eggs fertilized by different sperm and are genetically different.

DNA   Deoxyribonucleic acid, the macromolecule carrying the hereditary information in the chromosomes of cells.

DNA-DNA hybridization   A technique used to measure the similarity in base sequences between DNA molecules from different organisms.

DNA probe   A radioactive fragment of DNA carrying a particular cloned gene or sequence used to detect DNA fragments in a Southern blot.

dominant allele   The allele in a heterozygous organism that determines the phenotype.

Down syndrome   An inherited human disease caused by an extra chromosome number 21 in all of the person's cells.

duplication   Addition of one or more base pairs to DNA in a chromosome.

electron   A negatively charged particle that orbits the nucleus of an atom.

electrophoresis A technique used for the separation of nearly identical proteins based on differences in their electrical charge.

element A substance that cannot be broken down further by ordinary chemical or physical means.

endorphins Morphinelike substances synthesized in the human brain in response to stress or injury.

enhancer A sequence of bases in the chromosomes of eukaryoric cells that increases the expression of a gene at some distant location from it.

environment All nongenetic factors that affect the phenotypes of organisms.

enzyme A protein that increases the rate of a chemical reaction in a cell.

epidemiology A branch of science that studies the frequency and distribution of diseases in different populations and tries to identify disease causation.

epistasis The masking of the expression of a gene by a gene located elsewhere in the genome.

equivalence rule The amount of adenine in DNA equals the amount of thymine; the amount of guanine equals the amount of cytosine.

erythrocyte A red blood cell.

eugenics The science of improving people by breeding.

eukaryotes The superkingdom of all organisms other than bacteria; all eukaryotes have a true nucleus and undergo mitosis and meiosis.

evolution In general, a process of development and change through time; biological evolution refers more specifically to changes in gene frequencies that accumulate in populations of organisms through time.

exons The discontinuous segments of DNA in eukaryotic genes that carry information that is translated into the sequence of amino acids in proteins.

family pedigree The pattern of a trait's occurence in family members from generation to generation.

F$^+$ E. coli Bacteria that contain an autonomous F-plasmid in the cytoplasm.

F′ E. coli Bacteria that contain an autonomous F-plasmid that carries chromosomal genes in addition to plasmid genes.

feedback inhibition A mechanism by which the final product of a biosynthetic pathway (say, an amino acid) inhibits the activity of the first enzyme in that pathway.

fertility factor A plasmid such as the F-plasmid that allows bacteria to conjugate and exchange DNA molecules.

fertilization The union of male and female gametes to produce the first cell (zygote) of a new organism.

fetoscopy Insertion of an instrument (the fetoscope) into the amniotic sac to visualize the fetus.

fitness The reproductive contribution of an individual organism to subsequent generations.

founder effect Demonstration of a gene whose frequency is much higher in a particular population than it is in other populations.

F-plasmid Small, circular DNA molecule in E. coli bacteria that causes cell–cell conjugation. The F-plasmid can be transferred from a donor to a recipient bacterium or it can facilitate the transfer of chromosomal genes.

frameshift mutation A mutation that results from the insertion or deletion of one or two base pairs in DNA.

gamete Male or female reproductive cell; in animals, it is called sperm or egg.

gene A discrete hereditary unit located at a specific position (locus) on a chromosome; also a sequence of bases in DNA that codes for a protein.

gene cloning The insertion of a particular gene into bacteria or other microorganisms, where it multiplies.

gene library Millions of bacteria containing all of the genetic information from another organism whose DNA has been cloned and propagated in the bacteria.

generation time The time required to double the number of cells in a population.

genetic complementation In somatic-cell hybrids, the provision of essential functions by genes on different chromosomes from different cell lines. More generally, a test for whether mutations occur in different functional genes or on two different chromosomes.

genetic counseling A process of communicating the risks of having a defective child to prospective parents.

genetic drift Variation in gene frequencies in a population from one generation to the next as a result of chance.

genetic engineering The construction and utilization of novel DNA molecules that have been engineered by recombinant DNA techniques.

genetic load The average number of lethal (or deleterious) genes per member of a population.

genetic map The assignment of genes to specific locations on chromosomes.

genetic screening Procedures that detect individual members of a population who carry a defective gene.

genome The total amount of genetic information (haploid number of chromosomes) in an organism.

genotype The particular set of genes present in the chromosomes of an organism.

germ cell therapy Correction of a mutant gene in sperm or eggs.

germinal mutation One that arises in the chromosomes of reproductive (sex) cells.

glycogen A long chain of sugar (glucose) molecules found in animals.

half-life The time that it takes for one-half of the nuclei of a radioactive isotope to decay and be converted into some other element.

haploid The chromosomal state of a cell in which each different chromosome is present in one copy; sperm and eggs are haploid cells in animals.

Hardy-Weinberg law A mathematical expression of the relation between allele frequencies and genotype frequencies in an idealized population.

hemoglobin The oxygen-carrying molecule in red blood cells.

hemophilia A recessive, X-linked disorder that causes excessive bleeding due to lack of an essential clotting factor.

hereditary (genetic) disorder or disease A disorder or disease due to a defect in a gene or chromosome inherited from one or both parents.

heritability The amount of phenotypic variation in a population that is caused by genetic differences between individual organisms.

hermaphrodite A human being in whom both the male and female reproductive organs are present; however, neither organ is functional, and so hermaphrodites are sterile.

heterogametic sex The sex of a species that produces gametes with two types of sex chromosomes.

heterokaryon A cell containing two nuclei that have not fused into one.

heterozygous Having different alleles at the same locus on homologous chromosomes.

Hfr *E. coli* Bacteria that contain the F-plasmid integrated into the bacterial chromosome.

histocompatibility The antigens that determine whether tissue transplanted from one animal to another will be accepted or rejected.

histone A small DNA-binding protein thought to regulate gene expression and chromosomal structure in eukaryotic cells.

HLA genes Determine antigens on white blood cells (leukocytes) and other tissues. These antigens determine whether foreign tissue will be accepted or rejected.

hominid A primate whose characteristics are more humanlike than apelike.

homogametic sex The sex of a species that produces gametes containing only one type of sex chromosome.

homologous chromosomes The two members of a pair of chromosomes, one member from each parent. Homologous chromosomes pair up (synapse) at meiosis.

homozygous Having identical alleles at the same locus on homologous chromosomes.

homunculus A little individual that early Greeks believed was contained in the head of each sperm.

hormones A large class of different molecules in plants and animals that regulate gene expression and other physiological processes.

Huntington disease (Huntington chorea) An autosomal dominant disease that is expressed late in life, producing neurological symptoms and death.

H-Y antigen A protein that participates either in testes development or in sperm production in males.

hybridoma A somatic-cell hybrid between a B lymphocyte and a cancer cell; it produces monoclonal antibodies.

hybrid plant A progeny plant with a different combination of traits from those observable in either parent.

hydrocephalus A congential defect caused by excess cerebrospinal fluid in the brain.

hydrogen bond The sharing of a hydrogen atom between two other atoms. Weaker than a covalent or ionic bond.

hypercholesterolemia Elevated levels of cholesterol in the blood.

hyperlipidemia Elevated levels of lipids in the blood.

hypermutability An unusually high rate of mutation.

immune system The numerous mechanisms in vertebrates that recognize and eliminate foreign substances and alien cells.

immunoglobulin The scientific term for antibody.

inbreeding Mating between closely related individuals.

incomplete dominance Neither of two alleles is completely dominant or recessive.

incomplete penetrance The failure of a gene to produce the same effects in different individuals.

inducer Any small molecule (for example, lactose) that is able to switch on the expression of one or more genes.

industrial melanism The natural selection of dark-colored (melanic) moths in industrial areas of England.

inorganic Molecules or reactions in which carbon atoms do not participate.

insertion sequences (IS) Small pieces of DNA at the ends of transposons that have identical base sequences. These sequences pair with identical sequences in other DNAs and are responsible for transposon movement.

insulin A protein hormone produced in the pancreas that regulates the level of glucose in the blood.

interphase The interval in the cell cycle between mitoses.

intron An untranslated segment of DNA in a eukaryotic gene that separates exons.

inversion A chromosome segment that has been rotated by 180 degrees such that the order of genes is inverted with respect to the rest of the chromosome.

in vitro **fertilization** The formation of a zygote in a test tube by mixing sperm and eggs together.

ion A positively or negatively charged atom.

ionic bond A bond caused by the electrical attraction between two atoms; weaker than a covalent bond.

ionizing radiation High-energy radiation such as gamma rays, x-rays, ultraviolet light, and particles emitted by radioactive materials.

isotope A different form of an element; its chemical properties are identical with those of the element, but its weight differs because of differences in the number of neutrons in the nucleus.

karyotype Visual arrangement of all of the chromosomes from a single cell so that they can be identified and counted.

kin selection Selection for altruistic behavior in a group of individuals that are genetically related.

Klinefelter syndrome A syndrome characteristic of men who have an XXY chromosome constitution.

Lamarckianism The inheritance of acquired traits; an idea named after the French scientist Jean Baptiste Lamarck.

leukemia Cancer of immature white blood cells.

leukocyte A white blood cell.

ligase An enzyme that joins a sugar to a phosphate in a DNA strand.

linkage The joint inheritance of two or more nonallelic genes because they are located close together (linked) on a chromosome.

lipids Large molecules, some of which contain fatty acids, that are insoluble in water.

lymphocyte A specialized white blood cell present in lymph nodes, spleen, thymus, bone marrow, and blood; lymphocytes mature into B lymphocytes and T lymphocytes (also called B cells and T cells).

lymphoma Cancer of white blood cells in the lymphatic system and the spleen.

Lyon hypothesis Only one X chromosome is expressed in each female cell; the other X chromosome is inactivated.

lysis The bursting open and death of a cell.

lysogenic bacterium A bacterium that carries a prophage (unexpressed virus) in its DNA.

lysogeny A process by which phage DNA becomes integrated into a bacterial chromosome.

macrophage A large leukocyte that is able to destroy foreign cells by phagocytosis.

malignant tumor Cancer that can spread through the body and cause death.

map unit A measure of linkage between genetic loci; 1 percent recombinants equals one map unit; human genetic maps are calibrated in map units called centimorgans.

meiosis Process by which the haploid set of chromosomes ends up in gametes (reproductive cells).

messenger RNA (mRNA) An RNA molecule whose sequence of bases is translated into a specific sequence of amino acids (a polypeptide).

metabolism The sum of all the chemical processes in an organism.

metaphase The phase of mitosis following prophase; chromosomes line up along the equatorial plane of the cell.

metastasis The spread of cancer cells from the original tumor to other sites in the body.

migration Flow of genes from one population to another due to movement of individual organisms or gametes into or out of a population.

mimicry A resemblance of one organsim to another, which often protects it from its predators.

miscegenation Marriage or cohabitation between a white person and a person of another race.

missense mutation A single base-pair change that causes the substitution of one amino acid for another in the protein.

mitochondrion (pl. mitochondria) A self-reproducing organelle in all eukaryotic cells. The primary site for synthesis of ATP, which supplies energy to the cell.

mitosis Process of chromosome segregation and cell division.

modern synthesis of evolutionary theory Unites Darwin's ideas of natural and sexual selection with what is now known about the inheritance and chemical properties of chromosomes, DNA, and genes.

modification enzymes Enzymes that recognize specific base sequences and attach methyl (CH_3) groups to DNA at specific sites.

monoclonal antibody An antibody of a single type that is produced by genetically identical somatic hybrid cells (clones).

monosaccharide A single sugar molecule.

monozygotic (MZ) twins Twins that develop from a single fertilized egg and are genetically identical.

mosaicism A condition in which cell lines in an individual organism have different genetic expressions or chromosomal constitutions; women are mosaic for heterozygous loci on the X chromosome.

moveable genetic element Pieces of DNA such as viruses, plasmids, and transposons that can "jump" from one DNA molecule to another and move from one cell to another.

mutagen Any environmental agent that increases the frequency of mutations.

mutant An organism that carries a mutation that causes a phenotypic difference from a wild-type organism.

mutation A heritable change in the genetic information.

mutation frequency The number of mutant organisms in a population.

mutation rate The number of mutations in a gene per cell generation.

natural selection The differential reproduction of genetically different individual organisms as a result of their adaptation to a particular environment.

neural-tube defect A defect in the development of the brain or spinal cord that can result in anencephaly, hydrocephalus, or spina bifida.

neutral-gene theory Proposes that mutations arise that change the amino acid composition of a protein without any change in the protein's function or in the survival and reproduction of the organism; these "neutral mutations" can spread through a population purely by chance because a small number of all gametes produced contribute to individuals in the next generation.

neutral mutation A single base-pair change that does not change the amino acid in a protein and, hence, does not change the protein's function or the organism's phenotype.

neutron A particle in the nucleus of an atom that is identical with a proton but does not carry an electrical charge.

nondisjunction Failure of homologous chromosomes (or chromatids) to separate properly at meiosis, resulting in gametes with too few or too many chromosomes.

nonsense mutation A single base-pair change that generates a stop codon and terminates the polypeptide chain.

nucleic acid A chain of nucleotides; RNA and DNA.

nucleosome A beadlike structure observed in eukaryotic chromosomes that consists of DNA wrapped around histones.

nucleotide A small molecule consisting of a base, a sugar, and phosphate.

oncogene A cancer-causing gene.

oncogene hypothesis The idea that viral genes became integrated into the chromosomes of animal cells millions of years ago. These genes are normally unexpressed but if induced may convert normal cells into tumor cells.

oncogenic virus A virus capable of causing cancer in animals when the viral genes are expressed.

operator (o site) A sequence of base pairs in DNA to which repressor proteins attach, preventing transcription of structural genes of the operon.

operon A segment of DNA consisting of two or more adjacent genes capable of synthesizing polypeptides, along with the regulatory sites that govern their expression. The genes in an operon are transcribed together into continuous mRNA molecules.

organic Molecules or reactions that include carbon atoms.

origin of replication A site in a chromosome where replication of a new DNA molecule is initiated.

overdominance Also called hybrid vigor; heterozygous plants have a more vigorous phenotype than do homozygous plants.

paleontology The study of the life forms of past geological periods by means of the fossil record.

pangenesis The concept that each part of an adult organism produces a tiny replica of itself that is collected in the "seed" of the organism and then transmitted to offspring.

parthenogenesis Reproduction by females without fertilization by male gametes. This form of reproduction produces only female offspring.

pellagra A vitamin B deficiency disease in human beings.

peptide bond The specific covalent bond that joins amino acids together in polypeptide chains.

phage (bacteriophage) A virus containing either RNA or DNA that infects and destroys (lyses) bacteria.

phenotype The observable characteristics (traits) of an organism that result from the interaction of its genotype with the environment.

phenylketonuria (PKU) An inherited, autosomal recessive human disease caused by a defect in the gene that converts phenylalanine into tyrosine.

photosynthesis The cellular process of converting light energy into chemical energy.

phylogeny The relationships between groups of organisms that are revealed by their evolutionary history; a family tree.

placenta The organ in pregnant women (and other mammals) that regulates the exchange of nutrients and chemicals between the mother's blood and the fetus.

plasma cell A specialized cell derived from B lymphocytes that synthesizes antibodies.

plasmid An extrachromosomal, circular DNA molecule found in many kinds of bacteria. Plasmids are self-replicating and may exist in many copies.

pleiotropy The ability of a gene to affect several unrelated traits in an individual organism.

point mutation A change in only one base pair in DNA.

polar molecule One in which the charges are distributed unequally among the atoms such that some are positive and others negative; water is a polar molecule.

polygenic multifactorial inheritance Any trait caused by an undetermined number of genes and environmental factors, all of which interact together.

polymerase An enzyme that replicates a single strand of either DNA or RNA.

polymorphism The presence of several different forms of a trait, gene, or restriction site among individual members of a population.

polypeptide A chain of amino acids; one chain of a protein that has several chains, such as hemoglobin.

polyploidy The characteristic of having a multiple of the normal diploid chromosome number.

polysaccharide A chain of sugar molecules; another name for carbohydrate.

population genetics A branch of biology dealing with the measurement and calculation of allele frequencies in populations and with how gene frequencies change in successive generations over time.

porphyria An autosomal dominant disorder caused by abnormal metabolism of porphyrin molecules.

primary sexual differentiation Chromosomal determination of male or female development.

primary structure The sequence of amino acids in a polypeptide chain.

primase A special RNA polymerase that initiates replication of a DNA strand.

primates A mammalian order comprising human beings, apes, monkeys, and certain more-primitive animals, such as the lemur.

prokaryotes The superkingdom of all forms of bacteria.

promoter (*p* site) A sequence of base pairs in DNA to which RNA polymerase enzymes attach to initiate transcription of structural genes.

prophage An unexpressed phage carried in the DNA of bacteria.

prophase The first phase of mitosis following interphase; chromosomes contract and become visible.

protein A macromolecule consisting of one or more chains of amino acids that performs a catalytic or a structural function in cells.

proton A positively charged particle in the nucleus of an atom.

protooncogene A normal cellular gene that may give rise to an oncogene by mutation.

provirus An unexpressed viral genome carried in one of the chromosomes of animal cells.

punctuated-equilibrium theory A model of evolution suggesting that populations remain stable for long periods of time and change abruptly rather than gradually and continuously.

Punnett square A checkerboard method devised by R.C. Punnett showing the types of gametes in a cross and the phenotypes produced.

quaternary structure The folding together of two or more polypeptide chains to produce a functional protein.

race An arbitrary subclassification of a species based on physical or genetic differences.

radioactivity The release of energy from an atom's nucleus in the form of particles or radiation.

radioisotope An isotope whose nucleus is unstable and decays spontaneously, emitting energy.

random mating Mating in which any organism can mate with any other in a population.

recessive allele The allele in a heterozygous individual organism that is not observable in the phenotype.

recombinant DNA The construction and cloning of novel combinations of DNA molecules not found in nature.

recombination The breaking and rejoining of genetically different DNA molecules; the appearance of traits in progeny that were not observed in parents.

replica plating A technique in which bacteria are transferred to a series of Petri plates from an original plate by means of a velvet pad.

replicase An enzyme that replicates DNA by adding complementary bases.

replication fork A region in DNA where duplication of both strands is taking place.

repressor A protein that "turns off" or prevents a gene or a group of genes in DNA from being expressed.

reproductive isolation The inability of organisms to interbreed because of biological differences.

restriction enzymes Enzymes that recognize specific base sequences and cut DNA molecules at specific sites.

restriction fragment length polymorphism (RFLP) Variation among individual members of a population in the location or number of restriction-enzyme sites in their DNA. When DNA from individual members is digested and analyzed, different patterns of DNA fragments (RFLPs) are observed.

restriction map A genetic map constructed from the localization of restriction-enzyme sites in DNA molecules.

retrovirus An RNA virus whose RNA is copied (reverse transcription) into DNA in an animal cell; the DNA then becomes integrated into the DNA of a chromosome in the nucleus.

reverse transcriptase An enzyme present in certain RNA viruses that transcribes the information in the RNA molecule into DNA after the RNA has infected an animal cell.

Rh factor A red blood cell antigen that can cause anemia in a fetus.

ribosomal RNA (rRNA) The RNA molecule that is the main structural component of ribosomes.

ribosome Structure in the cytoplasm of cells on which proteins are synthesized.

RNA polymerase The enzyme used to synthesize rRNAs, tRNAs, and mRNAs from genes in DNA.

RNA splicing enzyme An enzyme that removes introns from eukaryotic mRNA and that splices the exons together in mRNA before translation.

saltation A sudden, large phenotypic change (jump) in a group of organisms.

sarcoma Cancer of bone and muscle cells

secondary sexual differentiation Hormonal determination of male or female development.

secondary structure The helical configuration of a polypeptide chain.

semiconservative replication Refers to the replication of a double-stranded DNA molecule consisting of one old polynucleotide strand and one newly synthesized strand.

sex chromosome The X or Y chromosome in human beings.

sex-linked trait A trait determined by a gene in an X chromosome.

sex pilus The physical appendage on bacteria that carry an F-plasmid.

sexual differentiation Expression of a male or female phenotype through development of sexual organs and characteristics.

sexual dimorphism The distinctly different physical forms of males and females in a species; for example, male birds are generally larger and more colorful than their female counterparts.

silencer A sequence of bases in the chromosomes of eukaryotic cells that decreases the expression of a gene at some distant location from it.

sociobiology The study of the biological (genetic) basis of all forms of social behaviors in all organisms, including human beings.

somatic-cell hybrid A somatic cell with a single nucleus containing chromosomes from two genetically different cells that fused to form a single one.

somatic cells All cells other than reproductive cells in plants and animals.

somatic gene therapy Correction of a genetic disease by inserting a normal gene into somatic cells of an individual.

somatic mutation Any mutation arising in a somatic cell.

Southern blot The digestion of DNA by several different restriction enzymes followed by separation of the DNA fragments according to size by use of agarose gel electrophoresis. The DNA fragments are transferred from the gel to a cellulose filter. Specific DNA fragments can be detected on the filter by using radioactive DNA probes carrying a known gene or DNA sequence.

species A population of organisms that interbreed with one another in nature but do not interbreed with members of other populations, from which they are reproductively isolated.

spina bifida A congenital defect caused by abnormal development of the spinal column.

split gene A eukaryotic gene in which genetic information is encoded in the DNA in discontinuous segments.

starch A carbohydrate in plants.

stromatolite A large, ancient, rocklike structure formed by bacteria and the minerals trapped by the bacteria.

structural gene One that codes for the synthesis of an enzyme or structural protein.

synapsis The pairing of homologous chromosomes in meiosis.

syndrome A group of symptoms used to characterize a disease.

telophase The phase of mitosis following anaphase; the chromosomes unwind and begin to return to their interphase condition.

teratogen Any agent that causes defects in a developing embryo.

terminator site A sequence of bases in DNA where transcription of RNA molecules is terminated.

terminus of replication A site in a chromosome where replication of DNA is terminated.

tertiary sexual differentiation Changes that occur at puberty in male and female human beings.

tertiary structure The overall three-dimensional configuration of a polypeptide chain.

testcross The mating of male and female organisms of known genotypes; used to determine linkage and distance between loci.

tetrad Four homologous chromatids (two in each bivalent).

thalidomide A tranquilizer drug that interferes with limb development in human embryos; a teratogen.

thymine dimer Adjacent thymine bases in DNA that are covalently linked by ultraviolet light.

T lymphocyte A white blood cell that regulates responses by other cells of the immune system.

topoisomerase An enzyme that can twist and untwist a DNA helix or circle.

totipotency The ability of a cell to proceed through all the stages of development, producing a normal adult organism. The nucleus from a single cell, in principle, contains all of the information for reconstructing the complete organism.

transcription The process of synthesizing RNA molecules from specific segments of DNA molecules.

transfer RNA (tRNA) A small RNA molecule that helps line up amino acids in the proper sequence by serving as a link between an mRNA codon and the amino acid for which it codes.

transformation The stable insertion of a fragment of DNA into a cell, causing its genotype and phenotype to be changed.

transgenic plant or animal An organism into which DNA from another species has been stably inserted.

translation The process of converting the information in the sequence of bases in a messenger RNA molecule into a sequence of amino acids in a polypeptide.

translation initiation site A sequence of bases in mRNA where polypeptide synthesis begins.

translocation The movement of a chromosomal segment from one chromosomal location to another, either in the same chromosome or to a different one.

transposase An enzyme that catalyzes the movement of a transposon.

transposon   A group of genes (often those conferring antibiotic resistance in bacteria) that move as a copied unit from one DNA molecule to another.

tumor   A mass of cells that accumulates at a particular site. If the cells spread to other parts of the body causing disease or death, the tumor is malignant. Benign tumor cells remain at the original site and do not usually cause disease.

Turner syndrome   A syndrome characteristic of women who have an XO chromosome constitution.

ultrasound scanning   Production of a visible image of a fetus in the uterus by using sound waves.

universal code   Refers to the fact that each codon specifies the identical amino acid in all organisms from bacteria to human beings.

variable expressivity   The range of phenotypes expressed by a given genotype in a defined environment or range of environments.

wild-type   An organism normally found in nature in contrast with a mutant organism.

X chromosome   The sex chromosome normally present in two copies in women and one copy in men.

xeroderma pigmentosum   An inherited disorder affecting the ability of cells to repair damage to DNA caused by ultraviolet light.

Y chromosome   The sex chromosome normally present in one copy in men.

zona pellucida   The gelatinous outer covering of an egg that allows sperm of the same species to attach and that facilitates fertilization by a single sperm.

zygote   A fertilized egg formed by the fusion of male and female gametes; the first cell of a new organism.

Index